DATE DUE

BRODART, CO.   Cat. No. 23-221-003

AMERICAN BOUNDARIES

AMERICAN BOUNDARIES

THE NATION, THE STATES, THE RECTANGULAR SURVEY

Bill Hubbard Jr.

THE UNIVERSITY OF CHICAGO PRESS
Chicago and London

Bill Hubbard Jr. teaches the introductory design studio for undergraduates in the Department of Architecture at MIT. He is the author of two previous books, *Complicity and Conviction: Steps Toward an Architecture of Convention* and *A Theory for Practice: Architecture in Three Discourses*.

The University of Chicago Press, Chicago 60637
The University of Chicago Press, Ltd., London
© 2009 by The University of Chicago
All rights reserved. Published 2009
Printed in the United States of America

18 17 16 15 14 13 12 11 10 09   1 2 3 4 5

ISBN-13: 978-0-226-35591-7 (cloth)
ISBN-10: 0-226-35591-8 (cloth)

Library of Congress Cataloging-in-Publication Data

Hubbard, Bill, 1947–
     American boundaries : the nation, the states, the rectangular survey / Bill Hubbard, Jr.
          p.   cm.
     Includes bibliographical references and index.
     ISBN-13: 978-0-226-35591-7 (hardcover : alk. paper)
     ISBN-10 : 0-226-35591-8 (hardcover : alk. paper)
     1. Surveying—Public lands—United States—History. 2. Surveying—Public lands—West (U.S.)—History. 3. U.S. states—History. 4. West (U.S.)—Geography. 5. United States—Boundaries—History. 6. West (U.S.)—Boundaries—History. 7. United States—Geography. 8. United States—Territorial expansion.
     I. Title.
     E179.5 .H83 2009
     320.1/20973 22
                                                     2007030995

♾ The paper used in this publication meets the minimum requirements of the American National Standard for Information Sciences—Permanence of Paper for Printed Library Materials, ANSI Z39.48-1992.

To the Surveying Profession

I am, by profession, an architect. All professions, including mine, have been changed by digital technology, but none so radically as the practice of land surveying. I may have seen my hard-won ability to draw precise lines rendered superfluous by software, but that skill was never really central to the work of designing buildings. For surveyors, though, precision is the very focus of their profession. Once, surveyors were those magicians who could lay hands on blunt mechanical instruments and with sharp eyes and skilled fingers conjure up precise readings of distance, angle, and location. Now, satellites pinpoint your location, you project angles by punching in numbers, and distance is given to you by a device that times the return of a reflected laser burst.

Surveyors tell themselves that those finger skills were only a means to achieve the real purpose of the profession: the quasi-judicial determination of correct boundaries. That is indeed true, but it is also true that the new instruments will call a different sort of person into the profession. The work of surveying once demanded a person who carried a jeweler's eyes and an accountant's mind in the body of a wilderness explorer. Such people tended to be bluff but persnickety, easiest perhaps with their own company, demanding of themselves, confident in their abilities but never prideful. They were a type not much seen these days, but only people possessing their qualities could have laid a million miles of boundaries across the face of the United States.

"Surveyor's surveyor" C. Albert White (1926–2007) at the monument, in Ohio, commemorating the beginning of the Rectangular Survey. In the course of compiling his history of the Initial Points from which the Survey spread across the continent, White and his wife located and visited all 38 points, including the five in the wilds of Alaska.

Courtesy of C. Albert White.

Contents

Boundaries in the United States as the Manifest Division of the Nation's Lands

If your travels ever take you near Lansing, Michigan, here's something that might tempt you off the interstate. Just west of the beautifully restored state capitol is the Michigan Museum of Surveying, and if you're at all interested in the evolution of the American landscape, the museum will introduce you to the unheralded man who worked out many of the techniques by which the American West was surveyed.

That man, William Austin Burt, invented a compass that could find true north not by the unreliable magnetic poles but by the invariant passage of the sun across the sky. In the museum you can see some of the actual instruments and tools by which Burt and his fellow surveyors apportioned the surface of Michigan into rectangular tracts that settlers could buy or claim as their own farmsteads.

Be certain to stop by the gift shop: among the items for sale, you'll find a poster that is both an inside joke among surveyors and a wry comment on the nature of American nationhood. The poster features a deadpan photograph of Mount Rushmore, but beneath the picture, the familiar image of the four presidents is recast by an unexpected caption:

Three Surveyors and the Other Guy

—by which the Michigan Museum of Surveying reminds us that Washington, Jefferson, and Lincoln were, all of them, surveyors when they were young men on their respective frontiers, long before they had (or, in the cases of Washington

FIG 1 Courtesy of Joseph Nonneman.

and Jefferson, could have had) any idea of being president. The poster hints, with only a modicum of hyperbole, that the United States is a nation formed *by* surveyors, people whose chief task is marking borderlines upon the land that cleanly apportion its surface into discrete parcels, each destined to be the sole property of some identifiable person, entity, or government.

Surveyors call such an apportionment a *cadastre*, and the achievement of such a perfect, no-conflicts allotment of the earth's surface—not just on maps but upon the land—is the great aspiration of the surveying profession. That our three greatest presidents held to that idea in their formative years must say something about the character of our nation in the first century after its independence, when the continent seemed a *tabula rasa* upon which any pattern of boundary lines could be written. That idea of unfettered mastery is inaccessible to us now, and indeed repugnant; but at the time of independence, it seemed correct—even manifest—to Americans recently or "anciently" arrived from a Europe where boundaries had seemed as inevitable as facts of nature but drawn to the advantage of the powerful.

In Europe, those boundaries were sometimes fuzzy, with landowners on either side sharing the disputed land or occupying it in sequence. But such arrangements have always struck us Americans as insufficiently precise. From the first we felt impelled to know, with certainty, the exact location of the boundary between "what is mine" and "what is somebody else's." To an American it doesn't really matter, too much, what the provenance of a boundary is. Where else but in the United States could citizens feel an intense identification with entities called Colorado and Wyoming, whose boundaries are mere rectangles 7° by 4° in longitude by latitude, drawn indiscriminately across mountains and plains?

And yet we do feel such a identification. We drive on our great interstates across an arbitrarily designated line into another state and feel that we have somehow crossed a crucial border. In spotting that "Welcome to . . ." sign on the verge of the highway, we believe that we have crossed over into a different reality. We suddenly notice subtle changes: license plates take on different colorations and mottoes; the land seems cultivated in different patterns; people seem to have different accents, live in different kinds of houses, eat different prepackaged candies and breads, call familiar sandwiches by unfamiliar names.

Visitors from other countries don't notice these fine distinctions. For them, America gradates slowly as they pass across it in rental cars. Only we Americans are so explicitly attuned to boundaries that we feel a difference when crossing the state line from, say, Kansas into Nebraska. We feel that we know in some sense what Nebraska is (that quirky unicameral legislature, that dichotomy between corn in the east and cattle in the west—bridged along I-80 by an intense shared loyalty to Big Red football). And when we drive across that invisible boundary, we expect things to be different; we seek out differences, collect them in a mental diary, and deploy them as evidence in comments like, "Well, you wouldn't see that in Kansas!"

We are people who find our way, know who and where we are, by boundaries. We take comfort in imagining that the essence of a bounded place pervades that place, undiluted, right up to its borders. Nebraska is everywhere within its borders "Nebraska," just as Kansas is "Kansas." And so are all the states, and the cities and towns, and my yard and the plants I grow and the way I live versus the plants you grow in your yard and the way you barbecue.

Do Italians feel such sharp distinctions when they drive on their own interstates, those *autostrade*, from the Latium region around Rome and across the border into Tuscany? There is a border there, for administrative purposes, but what is real to Italians on the road are the subtly accumulating differences in geography and culture as one region gradates into another.

We Americans are far from blind to such gradations. We feel them perhaps most strongly within the state where we reside, and especially if we have lived there long. But an awareness of clear-cut borders is never far from our consciousness. We see our American world through the lens of boundaries, even though we know, deep down, that these lines were drawn across the land in the most capricious ways. Nonetheless, we accept, even embrace those lines as proper, right, manifest—as if this apportionment of our geography had been determined by some ineluctable principle, as if these borders could not have happened in any other way.

Surely part of what makes the border between Tuscany and Latium different from that between Kansas and Nebraska is historical sequence. There were distinctive cultures centered in the lands around Florence and Rome long before there were definitive borders between them. When the need arose for boundaries to enclose regions, those borders were drawn, with difficulty and imprecision, to encompass cultural patterns felt to be similar enough to constitute entities. But in America, in almost all cases, the boundaries were drawn before the cultural patterns arose. Before anybody had a sense of what it meant to be a Nebraskan and what Nebraskans might share in common, there was a place marked on maps with clear borders called Nebraska.

As settlement progressed, cultural patterns arose organically out of geography, history, and the habits of the immigrant populations. But if people came naturally to grow corn around Omaha and raise cattle farther west, they nevertheless did so together, inside a single bounded entity. As settlers elected officials and representatives, and dealt in common with taxes and regulations, and aspired for their children to attend the state university, a thought must have occurred to

them, with increasing urgency: "Surely it must mean something to be a Nebraskan? Surely there is something distinctive and shared about living within these borders?"

Eventually all the states found, and continue to find, ways of looking at their boundaries and believing that consequential things happen within those borders that don't happen, or happen differently, the instant you pass beyond them. Sometimes the differences are ginned up, and sometimes they exist only in fond memory; but these distinctions are nonetheless felt, and they form part of our identity as Americans. I lament the passing of the roll call of the states at the national political conventions ("Madam Chairman, the Great State of ——, home of ——, proudly casts its votes for ——") where you could hear those attributes proclaimed in distinctive accents, often with self-conscious irony, knowingly overboastful, but always confirming how very much our capriciously drawn state borders matter to us Americans.

There is another set of arbitrarily drawn boundary lines that Americans have filled up with meaning, and those are the rectangles into which so much of the United States west and south of the Appalachians is apportioned. You've doubtless viewed the great gridded landscape from the air, either from your seat during a flight or in photographs. And if, when driving rather than flying across the landscape, something drew you off the interstate into that western and southern three-quarters of the nation, you will have traveled along those arrow-straight perpendicular lines.

It is the quintessential American landscape, characteristic of us, emblematic of qualities we value and aspire to: plainness, sobriety, even-handedness, evincing "good sense" as we reckon it. It is a landscape daunting in its implacability but, to some of us, both deeply moving and ineffably beautiful.

We Americans decided, even before we were a nation, that unsettled land would not simply be thrown open to claimants to divide up as they chose, nor would the land be offered up for sale to the highest bidder. Instead, the national government would hold the land in trust for all the people, survey it into rectangular parcels, and make those parcels available to individuals and families.

To the originators of this Rectangular Survey, the system was merely an efficient, foolproof way of apportioning the public lands. But once people began to live their lives within those rectangular boundaries, those borders—like those of the states—came to acquire meaning, significance. And the rectangular way of drawing those borders came to seem—as with the boundaries of the states—manifestly right, the only imaginable way to apportion land.

This book tells the story of how all of these borders came to be: how the national borders enclosing the public lands attained their continental extent, how the states carved from those lands got the shapes they now have, and how those states were apportioned into the rectangles of the Public Land Survey. But before those things could happen, several other things had to happen first, and those events form the armature of this book.

Underlying the whole story is the very idea of lands being held in trust as a public domain. At the time of our nation's independence, all of its land, from the Atlantic west to the Mississippi, was under the often conflicting jurisdiction of the thirteen new states, their claims based on the royal grants that had established them as colonies. Gradually the idea arose that lands not yet settled should be held by the national government, parceled out to settlers, and eventually formed into new states—an idea wholly unprecedented. By 1786 the states with claims north and west of the Ohio River had ceded their lands to the Continental Congress, and in 1802 the federal Congress added to the Public Domain the territory from Georgia to the Mississippi. When Louisiana was acquired from France, the principle of land held in trust was extended, and extended again as the nation expanded to the Pacific and then to Alaska and Hawaii. All of these stories are told in part 1 of this book.

Once it was determined that this domain would be apportioned into new states, a method had be found for accomplishing that. In 1784 the Continental Congress adopted a plan by Thomas Jefferson for establishing new states in the entire territory west of the Appalachians. Three years later, Congress reconsidered Jefferson's plan and adopted a more modest scheme for only the area between the Ohio and Mississippi, the Northwest Ordinance of 1787. Readopted by the new federal Congress, this act became a model for drawing state boundaries and establishing state governments (a model not always followed). When it came time to form states in territory far from Washington, Congress drew some rather capricious borders. This story is told in part 2.

Before the concept of apportioning the states into rectangular parcels arose, several competing ideas about how to apportion the public lands were in the air. Would contiguous townships be granted to proprietors, as had been the habit in New England? Would individual settlers be allowed to stake out their own boundaries around claims, as had been the pattern in the South? Would land companies be offered large tracts from the government for later resale to settlers?

In 1784 Thomas Jefferson proposed the unprecedented idea of surveying the public lands into tracts 10 nautical miles square. Congress rejected the idea but a year later adopted the Ordinance of 1785, mandating the 6-mile-square Townships that became the basis of the Public Land Survey System. But at first neither the surveying nor the land sales went well, and Congress flirted with selling huge tracts to land companies—also without success. There matters stood until 1796, when the federal Congress adopted the Rectangular Survey in essentially its present form. What remained was to carry this system into the field and refine both the instruments and the techniques surveyors used, a process completed by about 1855. All of these stories, and a lot of surveying lore and technique, are covered in part 3.

In part 3 you'll get a sense of the "ideal" system of survey as officials in Washington envisioned it, but to get a feel for how those millions of rectangles were determined and delineated onto the ground, you need to go out into the field and see how it actually happened. The survey of Montana can serve as a typical instance. In the final chapter of part 3, you'll meet an 1860s surveying team and then, through a series of yearly maps, track the progress of the survey from its inception in 1867 up to 1884.

FIG 3

When in 1900 the Government Printing Office published a report about the just-completed survey of the northern line on the boundary between Montana and Idaho, the cover featured this image. Heavily retouched in the mode of that era, it nevertheless tells us something of the difficulty of marking a perfectly straight line through rugged mountain terrain.

U.S. Government Printing Office.

Even as unabashed an aficionado as I am can't claim that the Rectangular Survey has been an unmixed blessing. In the epilogue I offer some thoughts about the survey's impact on the public domain, and acquaint you with one poignant "might-have-been" alternative to the system.

HOW THE BOOK'S TEXT IS ORGANIZED

The book's two main stories—the idea of a national domain and the technique of a rectangular survey—are intimately entwined. To help you through the thicket (and to make long chapters seem shorter), the text has been divided into sections, each with a summarizing title like the one above. You'll thus be able to skip sections if you like, but be aware that later sections might refer to ideas or phrases broached in earlier ones.

This backward and forward referencing will *not* be happening in . . .

SIDEBARS

In any research endeavor, you uncover facts and ideas that don't fit seamlessly into the narrative but are just too delicious to leave out. That's what you'll find in the sidebars. Nothing in them will be referenced in the main text, so you can safely skip over them—but then you'd miss all the fun.

Sidebars will be your signal throughout the book that although some text is nonessential, perhaps you'll find it interesting. Also set apart from the main text are the sections on . . .

THE TASK AT HAND

As you read about how people at the centers of power decide where boundaries are to go, you might want to have an image in your mind of surveyors out in the wilderness doing the work of marking those boundaries on the land. The concepts behind the work are complex, but they—along with the most basic techniques—can be summarized, and that's what the Task at Hand sections will do.

The main text will not refer to facts and ideas laid out in the Task at Hand sections, so if the subject doesn't interest you, just skip on to the next section of the main text.

PART I

Assembling a National Domain

OVERVIEW

We Americans came by our preoccupation with boundaries naturally, and right from our beginnings. Boundaries were very much on the minds of Englishmen when, a century after Spain and Portugal had done so, they cast their eyes seriously toward the New World. Boundaries were at the very center of two issues facing England around 1600—the planting of estates in Ireland and the enclosure of fields at home.

England's struggle to control Ireland lasted, incredible as it may seem, for nine hundred years, ending (if that is the word) only in the 1920s. But in the decades before and after 1600, England's hold on at least the northern and western part of the island seemed secure enough that estates could be set up in the territory, to reward nobles who had fought in or helped finance the wars for Irish conquest. As would be the case in most of colonial history, the indigenous patterns of occupancy were not deemed worthy of following, and so the problem arose of how to draw up brand-new boundaries across a perceived *tabula rasa*. Interestingly, many of the new boundaries were rectangular,[1] but the important factor for us in this story of American borders is not so much the shape of the boundaries but the fact of them for the English people. For five centuries, ever since the redistribution of land following the Norman Conquest, Britons had experienced lives lived pretty much within the boundaries they had been born into. Now, though, and quite suddenly, there was the spectacle of lines again being drawn across land.

And not just in Ireland: all across England, the plots on which people had farmed were being reorganized and enclosed, by fences and hedgerows, into wholly new configurations. For as long as anyone could remember, agriculture on English feudal manors had been conducted on an *open field* system under which tenants raised food crops on common fields, paying the manor's lord with a share of their produce and with work in his private lands. But as the sixteenth century drew to a close, nobles found that they could make actual *money* if they sowed their fields not in the crops humans eat but in those sheep graze upon. At first the sheep's wool would be traded to Holland, then later sold directly to English factories; but once begun, the movement toward pasturage only accelerated through the sixteenth and seventeenth centuries.

There had always been livestock on English farms, though limited to those animals raised either for milk or food or for plowing—few enough that they could be kept out of the crop fields without too much trouble. Now, though, big herds of sheep would have to be moved across the landscape, and that required containment, both to keep the sheep out of the remaining crop fields, and to keep them from wandering onto adjoining estates. *The enclosure movement*, as it came to be called, was of course a disaster for the tenant farmers forced off the land. But in the minds of both nobles and former tenants, a proper farm now became a bounded thing— not merely the right to use commonly held land, but a marked-off plot of ground, owned exclusively and indisputably by one person alone. So whether noble or middle-class or peasant, when Englishmen came to America, they came with boundaries on their minds.

Other matters were on their minds as well, and it's important to remind ourselves of some of them. We Americans, in our solipsistic focus on only our own history, sometimes slip into imagining that all that was happening in the world during the 1600s was the expansion of English colonies west from the Atlantic seaboard. In reality, during that century England was undergoing perhaps the greatest upheavals it would experience as a nation.[2]

In addition to the Irish conquest, Britons of the previous century had faced the conflict between Protestants and Catholics, and that problem too was unresolved as the seventeenth century opened. In 1604 James I succeeded Elizabeth I to the throne, to be followed in 1625 by his son Charles I. Charles brought a third problem to a head by refusing to convene a parliament from 1629 to 1640. Out of that crisis came the English civil war, the beheading of Charles in 1649, and rule by Parliament and Oliver Cromwell.

With the collapse of Cromwell's regime, the son of Charles I was placed on the throne in 1660 as Charles II. Charles ruled until his death in 1685, and was succeeded by his brother the Duke of York, who was crowned James II. James's rule brought such turmoil that he was overthrown in 1688 and replaced by his sister Mary and her Dutch husband, William of Orange. William and Mary's rule at last restored order, Mary ruling until 1694 and William until 1702, to be followed by James's daughter Anne. It was during Anne's reign in 1707 that true union with Scotland was finally achieved. With her death in 1714 came the first of the succession of Georges that led to George III (1760 to 1820), from whom the colonies secured their independence.

Such radical changes in governance could not but have effects on policy toward the colonies in America. The effects are many, but I want to emphasize three that most affect our story:

- *As might be expected from kings whose reigns could precipitate revolts, the colonial policies of the two Charleses and the two Jameses were often capricious, even erratic. They made grants of land in America that contradicted grants previously made, sometimes even by themselves.*
- *Everyone is aware of the long, gradual assertion of parliamentary authority over royal prerogative, but that story didn't reach its full climax until after America's independence. The story that most affects American colonial history is the gradual rise in the power of the King's Council of advisors. Up until the nineteenth century, these advisors were noblemen, in the tradition of the courtiers that rulers like Henry VIII and Elizabeth I drew around themselves. Thus, when a little later I speak of "king and council" doing something, I'll be acknowledging that English monarchs in the colonial period didn't act entirely alone.*
- *As the seventeenth century wore on, the King's Council evolved from a circle of "generalist" advisors to a group in which each member had a defined area of responsibility (culminating in the nineteenth century with full professionalization into ministries of government). It's not surprising that one of these councilors, the First Lord of the Treasury, should emerge as first among equals and*

become, again gradually, the equivalent of prime minister. When later you hear about Lord North, for example, what he was lord of was the Treasury, but he functioned, for George III, as a prime minister.

- *The third significant change lay in the manner with which colonial affairs were dealt. Before Cromwell, the King's Council handled such matters through a subcommittee drawn from its members. During the time of Commonwealth, Parliament administered colonial policy through its own special committee. With the restoration, Charles II set up the Lords of Trade in 1675 to supervise colonial affairs. In 1696 William III greatly expanded the powers of the Lords of Trade and renamed it the Lords Commissioners of Trade and Plantations, usually shortened, helpfully, to the Board of Trade. It was this group that nominated governors for the colonies, reviewed laws passed by their legislatures, and generally stood between the colonies and the king and his council. Any complaints from the colonies had to go through the Board of Trade first; any edicts from the king or council or Parliament would be interpreted to the colonies by the Board of Trade. The membership of the board exemplifies the movement of the English government toward a ministry system. Day-to-day affairs were conducted by eight salaried commissioners, who were joined for major decisions by eight unpaid, ex-officio members chosen from the King's Council. The Board of Trade would handle colonial policy until 1779.*

One primary outcome of these changes affected how colonial grants were made and how the resultant colonies would be governed. At the start of the seventeenth century, the king and his council had basically four ways of founding a colony, listed here in roughly ascending order of royal control:

- *The king could grant a feudal patent. Under such an arrangement the grantee would have almost absolute control over the land granted him. The legalistic construction sometimes used in such grants was that the grantee would have "all the rights and privileges of the bishop of Durham"—meaning that, by analogy, the grantee would own all the land, all the profits of the land would go to him, and he would be, in effect, the government of the grant. Charles I made such a grant to George Calvert, giving him the title Lord Baltimore, owner of Maryland. Charles II made such a grant to his brother the Duke of York for the conquered New Netherlands and for New Jersey, and to eight favored courtiers for the Carolina colony south of Virginia.*

- *Alternatively, the king could grant a boon to a favorite subject by definitively spelling out the rights and privileges in the form of a proprietorship. This is how William Penn acquired Pennsylvania and Delaware, and John Mason initially acquired New Hampshire.*

- *Or the king could accede to the formation of a joint-stock company and grant its members sole right to a specified tract of land. Clearly under such a setup, the terms of the company's charter were a matter for negotiation, allowing the king to extract more profit or assert more control, depending on the circumstances. Virginia and Massachusetts were originally founded by companies such as these.*

- *Finally, a group could receive a charter directly from the king, in which he could set the condi-*

tions of the agreement. Georgia was founded on this basis. Independent of royal sanction, people established settlements in Connecticut and Rhode Island, and at Plymouth: all three were later "legalized" by the granting of a charter from the king. Under such a royal charter, a colony would be ruled in whatever manner its charter set out, but it would have a direct relationship with the Board of Trade. In theory at least, the Board of Trade couldn't speak directly to the colonies granted as feudal manors, proprietorships, or corporate monopolies; it would need to speak through the hereditary lord, the proprietor, or the company's governing committee. When a colony had a royal charter, nothing stood between it and the Board of Trade.

Not surprisingly, by the eve of the Revolution, the British government had brought about a situation in which it had royal charters with nearly all the colonies. Kings, councils, parliaments, and boards of trade realized that central control could be asserted most effectively through such charters, and gradually—often against resistance from feudal lords, proprietors, corporations, and even colonists—London moved to "rationalize" its governance of the colonies under that model.

That rationalization, though, would come only as the seventeenth century neared its end. At the century's opening, the planting of English colonies on North America's Atlantic coast seemed such a risky proposition that king and council were prepared to entertain proposals from almost any group or person offering to undertake the venture.

The Maps in the Book

The great problem of mapmaking has always been how to represent the spherical earth on a flat sheet of paper. Dozens of projections of the globe have been devised, the most familiar, perhaps, being the Mercator projection—that map in which Greenland is the size of South America.

The map of "the Lower 48" in figure 1.01 is produced by a variant of the Mercator projection called the *plate carée*. This shape is certainly recognizable as the contiguous states, but doesn't it look a little funny to you? If your visual experience has been like mine, the shape in figure 1.02 is more familiar. This is the map produced by a conic projection, and somehow that swaybacked shape, with Maine holding its head up proudly, resembles the iconic "U.S. Map."

But look at the two versions of Oklahoma: the *plate carée* produces the flat-topped shape of figure 1.03; the conic projection gives us the swaybacked version in figure 1.04. Which shape seems more "right" to you? Here the answer is not so obvious, but I'll bet that in your mind the iconic Oklahoma has a perfectly straight Panhandle, not one that curves upward.

So since I want you to be able to compare "might-have-been" map shapes with those in your visual memory, I'll generally be using the conic projection when I'm depicting large areas of the United States, and using the *plate carée* when showing states or small regions.

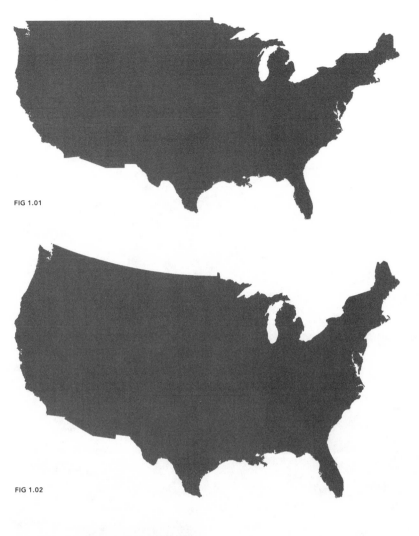

FIG 1.01

FIG 1.02

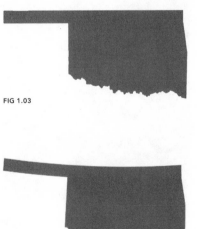

FIG 1.03

FIG 1.04

ONE

The Colonies Stake Claims to Land in the West

DECIDING BOUNDARIES FOR COLONIAL GRANTS

In 1606 two companies of "Knights, Gentlemen, Merchants, and other Adventurers" advanced proposals to James I of Great Britain to plant settlements on the Atlantic coast. In truth, all the men in the companies were adventurers, since the word at that time carried the sense of "investor." But in that era, joint-stock syndicates such as these were so risky that the term must have carried some of its present-day sense as well. What the companies sought, in return for risking their investments, were monopolies on settlement in, respectively, the northern and southern parts of the coast of His Majesty's new American dominion of Virginia. (The part of Virginia we now call New England would be given its separate name by John Smith when he explored there in 1619.[1])

Both companies had grandiloquent Jacobean names, but historians have helpfully shortened them to the two English cities from which each drew the bulk of its investors. The Plymouth Company sought to control settlement in the north, the London Company in the south. The question facing the King's Council was how to apportion the territory between the two companies.

Apart from some inexact mapping of the Atlantic coast, the England of 1600 possessed very little knowledge of the geography of America. How, then, to declare definitive boundaries for the land grants? The strategy adopted by kings and their councils was that the charters would designate points and lines that could be named in England and then later found, mapped, and finally traced on the ground in America by surveyors.

(This was to be the strategy adopted all through the settlement period whenever "the center" had to legislate boundaries for "the periphery." And from the seventeenth century to the early twentieth, officials at the center were continually surprised by the consequences on the ground when their projected boundaries were ultimately traced.)

Lines of longitude and latitude could provide definitive boundaries once they were located and then traced, and such lines were often used in the language of the grants. But in 1600 very few geographic features in Atlantic America had had their positions located within that global net of lines. There was always the risk that if a grant cited a specific latitude or longitude, the final tracing of those lines might reveal a vital river or harbor to be on the wrong side of the line, outside the bounds of the grant. So the grants often specified river courses as boundaries, especially if the intention was that two adjacent grants would be able to use the river as an outlet to the sea (as was to be the case when the Delaware River was made the common border of Pennsylvania and New Jersey). If the intention was to give one grant exclusive rights to a river, the charter might specify that the borderline of the grant head inland from the point where one bank of the river's mouth intersected the ocean—or, to make those rights absolutely certain, that the border be set at some distance back from the river's edge.

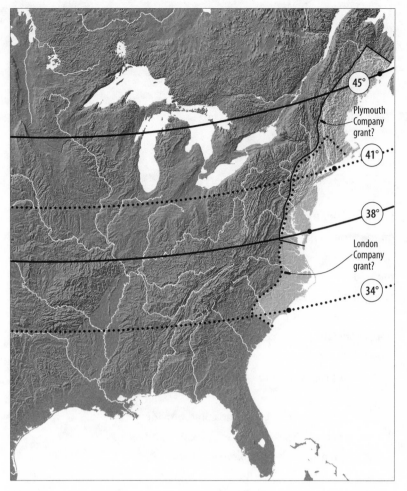

FIG 1.05

(Remember that in 1600 the Pacific coast had barely been sailed by Europeans, much less mapped. Plus, the western coast's longitudinal position was not well fixed: people weren't sure how "thick" the North American continent was. There was still hope that some estuary or bay might extend partway—or even all the way—across the continent. A grant that contained such a passage to China might become prosperous beyond anyone's imagining.)

In its charters for the London and Plymouth companies, the King's Council boldly used lines of latitude to apportion the grants, but hedged its bets. I will subject you to reading only the London Company's charter. It is written, like all the royal grants, in an embroidered legalese that obfuscates as much as it clarifies:

And they shall and may begin their said first Plantation and Habitation at any Place upon the said coast of Virginia or America, where they shall think fit and convenient, between the said four and thirty, and one and forty degrees of said Latitude; and that they shall have all the Lands . . . from the said first seat of their Plantation and Habitation by the Space of fifty Miles of English Statute Measure, all along the said Coast of Virginia and America, towards the West and Southwest, as the Coast lyeth . . . and also all the lands . . . from the said place of their first Plantation and Habitation for the space of fifty like English Miles, all along the said Coast of Virginia and America, towards the East and Northeast or towards the North, as the coast lyeth . . . and also all the Lands . . . from the same fifty Miles every way on the Sea Coast, directly into the main Land by the Space of one hundred like English Miles.[2]

In essence, the London Company was empowered to establish its settlement anywhere on the

The long courses of rivers could also be used to specify how far inland a grant would extend. The charters often cited the source of a river as a definitive point from which to project a borderline. The source of a river is indeed a unique point in the terrain—once everyone has agreed which of the innumerable branches of the river is its true main course (no small matter, as we shall see repeatedly).

On the other hand, if the king and his royal council were in a particularly magnanimous frame of mind, they might make the grant extend inland from the Atlantic coast as far west as there was land. The phrase often used was "to the South Sea," meaning all the way to the Pacific Ocean.

coast between the 41st and 34th parallels. The grant to the Plymouth Company was analogous in every way except that its two bounding parallels were the 45th and the 38th.[3]

Throughout the colonial period, there would be continual wrangling among the colonies about just where their boundaries lay. It was language like this that kept the lawyers busy. First and most obviously, the two grants overlap between the 38th and 41st parallels. Second, each grant is itself ambiguous in its own provisions.

The London Company's grant could mean what is shown in figure 1.05: its lands extend all the way between the 41st and the 34th parallels, for 100 miles in from the shore; they can plant a settlement anywhere within those lands; and if the settlement happens to fall on either extreme parallel, they can extend their hundred-mile-deep claim a further 50 miles beyond the parallel, north or south.

Or it could mean what's shown in figure 1.06: the company can plant its settlement anywhere between the 41st and 34th parallels; the company's grant will then extend 100 miles inland, and 50 miles up and down the coast from that settlement, even if the company plants it on one of the extreme parallels.

Quite some difference. The first is a grant of a huge swath of territory, the second is the grant of a compact square of land measuring 100 miles on a side. Clearly, in the first interpretation the grants are mutually contradictory, whereas in the second reading no inconsistency would arise unless the companies settled closer than 100 miles apart (on, say, the coast of present-day New Jersey). But ask yourself: if you were to find that a contract you had signed could be interpreted so as to dump a windfall on you, wouldn't you claim—with a straight face—that that was

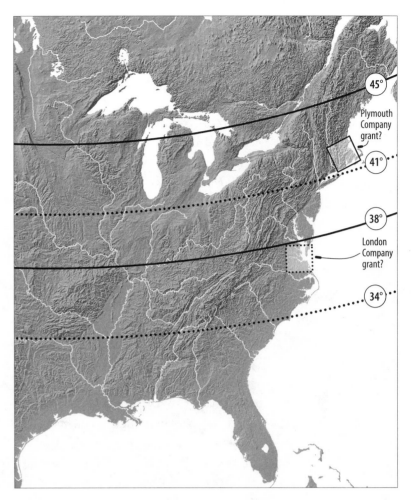

FIG 1.06

how you had read the contract all along? That's what both the London and Plymouth companies did, and who can really blame them?

Thus in 1606, before a single permanent English settlement had been established in North America, the powers at the center had set up conditions for conflict at the periphery by their cavalier declaration of ambiguous boundaries. In the nineteenth century, Washington, DC, would succeed London as "the center," and it would continue the pattern of dividing up the periphery with arbitrary boundaries.

The conflict between the two 1606 charters could have become moot when the two companies' grants were redrawn, in 1609 and 1620, respectively. But in the new grants, the conflicts in the claims were merely displaced far to the west, and would arise only in the 1780s.

In 1607 the London Company established its beachhead at Jamestown, an upriver spot perilously close to the hundred-mile limit of its inland boundary. The company sought to have its grant redefined to reflect this new center of settlement, and a fortuitous geographic fact gave them a point of negotiation. Coming into Chesapeake Bay during the reconnaissance of their grant, the settlers had spotted a prominent point of land which, remembering the rigors of their long voyage, they named Point Comfort. In the undulating western shore of the Chesapeake, Point Comfort pinches the broad James River estuary and thus marks the point at which it can be said that the river debouches into the bay—a natural landmark from which a new, more generous land grant might be laid out.

It is possible to sail east from what is now called *Old* Point Comfort and pass straight through the mouth of the bay, between Capes Charles and Henry (as shown in figure 1.07). Interestingly, though, the capes are less than 20 miles apart, so a sharp-eyed sailor in a crow's nest could see both capes simultaneously and thus know when his ship had passed from the bay into the open ocean. If the captain were to bring the ship about at that point and sail directly up or down the coast, he would be able to calculate pretty accurately how far north or south he had ventured from Point Comfort's latitude.

In 1609 King James I and his royal council used this happy fact to prescribe new boundaries for the Virginia colony. The grant would extend from that point at the mouth of the bay, north up the coast 200 miles, south down the coast 200 miles, and then from those two endpoints, "up into the Land, throughout from Sea to Sea, West and Northwest."[4]

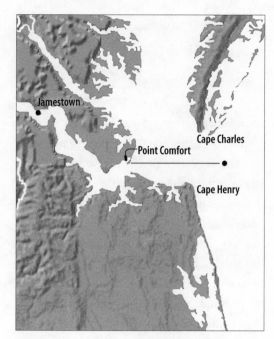

FIG 1.07

Here the drafters of the grant left yet another ambiguity to plague the future. Did they mean that the northern point should be projected west and the southern point projected northwest (the grant does describe the two points in that order), resulting in the wedge of land shown in figure 1.08? Or did they mean to reverse the projections—west from the southern point and northwest from the northern point—giving Virginia the broadening fan of territory shown in figure 1.09?

In the 1600s, with settlement close to the coast, there was no need to interpret the grant closely. By the late 1700s, however, when explorers and settlers began to push into present-day Kentucky and Ohio, Virginia quite naturally seized upon the more expansive reading of the grant. It was on the basis of this 1609 charter that Virginia would claim as its own almost all the land northwest of the Ohio River.

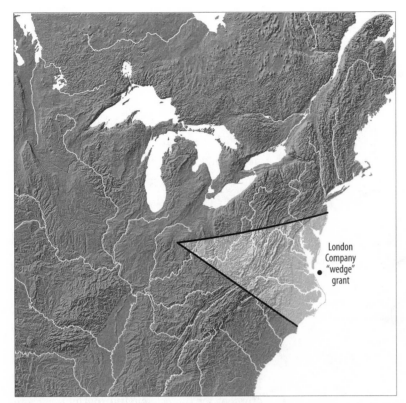

FIG 1.08

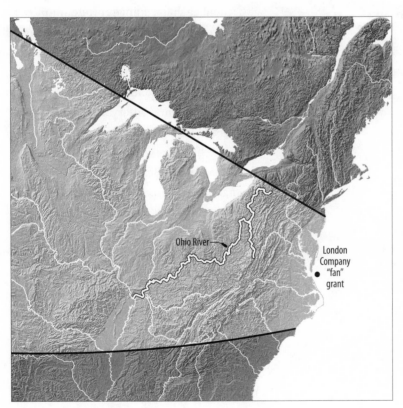

FIG 1.09

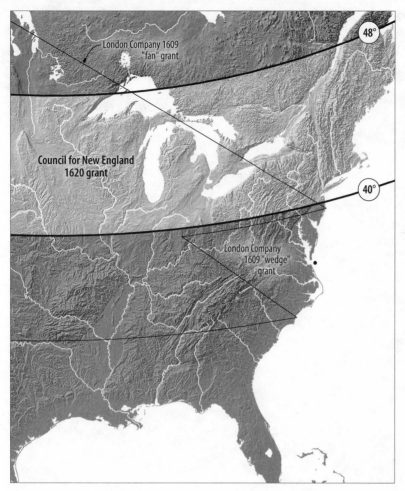

Council for New England
1620 grant

London Company 1609
"fan" grant

London Company
1609 "wedge"
grant

48°

40°

FIG 1.10

THE CLAIMS OF MASSACHUSETTS AND CONNECTICUT FOR LAND IN THE WEST

By 1620 the Plymouth Company had failed to establish a settlement within its grant, and a new company was brought into being, whose name historians have shortened to the Council for New England. To the Council a new grant was made by James I: all the land between the 40th and 48th parallels, from sea to sea (shown in figure 1.10).

This grant obviously intersects the "fan" definition of James's grant to the London Company just eleven years earlier (though it aligns closely with the "wedge" reading of that grant), but that would be a problem only later. Despite having a name that makes it sound like a constituted government, the Council for New England was more nearly intended as a vehicle by which royal favorites would be granted huge estates in America. The prime force behind the creation of the Council was one such favorite, Sir Ferdinand Gorges (or Ferdinando, when not using the Anglicized version of his name), and so it is not surprising that the first grant made by the Council was to Gorges and his colleague Captain John Mason.

In 1622 they were given a strip of land 60 miles deep between the Merrimack and Kennebec rivers (fig. 1.11), thus initiating the practice of using rivers as boundaries for land grants. Later each of the two parties wanted his own land, and so the grant was redrawn. In 1629 the Council gave Mason a 60-mile strip between the Merrimack and the Piscataqua rivers—roughly the seacoast portion of present-day New Hampshire, making Mason and his heirs a force in the early years of that colony. In 1639 Gorges was given land just up the coast, between the Piscataqua and Kennebec rivers, this time to a depth of 120 miles inland.

Between the times of these two grants, in 1635, came a grant to Lord William Alexander of a parcel still farther up the coast: the land between the Kennebec and St. Croix rivers (the St. Croix being the eastern border of present-day Maine). His grant extended farther inland—from the heads of both rivers all the way to the St. Lawrence (fig. 1.12). The intention was that this grant would abut the one made to Alexander's father, Sir William Alexander, by James I in 1621—essentially, all the rest of the peninsula south of the St. Lawrence River, which now contains Nova Scotia, New Brunswick, Prince Edward Island, and the Gaspé Peninsula of Quebec.

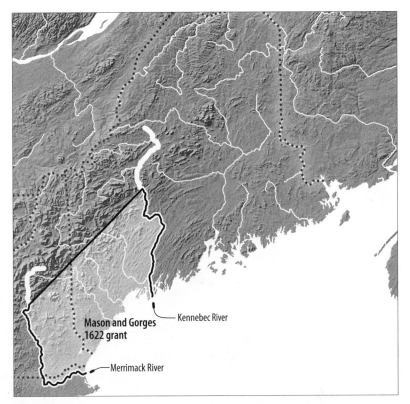

Mason and Gorges 1622 grant

Kennebec River

Merrimack River

FIG 1.11

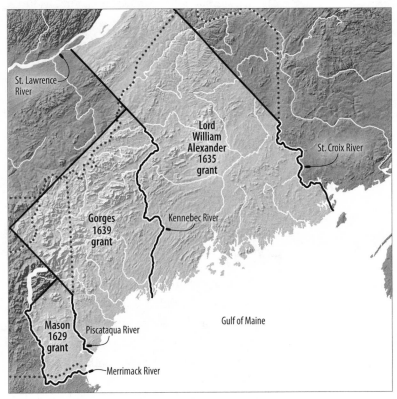

St. Lawrence River

Lord William Alexander 1635 grant

St. Croix River

Gorges 1639 grant

Kennebec River

Mason 1629 grant

Piscataqua River

Gulf of Maine

Merrimack River

FIG 1.12

But before these grants to Gorges and Alexander—indeed before the end of the 1620s—it had become apparent that the Council for New England would not, itself, be able to found successful settlements. Consequently, the Council became something of a shell corporation, the designated vehicle through which grants of New England land would be made.

The Council grant that would have the greatest impact on American history came in 1629. This grant was made to the Massachusetts Bay Company and confirmed by the new king, Charles I. Its territory was all the land between two parallels, from sea to sea. This time, the grantors avoided the shot-in-the-dark process of naming whole-number latitudes and instead tied the bounding parallels to natural features. Specifically, they chose two rivers that they wanted to ensure would fall within the grant's boundaries. The parallels are indeed specifically described, but the descriptions only reveal London's ignorance of the territory being apportioned. (Figure 1.13 shows the unintended extent of the grant.)

The southern parallel was to be 3 miles south of either Massachusetts Bay or the southernmost bend of the Charles River, whichever was found to be farthest south. Luckily for the Massachusetts Bay Company, the Charles River (we now know) starts its course quite a few miles south of the southern curve of the bay, even obligingly flowing a ways southeast from its source point before turning north to meander its way to Boston.

A line of latitude 3 miles south of that bend in the Charles falls at approximately 42° 2' north latitude and forms the basis for the southern bor-

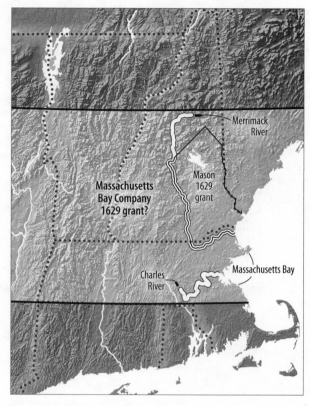

FIG 1.13

der of the later colony and state of Massachusetts. (The present-day border wiggles north and south to record subsequent adjudications with Connecticut and Rhode Island.) And since the grant awarded land north of the line all the way to the ocean, Massachusetts would, much later, invoke this grant to lay claim to land to the west.

The grant's northern boundary, when finally found, had equally far-reaching consequences. That border was to be a line of latitude 3 miles north of the northernmost point on the Merrimack River, but as with the Charles, explorers had not yet traced the full course of the river. Little did anyone suspect, in 1629, that the Merrimack actually starts its run from far north into present-day New Hampshire, then heads south,

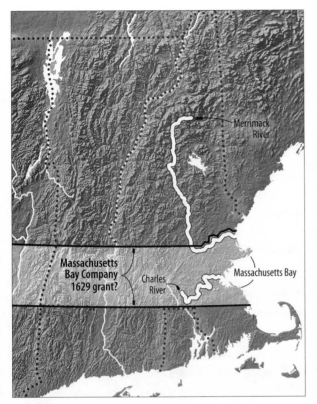

FIG 1.14

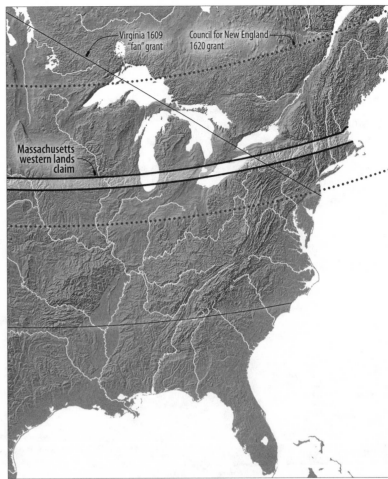

FIG 1.15

and only in its last few miles makes its abrupt turn to the northeast to find an outlet to the sea.

In 1740 the King's Council finally clarified the northern border of the Massachusetts colony. It would indeed be "three miles north of the Merrimack"; in fact it would be a line tracing the river's course, displaced 3 miles to the north. But that mirroring line would stop at the bend where the river makes its great turn. From that point a line of latitude would be projected west, as shown in figure 1.14. This amended grant would later become the basis for a claim by Massachusetts to land in what is now Michigan and Wisconsin (figure 1.15).

In 1620, before the Council for New England was up and running, a group of religious dissent-

ers had obtained permission from the London Company to found a settlement in the northern part of its 1609 "Virginia" grant (200 miles north and south of Point Comfort, you will remember). The Pilgrims, though, made landfall far to the north, on Cape Cod, and dug in for the winter at Plymouth, due west on the mainland. They were thus well outside the bounds of either the "fan" or the "wedge" interpretation of the London Company's 1609 grant, but they hoped that their status would be made official. In 1630—one year after the grant to the Massachusetts Bay Company—the colonists got their sanction.

The main part of their grant was defined by a line drawn from the inlet of the present-day town of Cohasset west to the place where the Narragan-

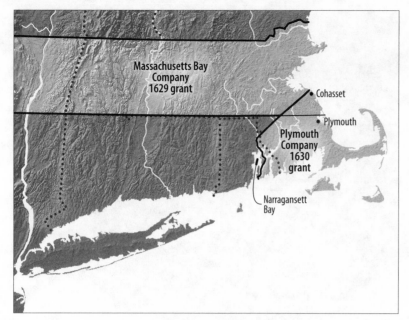

FIG 1.16

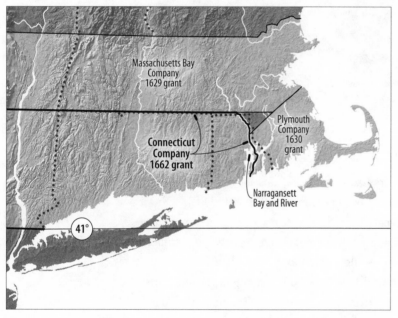

FIG 1.17

sett River widens into Narragansett Bay. (Cohasset marks the southern "corner" of Massachusetts Bay, as you can see in figure 1.16.)

It is a point of historical curiosity that the Pilgrims were also granted a tract in what is now Maine: a rectangle of land 15 miles to each side of the Kennebec River and between two falls of the river called Cobbiseconte and Mequamkie. Local officials were never able to agree on precisely which falls had these names until 1783, by which time the question was largely moot: the Plymouth and Massachusetts Bay colonies had been merged into the colony of Massachusetts in 1691, with jurisdiction over Maine awarded to the new colony.

By 1630, then, apart from the grant to the Pilgrims, all the land granted through the Council for New England had been north of that "bend-of-the Charles" line near latitude 42°, meaning there were still parts of New England that were undistributed—depending on how you define New England. Under the 1620 grant from Charles I to the Council, the southern border of New England was the 40th parallel. The distinction matters because in 1662 Charles II made a grant to the new Connecticut Company of "all that part of our Dominions in Newe England" west of Narragansett Bay and south of Massachusetts, all the way to the South Sea—implying that the Connecticut Company's lands were a wide band from the 42nd to the 40th parallel. (Connecticut's claim was later defined as extending only to the 41st parallel, as shown in figures 1.17 and 1.18. We'll see why in a moment, when we look at New York's claim of lands in the West.)

In 1643, at the height of the turmoil in England between Parliament and Charles I, a parliamentary commission had control of colonial affairs, and from that commission Roger Williams obtained a patent for governing the "Providence Plantations" he had established around Narragansett Bay. The patent described an area of jurisdiction—south of the grant to the Massachusetts Bay Company, from Narragansett Bay to a line 25 miles to the west—but it did not grant Williams and his followers actual control of the land. Twenty years later, in 1663, Charles II (on the throne after the Cromwell interregnum) rectified this anomaly with a grant of real land to the now renamed Rhode Island and Providence Plantations (still the official name of the state). The grant gave Rhode Island less land west of Narragansett Bay (from only the head of Pawcatuck Creek north to the Massachusetts border, about 20 miles west of the bay) but compensated by granting land on the east side of the bay, as shown in figure 1.19.

This new colony had thus been shoehorned between the claims of the Connecticut and Plymouth colonies, whose common border had previously been Narragansett Bay. Litigation among the colonies naturally ensued, with Rhode Island's border with Connecticut fixed by the King's Council in 1727. Its border with the Plymouth colony's successor, the state of Massachusetts, was not finally settled until 1862, by the U.S. Supreme Court.[5]

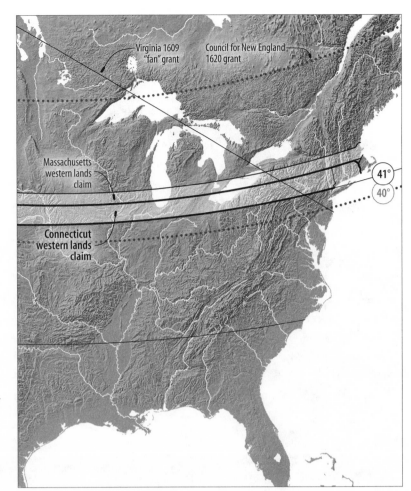

FIG 1.18

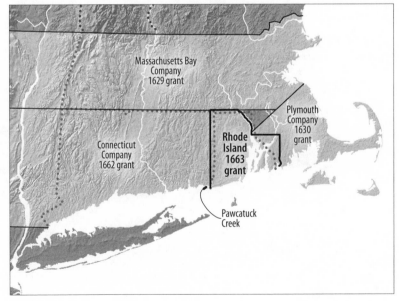

FIG 1.19

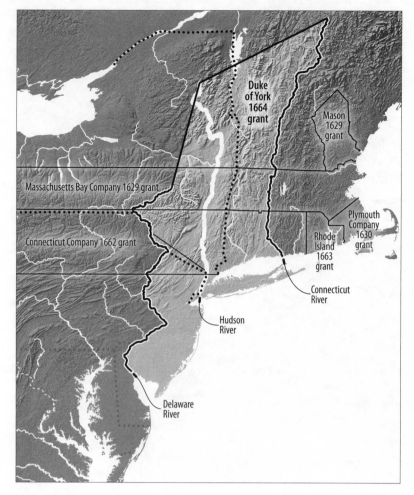

FIG 1.20

In 1674 Berkeley and Carteret divided York's grant between them, Berkeley taking the colony of West Jersey and Carteret, East Jersey. It was only in 1702 that the two colonies were united as New Jersey.

NEW YORK'S CLAIM FOR LAND IN THE WEST

New York also had claims to land in the west, the basis for its claims being as complex as those of all the other claimants. The story starts in 1664 when Charles II made a two-part grant to his brother the Duke of York (who would, you recall, succeed Charles in 1685 as King James II).

The first part of the grant was an almost exact duplicate of the grant made in 1635 to Lord William Alexander—all the land between the Kennebec and St. Croix rivers, north to the St. Lawrence. York never seriously claimed this land, but the second part of the grant he took very seriously, for it included the Dutch colony of New Amsterdam, which the English had seized earlier that year.

Charles's penchant for granting land already granted was repeated in this second boon to his brother—all the land between the Connecticut and Hudson rivers, and all the land from the Hudson to the Delaware River, a tract that can be outlined, as in figure 1.20, by drawing lines connecting the source points of the three rivers.[6]

Even with the geographic knowledge of 1664, it would have been plain that this grant cut straight across territory previously granted "from sea to sea" to the Connecticut and Massachusetts Bay companies. The result was a series of border disputes that plagued the three colonies for decades. In the discussions, the one point on which York and the later colony of New York would never yield was possession of both banks of the Hudson. Thus, during negotiations in the late 1600s, Connecticut agreed to a border about 20 miles east of the Hudson, but held fast to settlements on Long Island Sound, resulting in the "tail" that extends southward, just to the 41st parallel.[7] Massachusetts entered into similar negotiations, and in 1773 agreed to a borderline also about 20 miles east of the Hudson.[8]

The 1629 grant from the Council for New England to John Mason was the basis for the colony of New Hampshire, and even though that grant's 60-mile depth didn't extend to the Connecticut River, New Hampshire settlers reached the river (and beyond) long before people coming in the other direction from New York. The King's Council decreed in 1764 that New Hampshire's western border would be the Connecticut River, but by that time New Hampshire settlers had nearly reached as far west as the Hudson. Those settlers took advantage of the confusion of the Revolution and declared themselves the new state of Vermont in 1777—a declaration which New York naturally challenged. When the controversy was finally resolved with Vermont's admission as the fourteenth state in 1791, the precedent established farther south was followed, and both states accepted a border about 20 miles east of the Hudson.

Just three months after receiving his grant from Charles II, the Duke of York gave the portion of land that is now New Jersey to his friends Lord Berkeley and Sir George Carteret. The duke's grant gave the two men all the land between the Delaware River and the Atlantic, up to a precisely defined line—from the point on the Delaware at 41° 40' north latitude, then down a diagonal line to the point where the Hudson crosses latitude 41°. That precision, though, caused its own problems: the diagonal line from the 41° 40' point gave New Jersey an absurdly narrow shoreline on the northeast bank of the Delaware, as shown by the thin solid line in figure 1.20. A more sensible termination point would have been at the Delaware's abrupt eastward bulge, as depicted by the dotted line. Nonetheless, the later colonies of New York and New Jersey couldn't bring themselves to agree to such a border until 1773.[9]

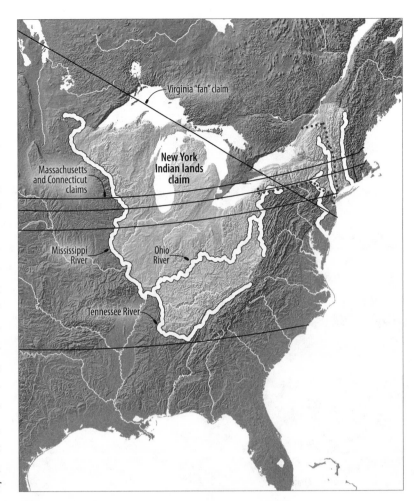

FIG 1.21

Even with New Jersey given away, the Duke of York's holdings were extensive, but they had a shape nothing like the New York State we now know. Where did that shape come from?

Unique among all the colonies, New York based its western territorial claims not only on royal grants but also on successive treaties it "negotiated" with Indian tribes, most especially with the Six Nations confederation that included the Iroquois. If one were to interpret these treaties expansively (and include within them the additional lands of all the tribes allied with the Six Nations), New York's jurisdiction could be said to have extended west from the Hudson to all the land above the Tennessee and Ohio rivers, all the way to the Mississippi, as shown in figure 1.21.

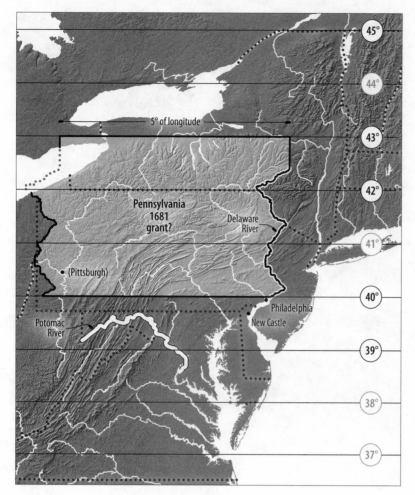

FIG 1.22

The shape of New York that we now know was carved out of this immense area by the actions of two different kings. To the north, the general realignment of borders that took place in 1763 after the French and Indian War placed New York's northern border at the 45th parallel (see figure 1.22). A grant that Charles II made to William Penn in 1681 curtailed New York's lands to the south. Their common border falls at the 42nd parallel, but that's not exactly what Penn's grant says.

Perhaps taking a leaf from his brother York's grant of New Jersey, Charles II drew Penn's grant with precision. First, a circle 12 miles in radius would be drawn around the town of New Castle on the Delaware River, and the land within the circle excluded from Penn's grant. The grant's southern border would start where that circle touched the river, then travel the circle's circumference counterclockwise until it hit "the beginning of the fortieth degree of Northern latitude." From that intersection point the border would follow that latitude until it reached a point 5° west of that line's projected intersection with the Delaware. The northern boundary of the grant was to be "the beginning of the three and fortieth degree of Northern latitude," from its intersection with the Delaware to a point 5° west.[10] Figure 1.22 depicts the shape this grant would have had if surveyed on the ground: on the south and north, two east-west lines, the 40th and 43rd parallels; on the east, the undulating course of the Delaware River; on the west, displaced westward by 5°, a copy of that undulating line.

But Pennsylvania's present border with New York is the 42nd parallel, not the 43rd. The key to understanding this apparent contradiction

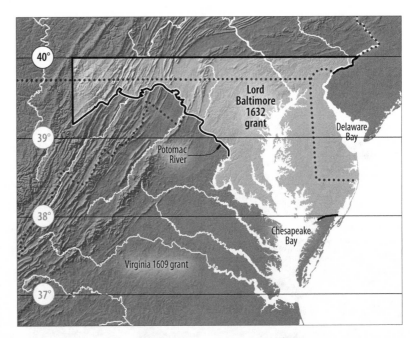

FIG 1.23

comes out of how one can interpret that phrase in Penn's grant, "the beginning of" the forty-third degree of latitude. "Degrees of latitude" can be thought of as lines, but they can also be visualized as a stack of bands circling the earth, each band 1 degree wide. Under that conception "the first degree of north latitude" is a 1-degree-wide band with its "bottom" resting on the equator—0 degrees of latitude. So if the *first degree* of latitude begins at *0 degrees*, then the *second degree* of latitude begins at the line *1 degree north* of the equator, and the *third degree* begins at the line *2 degrees north* of the equator—and so on, up the stack of 1-degree-wide bands to the beginning of the *forty-third degree* of latitude at the line *42 degrees* north of the equator (the line more conventionally called "42° north latitude").[11]

One can see why the New York colony was happy to accede to this interpretation of Penn's grant: a conventional reading of the grant would have pushed the border 1 degree farther north (see figure 1.22). A problem arose, however, with Maryland, the colony immediately to the south of Pennsylvania.

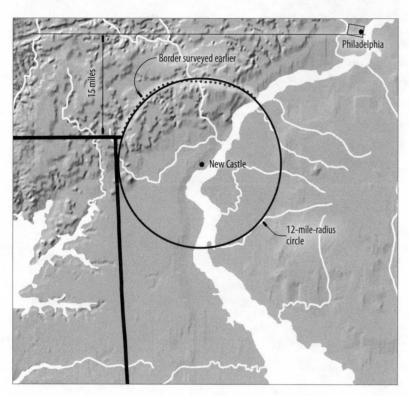

FIG 1.24

PENNSYLVANIA'S GRANT BOUNDS VIRGINIA'S CLAIM FOR LAND IN THE WEST

Almost fifty years before Penn's grant, in 1632, Charles I had given a boon to his subject George Calvert. Calvert (and a succession of sons) would henceforth be titled Lord Baltimore, the proprietor of the peninsula between the Chesapeake and Delaware bays, and the land north of the entire length of the Potomac River, up to "the Fortieth Degree of North latitude from the Aequinoctal, where New-England is terminated"[12] (thus conforming to the definition in the 1620 grant to the Council for New England). So if one adopts the conventional understanding of "degree of latitude," there is no conflict at all between the grants to Penn and Baltimore: Maryland and Pennsylvania would have had a common border at the line of the 40th parallel (see figure 1.23, on previous page).

But it's clear from the terms of the grant that Penn and Charles had assumed that the 40th parallel fell near enough north or south of New Castle that it would intersect the 12-mile-radius circle drawn around the town. So when Penn's emissary arrived on the scene in 1681 and broached that assumption to Lord Baltimore, Baltimore pointedly informed him otherwise. Not only did the 40th parallel fall north of the New Castle circle, it fell north of the site intended for Penn's green country town, Philadelphia.

Negotiations ensued, closing with letters back to England to Penn and King Charles. When Penn himself arrived late in 1682, the matter might have been settled, for Penn carried a proposal from Charles for a compromise borderline (about 4 miles south of the present line) which would have gone some way toward mollifying Baltimore. Unfortunately, in the interim between Penn's original grant and his arrival in America, he had been additionally granted what is now Delaware, the territory anointed the "three lower counties" of Penn's grant.[13]

Now there may have been legitimate confusion over the location of the 40th parallel, but Baltimore's grant had unequivocally given him the peninsula between Chesapeake and Delaware bays. Penn and Baltimore agreed to disagree, as would the descendants of the two proprietors, for the next hundred years.

Finally, in 1732 the descendants of William Penn and Lord Baltimore worked out a provisional agreement that, after more wrangling, was signed in 1760. For the peninsula border, a line would start on the Atlantic Ocean at Cape Henlopen (not the Cape Henlopen we know today at the mouth of Delaware Bay, but a bulge farther south that William Penn and his descendants insisted the grant had intended when it specified "Cape Henlopen" as the southern border of the three lower counties—see figure 1.25). That line would go due west to a point halfway across the peninsula, and from that point a second line would be projected up the peninsula (not necessarily due northward) so as to be tangent to the 12-mile-radius circle around New Castle. That would be Maryland's border with what had, by then, become the colony of Delaware.[14]

The border with Pennsylvania would then be a line of latitude 15 miles south of the southernmost point of Philadelphia. (The location of that point was based not on Penn's rectangular plan for the complete city, but on the southernmost house then built—the dot shown in figure 1.24.) As it turns out, this line hits the New Castle circle a little north of the tangent point specified for the Delaware boundary—which accounts for that

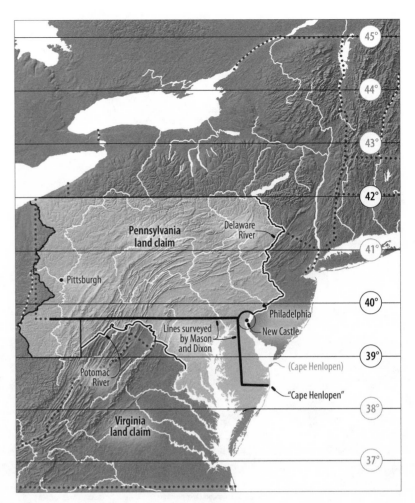

FIG 1.25

little bump in the state's border. (By rights, that tiny sliver could have gone to Maryland, but later negotiations awarded it to Delaware instead.)[15]

The work of surveying these three lines began in 1760 on the ocean at "Cape Henlopen." The work was suspended, and then resumed in 1763 by Jeremiah Mason and Charles Dixon. Four years later they had carried the line up the peninsula and then all the way to the western limit of Maryland's grant, a line traced in 1746 north from the source of the Potomac. The two surveyors even exceeded their mandate, projecting the latitude line 40 miles farther west, to about the point shown in figure 1.25.[16]

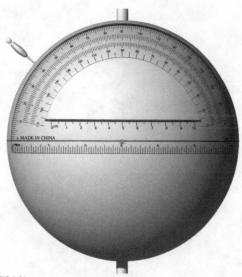

FIG 1.26

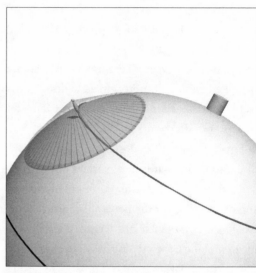

FIG 1.27

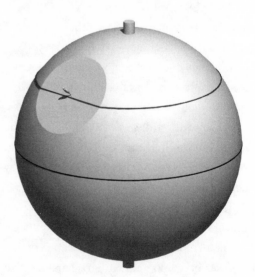

FIG 1.28

When Mason and Dixon found where a line 15 miles south of Philadelphia would cross the New Castle Circle, their next task was to determine and record the precise latitude of that point.[17] That way, as they projected their line westward, they could periodically stop at a point on the line, determine that point's latitude, and if necessary shift the line to put it back on the proper latitude. Surveyors will tell you that finding the latitude of some point on the earth is relatively easy. What's difficult is projecting a latitude line across the earth's surface. Here's why finding latitude is easy.

First, remind yourself that the latitude of a place is measured by how far—in degrees—that place lies north (or south) of the equator. (The grade-school protractor in figure 1.26 makes the concept plain.) Remember too that as you stand at, say, latitude 40° north, gravity puts your body on a line running straight to the center of the earth.

When you measure something, you often measure "away from" from some starting point, and the start point for latitude measurements is a horizon plane. In figures 1.27 and 1.28, I'm going to engage in a small surveying fiction to make the idea of a horizon plane easier to visualize.

Imagine that you are standing upright on the surface of a perfectly spherical earth. You look out to the point where your angled-down line of sight touches the curving-down surface of the sphere. Keeping your line of sight trained in that direction, you slowly turn yourself around in a complete circle. Can you see that the rays of your vision sweep out a very broad cone?

Then imagine that we run a knife directly across the base of that cone, removing the slice of

FIG 1.29

earth above it. We now have a perfectly flat circle, a "horizon" we can visualize and measure against. (I'll be using this image repeatedly, and I hope you'll soon appreciate its usefulness— and overlook its unfortunate resemblance to the Death Star in the first *Star Wars* movies!)

What we now want to measure is the angle, in degrees, between our horizon plane and the **north polar point**, the place in the night sky to which the axis of the earth's rotation points, that axis shown by the "spindle" on our sliced globe. We identify that north polar point with Polaris, the star at the end of the "handle" of the Little Dipper constellation, but that's not entirely accurate. A line through the earth's axis doesn't shoot directly "through" Polaris but about 1 degree to the side. To complicate matters still more, the pattern of stars in the night sky rotates around that polar point, so that Polaris

doesn't just stay in place but circles around the polar point, in an orbit 1 degree in radius. (Actually, of course, it's the daily rotation of the earth that makes the sky seem to spin—counterclockwise to us in the Northern Hemisphere.) Even though Polaris rotates around the polar point, its distance from the polar point is precisely known, and surveyors have a whole bag of tricks for finding that empty point in the sky.

Figure 1.29 shows a computer-generated image of how the night sky would have looked in April of 1784 from the 40th parallel of latitude. This is the sky Mason and Dixon would have seen as they set up their instruments outside New Castle. Figure 1.30 shows the geometric condition they were trying to achieve: they wanted to find a spot on the earth where, if they looked outward toward the horizon and then rotated their line of

sight upward until they were looking at the north polar point, the angle between those two sightings would be precisely 40 degrees; that spot would, by definition, be on the 40th parallel of latitude.

(In Figure 1.30, only the spindle of our spherical earth points directly to the polar point. Stars like Polaris, though, are so distant that the two sight lines are effectively identical. I think you can intuit that, as you mentally slide that "horizon plane" down toward the equator or up toward the pole, the two measured angles will change but always be equal to each other—the angle between the horizon and the north point always equaling your latitude.)

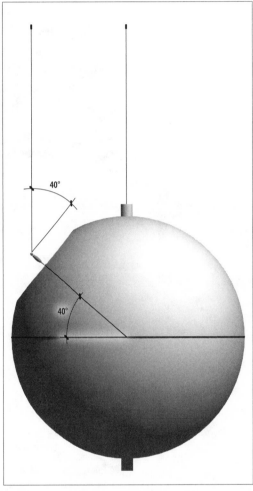

FIG 1.30

FIG 1.31

Courtesy of Physics Department,
University of California at Berkeley.

THE TASK AT HAND
Projecting a Line of Latitude

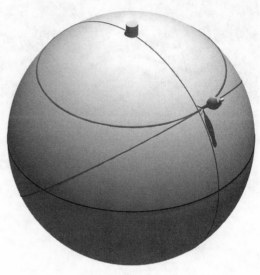

FIG 1.32

The instrument used to measure this angle would have been a smaller version of the one shown in figure 1.31. This is an astronomical transit from the turn of the last century, but it is pointing in a direction close to 40 degrees above the horizon. Mason and Dixon would have set the base of their instrument on a stout post and leveled it to achieve alignment with the horizon. You can see that the scope of the instrument can rotate up and down, and that the angle of rotation can be precisely measured on those large circular plates. Mason and Dixon would have had to move their instrument northward and southward several times to home in on a spot where the angle was exactly 40 degrees, but they accomplished the feat, and with great accuracy.

Having found and recorded their latitude, the next task facing Mason and Dixon was projecting a line westward that would stay continually at that latitude and not veer north or south. This is by far the more difficult of the two tasks, and here's why.

Sighting on Polaris gives you not just your latitude; with a few more tricky calculations, you can determine true north from that sighting. And having found true north, you can find the true westward direction by rotating your theodolite (a surveyor's instrument for measuring angles in a horizontal plane) precisely 90 degrees counterclockwise. (That's in fact the only way a surveyor can find west—or east.) Problem is, you can't just start walking in that westward direction: continue on that line, and you will soon find yourself veering ever more south of the true parallel. The image in figure 1.32 shows why: we are dealing

here with spherical geometry, not plane geometry.

Your problem as a surveyor is that you have only one way to mark a line on the ground: through your instrument you sight in a direction and then send someone forward on your line of sight; he faces back to you, and with hand signals you get him to move a marker pole side to side until it falls directly on your vertical crosshair, at which point you have him drive the pole into the earth to mark a point on the line. You next position your instrument over that point and repeat the process. With that as your only technique, how can you mark a curved line?

Remember first of all that you don't have to trace the full contour of the curve on the ground. Your job is merely to place markers at points along the curve—in the case of the Pennsylvania border, every mile or so. With your instru-

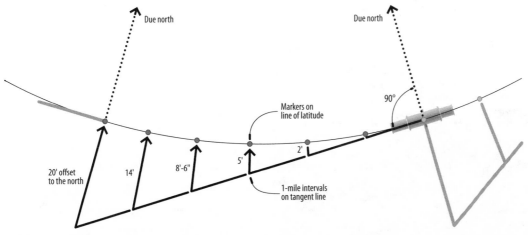

Due north

Due north

90°

Markers on
line of latitude

20' offset
to the north

14'

8'-6"

5'

2'

1-mile intervals
on tangent line

FIG 1.33

ment pointing west, you send your crew ahead to clear away brush and trees from your line of sight, making what surveyors in the eighteenth century called a *vista*. You make the vista as long as practicable, since each move of your instrument will be a source of possible error. You then send crew members down your line of sight to specified distances. You know that the true latitude-line is curving northward away from the line of sight, but from a book of tables you know how far northward your crew needs to offset their markers to have them fall on the latitude line. If you extend your line for 6 miles, for example, the offset needs to be 20 feet, with lesser offsets at the required one-mile intervals—as shown in figure 1.33.

That offset holds, of course, only for the 40th parallel.

Northward of that, the parallels "curve more," so that at the 49th parallel—the U.S.–Canada border—the offset after 6 miles is 28 feet. The 31st parallel is the northern border of Florida, and there, by contrast, the offset is only 14 feet.

Surveyors today call the technique used in figure 1.33 the *tangent method*, but Mason and Dixon seem to have used what is now known as the *secant method* of projecting a latitude line. Figure 1.34 shows how the secant method it works; what's striking is the accuracy the method requires. The first marker is set 32 inches north of where the scope is positioned. To project the correct secant line, the scope must be rotated a tiny fraction more than three-hundredths of a degree, and then markers are offset tiny distances south of the sight line.

Surveyors then and now have techniques for achieving such accuracy, and Mason and Dixon used them skillfully. In addition, they seem to have recalculated their latitude about every 12 miles. The result is that the Mason-Dixon line—and its extension to Pennsylvania's southwest corner—are revered by surveyors. Mason and Dixon determined that the latitude of the westernmost point of their line was 39° 43' 17.6". A resurvey in 1892 found the point to be at latitude 39° 43' 19.91"—meaning that the two surveyors were 180 feet off in their calculations. In an 1883 resurvey, the other end of their line, back at New Castle, was found to have a latitude of 39° 43' 18.2"—meaning that their line ended, after more than 250 miles, 3 feet north of where it started!

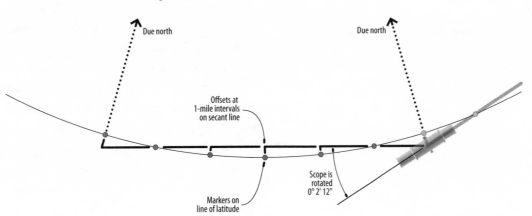

Due north

Due north

Offsets at
1-mile intervals
on secant line

Markers on
line of latitude

Scope is
rotated
0° 2' 12"

FIG 1.34

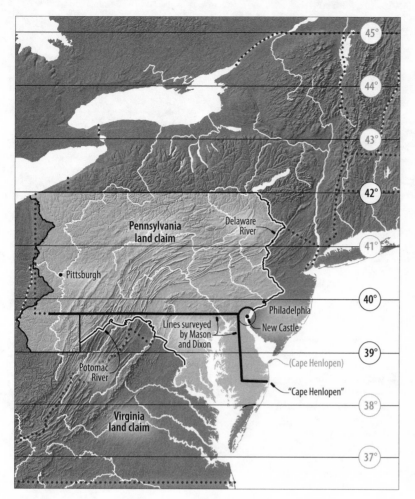

FIG 1.35

With Maryland's boundary run, Pennsylvania's only unresolved border was that with Virginia. You will recall that Virginia's grant had the shape of a fan, broadening out from the coast, into which the grant of Pennsylvania intruded. It was an accepted principle among the colonies that when royal fiat intervened, early grants could be diminished by later grants; so the only question was, how big a bite would Pennsylvania's grant take out of Virginia's great fan of territory?

When the commissioners from the two colonies sat down in 1774, both sides assumed that the intention of the grant to Penn was that Pennsylvania's western border would be a tracing of the course of the Delaware River, displaced 5 degrees westward.[18] By this time, however, Fort Pitt (today's Pittsburgh) had been established and had become an important center of settlement and trade. A look at a map of the Delaware River shows that above Delaware Bay, the river's course veers quite a way toward the east, a fact well known to the negotiators. Both sides suspected that when the surveyors retraced this course on the ground 5 degrees westward, the line would swing to the east of Fort Pitt—placing that important post in Virginia's claim.[19] (In point of fact, figure 1.35 shows how such a line would actually fall about 20 miles west of Pittsburgh.)

With that fear in mind, Governor John Penn adopted a position that was either a cynical bargaining ploy or a long-held family belief. He claimed that regardless of the agreement reached with Maryland, Pennsylvania's southern border was the 39th parallel—invoking that unconventional but plausible reasoning about the border of William Penn's grant being "the beginning of" the fortieth degree of latitude. With that position on the table, the negotiations went back and forth, each side proposing a series of boundary lines. Finally, five years later, in 1779, worn down by the debate and in the midst of the Revolutionary War, the commissioners found a compromise. Pennsylvania would give up the 39° line and accept, as its southern border, an extension of the previously agreed Maryland border to a point 5 degrees west of that line's intersection with the Delaware River. In return, Virginia (to compensate Pennsylvania for the territory it would "lose" south of the Maryland line) would accept, as Pennsylvania's western border, a meridian line drawn straight north from that western endpoint.[20] Figure 1.36 shows the western land claims resulting from the agreement, with Virginia's vast claim diminished by grants made later. (Strictly speaking, the map is anachronistic, because by the time of the Pennsylvania negotiations in 1779, Virginia's southern border had been pushed northward by earlier grants; we'll get to that story in a moment.)

According to the agreement, the extension of the line would be surveyed by a team headed by four commissioners from each state. Virginia's team included James Madison and Andrew Ellicott, who was later to help mark Pierre Charles L'Enfant's plan for Washington, DC, on the ground. On Pennsylvania's team were Thomas Hutchins, who, as the first Geographer of the United States, would later begin the Rectangular Survey; and David Rittenhouse, America's preeminent maker of precision instruments. By the fall of 1784 the surveying party had run the border to the southwest corner (see figure 1.37), marked it with a squared-up oak post set in a mound of stones, and retired from the field to await the next surveying season.

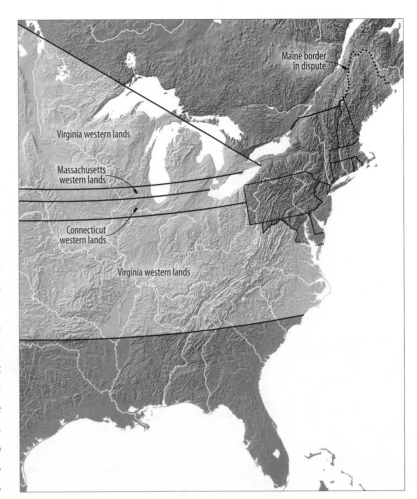

FIG 1.36

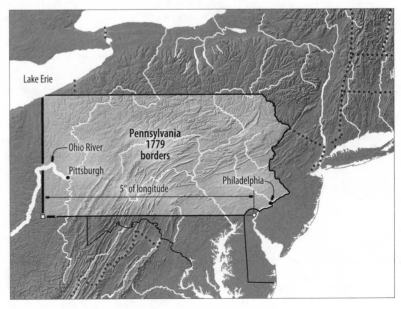

FIG 1.37

Finding Longitude

To know where to set that oaken post, the commissioners needed two pieces of information: the longitude of the west bank of the Delaware River where an extended Mason-Dixon line would touch it, and the location on that line of a point 5 longitude degrees to the west. We saw earlier how finding latitude is relatively easy, whereas projecting a line of latitude is quite difficult. In a curious paradox, projecting a line of longitude is pretty simple, but in the eighteenth century there was no surveying task more difficult than finding the longitude of a place.

To understand why, remind yourself that (in the United States) just as the latitude of a place is the number of degrees northward from the equator to that place, so the longitude of a place is the number of degrees westward from Greenwich to that place—*Greenwich* being shorthand for a precise line on the earth at the Royal Naval Observatory outside London.

Up until very recently, the only way to know the number of degrees between your position and that of Greenwich was to know the difference between the time at your position and the time in Greenwich. The earth rotates 360 degrees in twenty-four hours, so every hour it rotates 1/24th of 360 degrees, or 15 degrees. If it's noon where you are, and it's 3:00 p.m. in Greenwich, the earth will have turned through three hours of rotation since it was noon at Greenwich. There are thus 3 times 15 degrees of longitude between you and Greenwich: you are at longitude 45° west. Figure 1.38 illustrates the principle, with the translucent plane fixed in place as the earth rotates beneath it.

To make use of this handy equation about longitude, though, you need the answers to two questions: What time is it at Greenwich? And what time is it where you stand?

It was the business of the Royal Naval Observatory to let the world know "the time at Greenwich," and the way it did that was to calculate and publish tables listing the precise Greenwich time of future celestial events, a prime example being the eclipses of Jupiter's Galilean moons. Jupiter has four moons—Io, Europa, Ganymede, and Calisto—that are so large that in 1610 Galileo was able to view them through his new telescope. What makes them useful is not just their visibility but the speed with which they orbit Jupiter: Io once every 1½ earth days; Europa, once every 3½ days; Ganymede, once a week; and Callisto, once every 16½ days. What this means for us on earth is that a moon will frequently be eclipsed by Jupiter, either as it orbits behind the planet or moves into the shadow the planet casts out into space, and thus (to us) suddenly "blinking out."

So there you are, standing on the bank of the Delaware in 1784, looking up at Jupiter with your best telescope. From the Greenwich *ephemeris* (that's the name for a book of celestial tables; the plural is *ephemerides*) you know that tonight Io will be eclipsed by Jupiter. You aren't sure which of the two moons you can see is Io, but only Io is scheduled to be eclipsed tonight. So you watch both moons, and as soon as one "blinks out," you instantly look down at your clock (which you just yesterday set to solar time) and record the time of the eclipse. Your clock reads 1:05 a.m.; your ephemeris says that Io will be eclipsed when it's 6:07 a.m. in Greenwich. That's a difference of five hours and two minutes. Doing the math, that's 75½ degrees: Mason and Dixon's line touches the Delaware at longitude 75° 30' west of Greenwich. (To get greater accuracy for your readings, you would doubtless stay there on the Delaware, observing and recording as many other celestial events as your schedule would allow, then averaging the results.)

12:00 noon, Greenwich time

FIG 1.38

3:00 pm in Greenwich
12:00 noon on horizon plane at
Longitude 45° west

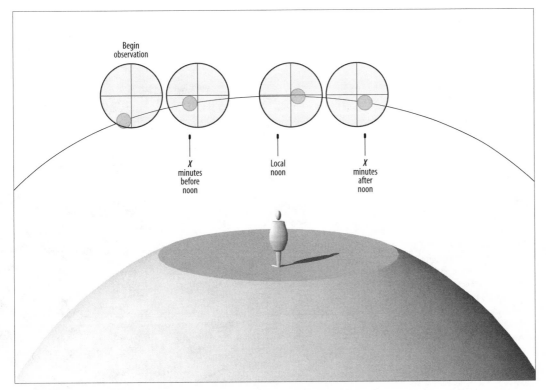

Begin
observation

X
minutes
before
noon

Local
noon

X
minutes
after
noon

FIG 1.39

How, though, do you determine the precise time for your position on the earth? You've got a timepiece, but it loses (or gains) fractions of a second every day it runs, so you'll want to reset it every time you make a celestial observation; you'll do that with the only accurate "time signal" at your disposal, solar noon.

Of all the moments the sun is in the sky, noon is unique (which is why it's used for time-keeping). Once you're well north of the equator, the sun's movement from east to west occurs in a long arc upward and then down across the sky. Noon is the peak of that arc, the moment when the sun stops rising in its arc and begins falling. It's this change from upward to downward that you use to find the moment of solar noon, and figure 1.39 illustrates a technique you can use.

You set up an instrument whose sighting telescope can swing back and forth horizontally; you precisely level its base, and then at around 11 a.m. by

your clock, you point the instrument at the sun, maneuvering the scope until the sun is in the lower left of the instrument's aperture, its right edge against the vertical crosshair. You then track the sun's movement across the sky by rotating the scope to the right, keeping the sun's edge tight against the crosshair. As the sun moves, it rises; and at some moment it will have risen along the vertical crosshair to also touch the horizontal crosshair. At that precise moment you record the time given on your timepiece.

You can now have a quick lunch, during which solar noon happens and the sun begins moving downward in its arc. As soon as you get back, you again rotate the scope, this time to get the sun's left edge against the crosshair. You again track the sun's movement across and (this time) down. At the moment when the sun touches both the vertical and horizontal crosshairs, you record the time your clock now gives you. By averaging the two times, you

can calculate what time your clock would have read at the moment of solar noon. Your clock will probably have struck noon some seconds early or late, but by setting your clock forward or back by that amount, you can restore it to accuracy for the coming night's observation—after you have made an adjustment called an *equation of time*.

To understand this concept, it's helpful to think of solar noon not as a time but as a condition—the one illustrated in figure 1.40, when a line projected from the center of the sun hits a line drawn between earth's two poles. The plane in the diagram stands for this "condition-called-noon" (and its converse, "midnight," of course), and the condition exists continually, since the plane keeps its edge pointed at the sun. The "time-called-noon" happens for you when your position on the planet rotates under this plane. By worldwide convention, a new day begins (on the opposite side of the earth) when Greenwich rotates under the plane, but in the course of each day, the earth has also moved along its orbit around the sun, which introduces a complication.

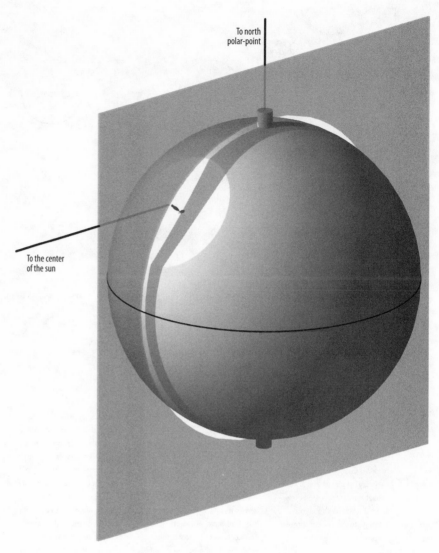

To north polar-point

To the center of the sun

FIG 1.40

In figure 1.41, we're looking down on the earth from above the North Pole, with the earth thus rotating counterclockwise and orbiting "downward." You can see that in the second, lower position, the earth has rotated a full 360 degrees, but you on the horizon-circle have not yet rotated back under the noon plane: you'll need, in fact, about four more minutes of rotation (on average) before your position moves back under the plane. That "day of one rotation" is called a **sidereal day**, but the "day of noon-to-noon at Greenwich" is called the **diurnal day**, and that is the "day" we divide up into twenty-four hours. But wait: there's more . . .

You may recall, from grade-school science, something called Kepler's law. It states that an object orbiting the sun moves not in a perfect circle but in an ellipse, with the sun at one of the two focuses of that ellipse. The law also states that an object orbits fastest when it approaches the sun, slower when it swings away.

The long elliptical orbit of Halley's comet is an extreme example of this: as it swings around the inner focus of its ellipse, at the sun, the comet traverses most of the solar system in a matter of months; yet it spends eighty-six years getting around the outer (empty) focus of its orbit.

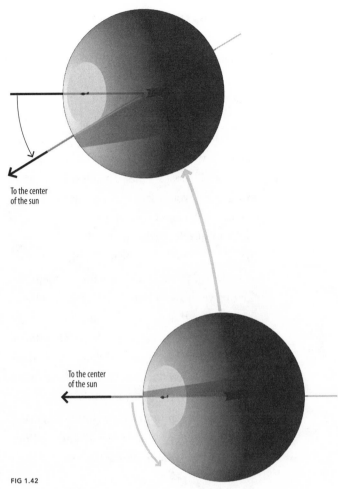

To the center
of the sun

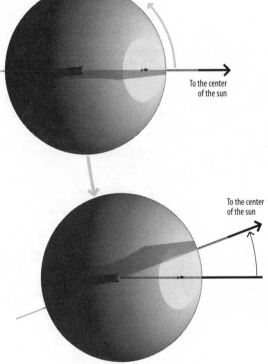

To the center
of the sun

To the center
of the sun

FIG 1.42

FIG 1.41

To the center
of the sun

The earth also speeds up and slows down—not as much, of course, but enough that it affects our clocks. Think of figure 1.41 as showing the earth farthest from the sun, in the "slow" part of its orbit. Figure 1.42 shows the earth nearest the sun, orbiting faster and thus moving farther along its orbit in the course of a day. You can see that in this part of the orbit, you'll need more minutes of rotation to get back under the noon plane.

Obviously, it would be unacceptable to have every succeeding day be shorter or longer than the one before it, so we average-out those variations. It's that "averaged" day that we divide up into twenty-four hours—and it's those "aver-

aged" hours that Greenwich uses when it publishes the times of future celestial events.

So after you set your clock to solar noon and use it to time the eclipse of Io, you have to convert the "local solar time" on your clock to the time scheme Greenwich uses. In surveying parlance you have to **do an equation of time**: you check the table in your handy ephemeris, and add or subtract the number of minutes and seconds specified for that day of the year. Once you've done that, you know—at last—both the time where you stand and the time at Greenwich.

Knowing that the start point of Mason and Dixon's line was at longitude 75° 30' west, the job of the bistate commission-

ers was to find the line's end at longitude 80° 30' west. This they did by choosing a place that, they calculated, was near the endpoint, and there constructing an "observatory" (their sighting instrument leveled on a stout post under a movable cover). After a series of observations, they determined their longitude, and thus the number of longitudinal minutes and seconds required to reach the 80° 30' endpoint. Converting those minutes and seconds into feet, they carefully measured out that distance along the line and set up their oak post in its pile of stones.

That's an example of how difficult it is to *find* longitude. Now watch as the bistate team easily *projects* a line of longitude.

Late in the spring of 1785, the Virginia-Pennsylvania surveying team took to the field, reconstituted with Pennsylvania's Rittenhouse and Virginia's Ellicott as members of smaller two-man teams for each state. On June 6 the axmen of the party began to clear a vista northward from the stout oak post set the previous fall. The vista through the woods would make the border visible to travelers and settlers, and allow the team to take full advantage of the "Transit Instrument" that Rittenhouse had brought with him. Rittenhouse's instrument very much resembled the one you saw in figure 1.31: it had a powerful telescope that could rotate upward to make sightings on the polar point, then swing downward to point directly north to guide the workers as they advanced the line. With the instrument the bistate team was able to traverse the 63 miles to the Ohio River in just over two months. The journal of Pennsylvania commissioner Andrew Porter gives an impression of their work:

- *17th [August]. Moved and fixed the Instrument on a high hill near the 60 ¹/₂ miles. Could find no stone to plant. Heavy rain about noon. Aaron Mille struck with a tree, his thigh broken, lay senseless for nearly an hour. Set his thigh and carried him to the Virginia Camp. Moved our Camp by a very circuitous course, towards the East and found a tolerably easy descent to the River. Encamped near the banks of the Ohio, at the mouth of Mill Creek. This place is called Raritons Bottom.*
- *18th. Heavy rain this morning and continued showry thro' the day—so as to prevent us going on the line.*
- *19th. Moved the Transit Instrument and fixed it on a very high hill about ³/₄ of a mile from our last post. Planted a stone marked P. & V. [Pennsylvania & Virginia] From this hill we have a very*

extensive view of the Vista to the South end of the Ridge to the North on the other side the Ohio. This Ridge is said to divide the waters of Little & Big Beaver Creeks. Moved & fixed the Instrument a short distance forward.
- *20th. This morning continued the Vista over the hill on the South side of the river and set a stake on it by the signals, about two miles in front of the Instrument, brought the Instrument forward and fixed it on a high post, opened the Vista down to the River and set a stake on the flat, the North side of the river.*
- *21st. Sunday. A number of the men were paid off and returned to their respective homes.*
- *22nd. Rained the greater part of the day—cut down a few trees that stood too near in the Vista— drew up and signed the Report with the Virginia Commissioners.*[21]

The Virginia team could then retire from the field, because by 1785, Virginia had ceded most of its lands northwest of the Ohio River to the Continental Congress. Thus, beyond the Ohio, the north-south line became the border between Pennsylvania and the new national domain. Since the weather was still good, the Pennsylvania team chose to push the line farther north, with Virginia's Ellicott accompanying Porter and Rittenhouse. Porter gives us some indication of the power of the telescope in the Transit Instrument when he tells us that from one high point, "we have a view of the Vista for upwards of 7 miles back."

The crew carried the line 40 or 50 miles north of the river, then retired for the season. The following August, Porter returned to the field with Alexander McClean, and together they and their crew advanced the line to Lake Erie by the end of September 1786.[22]

THE CLAIMS OF GEORGIA AND THE CAROLINAS FOR LANDS IN THE WEST

In 1629 Charles I made a grant to his friend Sir Robert Heath of all the land from Albemarle Sound, the great shoal-enclosed bay that forms the east coast of what is now North Carolina, south to the St. Mary's River, the northern border of present-day Florida (see figure 1.43). Heath, however, mounted no settlement, and so the grant was voided in 1663 and a new grant made (by Charles II) to a company headed by the Earl of Clarendon called The Lords Proprietors of Carolina. The grant's southern border was the St. Mary's River as before, but its northern limit was now defined more precisely as the 36th parallel. Two years later the grant was redefined: its southern border was moved south to the 29th parallel, its northern boundary adjusted to a parallel at 36° 30' north latitude. As with the previous Carolina grants, this one extended all the way to the Pacific Ocean.[23]

(The latitude 36° 30' is important in American history, for not only did it become the basis for the border between Virginia and North Carolina, and later between Kentucky and Tennessee; it became the southern border of Missouri and the demarcation line of the Missouri Compromise of 1820—the limit north of which there would be no slave states admitted to the Union after Missouri. Later we'll see a reason why this particular line was the one chosen for the compromise.)

As residents of the two Carolinas will tell you to this day, there are great differences between the territory around Albemarle Sound and the Low Country centered on present-day Charleston. The two regions attracted different kinds of settlers, grew different crops, and eventually felt themselves distinct enough to ask for separate

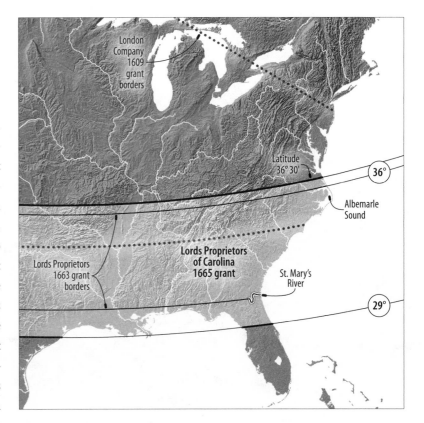

FIG 1.43

governments. In 1729 George II revoked the charter of the Lords Proprietors and established two separate colonies, each to have a governor appointed by the king.

The new charters, however, failed to specify a definitive border between the two colonies, and so a dispute immediately ensued. The controversy centered on which local tradition would determine where the "ancient boundary" between the colonies lay. In northern Carolina, tradition had declared the Santee River to be the dividing line, whereas to people around Charleston, the border had always been the Cape Fear River. You can see in figure 1.44, on the next page, how these two border claims overlap.

In 1730 the Board of Trade in London declared a compromise border that pleased no one, so in 1735 the two colonies took matters into their own hands and established their own boundary. It would start at the ocean at a point 30 miles

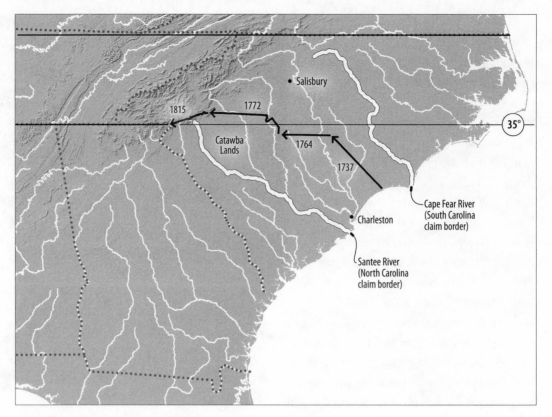

FIG 1.44

down the coast from the mouth of the Cape Fear River and then proceed northwest to the 35th parallel; from that point the border would extend westward to the Pacific. One exception was that if the line of the 35th parallel was found to cut through the lands of the Catawba Indians, the line would veer northward in an arc around those lands, then arc back to the 35th parallel and proceed on to the Pacific. All the land south of that line (including the Catawba lands) would go to South Carolina.

Between 1735 and 1737 a team of surveyors from both colonies ran the northwest line from the ocean to where they thought it touched the 35th parallel. (Their line actually stopped about 11 miles south of the parallel's true location, as shown in figure 1.44.) In 1764 a second team of surveyors began at the terminus of the northwest line and extended it west to an important road between Charleston on the coast and the inland town of Salisbury. The line was stopped there because just to the west, they knew, were the lands of the Catawbas which were to go to South Carolina.

How can boundaries be defined around the lands of people who don't think of land in those terms? In 1771 the Board of Trade issued complicated instructions that would carry the border up and around what it conceived to be the Catawba lands. The following year surveyors extended the previously marked line up to the northwest and then west. It was not until 1815 that the two Carolinas could agree on how to carry this line back down to their agreed-upon border at the 35th parallel, but they did, and the result is the portion of South Carolina that bulges northward into North Carolina.[24]

In 1732, three years after dividing the Carolinas, George II awarded a grant to The Trustees of Georgia, a syndicate headed by James

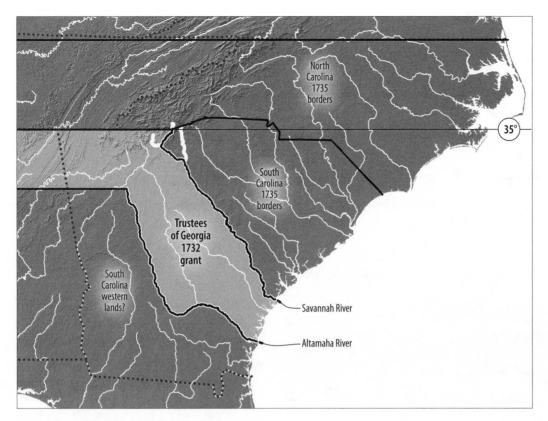

FIG 1.45

Oglethorpe, who hoped to found a refuge where paupers from England could lift themselves to prosperity in America. The borders of the grant were straightforward: all the land between the Savannah and Altamaha rivers, from the sea to the rivers' headwaters, and then from those headwaters west to the Pacific (fig. 1.45).

But nothing is truly straightforward when it comes to colonial boundaries. We've already seen instances of the principle that grants from the Crown could "bite into" grants made earlier. South Carolina's grant had stated that, west of the Catawba lands, it had a claim to all the land south of the 35th parallel; but the new Georgia colony would extend north as far as the source of the Savannah River. So if the source of the Savannah was south of the 35th parallel, South Carolina could lay claim to a strip of land (all the way to the Pacific) with a "height" equal to the distance from the river's source north to the 35th

parallel. But if the Savannah's source was north of the 35th parallel, Georgia's band of land would entirely overlap South Carolina's claims and thus supersede them. The controversy came to a head in the 1780s when settlement began to push into the region.

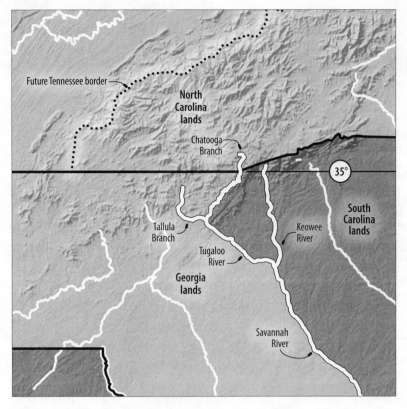

FIG 1.46

course of the river—the branch that also happened to throw the bulk of the new settlements into its territory.

Both states submitted their cases to the Continental Congress in 1785, but while the congress dithered, the two states sent delegates to the town of Beaufort in South Carolina, and there they reached their own compromise. The border would be a line of latitude drawn from whichever of the two branches of the Tugaloo (the western branch of the Savannah) reached farther north. By hard bargaining, South Carolina got the new settlements, and still held out hope that it would also get that narrow strip of land when the Tugaloo's branches were traced.

Before the question was settled, though, South Carolina surrendered its western claims to the United States—but that only meant that the controversy now lay between Georgia and North Carolina. Finally, in 1807 the states had surveyors locate the 35th parallel and mark the line on the ground. To Georgia's surprise the North Carolinians had been right: the line was indeed south of the source of the Chatooga Branch—in fact, some 18 miles farther south than Georgia had assumed.

Georgia's legislature refused to accept the results of the survey and so, in 1811 and 1812, they sent their own independent expert into the field. Their consultant, Andrew Ellicott (whom we met during the Pennsylvania border survey), found, however, that the 1807 line was correct, and Georgia consequently surrendered. But at least the location of the 35th parallel had been established—a fact important for us here, because as we will see in the next chapter, that parallel, west of Georgia, would mark the northern extent of the national domain in the southern states.[25]

Bear with me for a bit of regional geography, depicted in figure 1.46. By local nomenclature, the Savannah River was formed by the confluence of two rivers, the Keowee on the east and the Tugaloo on the west. The Tugaloo, in turn, was considered formed by the confluence of two branches, the eastern (and longest) branch today called the Chatooga.

South Carolina quite naturally asserted that the Savannah River "began" at the confluence of the Keowee and the Tugaloo (rather as the Ohio River "begins" with the confluence of the Monongahela and the Allegheny at Pittsburgh). Georgia took the more conventional position that a river begins not at the point where we start calling it by that particular name, but at the point where water comes out of the ground and starts flowing. But which of the many springs that converge to form a river is the true source of that river? Georgia understandably chose the easternmost branch of the Savannah, the Keowee, as the main

TWO

The Idea of a National Domain Emerges

By about 1760 all thirteen colonies had roughly established their present borders along the seacoast and up to the frontier of white settlement. After a long dispute, the Crown had awarded Maine to Massachusetts in 1691 and had set Maine's border with New Hampshire, although New Hampshire was still arguing with New York over the territory that is now Vermont. Pennsylvania had settled its boundary with Maryland, if not yet with Virginia. Virginia had acceded to the Carolinas' boundary at 36° 30', and the Carolinas had divided themselves, at least along the ocean, with a line of their own devising. And Georgia had a secure hold on the coastland, even if its western claim was still in dispute. But the ongoing disagreements about the colonies' western borders were to be interrupted by repercussions from the Treaty of Paris, which ended the French and Indian War.

THE FRENCH AND INDIAN WAR BRINGS NEW BOUNDARIES

That war was, of course, only the "North American theater of operations" of the much larger Seven Years' War—which was itself only one in a string of wars between Britain and France, beginning in the 1680s and ending only in 1815 at Waterloo. In each of the wars there were shifts in the countries allied with either of the two main combatants, but in the Seven Years' War, Spain was an ally of France. Quebec fell to the British in 1759 and Montreal surrendered the following

year, but before the final peace in 1763, France was able to draw Spain into the war on its side, making Spain a party (on the losing side) to the peace negotiations in Paris.

By the time of the negotiations, the British government that had pursued the war had been ousted, and the new government was less intent on crushing Britain's enemies than it was on erecting a stable balance of powers. Thus, France was forced to surrender only Canada (with the exception of a few islands off the Nova Scotia coast, as fishing resupply stations); it would keep the vast Louisiana territory that it had strung with trading posts. And although Spain had been defeated in Cuba and the Philippines, it would keep both—but it would surrender its possessions in Florida to Britain. (What Britain would not learn until later was that in November of 1762, France had agreed to cede its Louisiana territory to Spain.)

With its hold on the eastern half of the continent secure, Britain proceeded to rationalize the boundaries of its possessions with a royal proclamation on October 7, 1763. The western and southern boundaries of Britain's territories had been spelled out definitively in the Treaty of Paris. The borderline would begin at the source of the Mississippi and run down the center of that river to "the river Iberville" and then down the center of that river through Lakes Maurepas and Ponchartrain to the Gulf of Mexico. (You can trace this line in figure 2.01.) In the eighteenth century the River Iberville was a bayou—a water channel through a marsh—important according

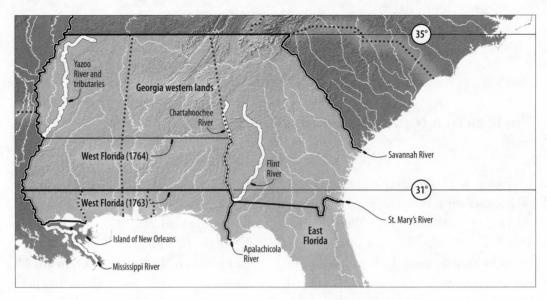

FIG 2.01

to the perceptions of people then, because at that time they couldn't get from New Orleans to the "mainland" without crossing either the Mississippi to the west or the channel of the Iberville and the lakes to the north. The eastern bank of the Mississippi south of this water passage was thus called "the Island of New Orleans," and the treaty language made it plain that the island was to be outside British territory.

East of the Island of New Orleans lay Britain's new possession of Florida—actually, it was the two territories Spain had called East and West Florida, and the royal proclamation defined their borders with more precision than Spain had done. West Florida would be bounded on the west by the Iberville-and-lakes watercourse, and then by the Mississippi River, up to the 31st parallel. Its northern border would run east on that parallel to the Apalachicola River, then down that river to the Gulf. East Florida would extend east from the Apalachicola, but only up to the point where the Chattahoochie and Flint rivers join to form the Apalachicola, a point some miles south of the 31st parallel. From that confluence, the border would be a line drawn straight to the source of the St. Mary's River, and then along the

S-shaped course of that river to the Atlantic (that line is the present Georgia-Florida border).

The same Proclamation of 1763 also enlarged the boundaries of the Georgia colony immediately to the north. Its northern border would remain the parallel touched by the source of the Savannah River, but its southern boundary would now be the northern borders of East and West Florida, from the ocean "westward as far as our territories extend"—that is, to the Mississippi.[1]

The proclamation thus definitively set the southern boundary of the thirteen colonies—for a time. In 1764 the governor of West Florida moved his border northward, from the 31st parallel up to the parallel where the Yazoo River flows into the Mississippi (just above Vicksburg), and along that parallel to the Chattahoochee. Since the Georgia colony's southern border had been defined as that of East and West Florida, this new border bit into the area of Georgia's claim of land to the Mississippi. The change was only "on paper"—Georgia didn't acknowledge it—but it would figure in the border negotiations following the Revolutionary War.

Farther to the north, the proclamation set the boundaries between New York and New England,

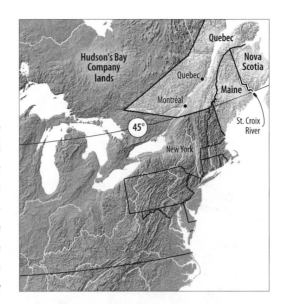

FIG 2.02

on one side, and Quebec and Nova Scotia on the other. Quebec's border with the colonies would be the 45th parallel, east from the St. Lawrence River to the watershed line that divides the rivers flowing into the St. Lawrence from those flowing into the Atlantic (the curving line in figure 2.02). The border would follow that line to the inlet that bounds the Gaspé Peninsula on the south. Moreover, the proclamation notoriously set up the new colony of Nova Scotia to encompass the colony of Acadia—whose French settlers would be exiled to Louisiana. Nova Scotia's border was to be the St. Croix River, from the ocean up to its source, and from that source point due north to the St. Lawrence watershed line.[2] (Confusion about these two borders—the watershed line and the St. Croix line—would cause Britain and the United States to dispute the border of Maine until 1842.)

From the point of view of the thirteen colonies, all these newly precise borders were either of little consequence or confirmed boundaries already presumed (Nova Scotia and Quebec where they touched New England), or else they were enlargements of previous borders (Georgia)—new conditions that were, by and large, a good deal for the colonies. But the boundary that shocked the colonists—and which they never fully accepted— was what came to be called the Proclamation Line of 1763.

With the exceptions of Georgia and the territory of the Hudson's Bay Company, all the land under British jurisdiction "lying westward of the sources of the rivers which fall into the sea" was declared to be an Indian reserve. This meant that all the land east of the Mississippi, north of the 35th parallel, and west of the Atlantic watershed line in the eastern mountain chains was off-limits to white settlement. By the 1760s settlers from the colonies had already pushed over the crest of the mountains, and in the case of Kentucky, deep into the land beyond. The Proclamation Line thus not only abrogated the "sea-to-sea" land claims of the colonial governments, it challenged the colonists' historical patterns of settlement and land possession.

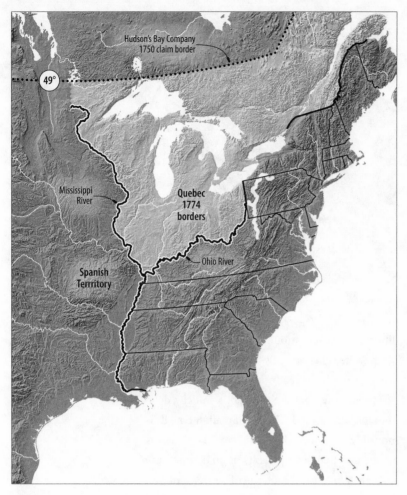

FIG 2.03

Britain only compounded the affront with the Quebec Act of 1774. Under that act Quebec was enlarged to the north and east, but its borders were also extended southward, in the manner shown in figure 2.03. Quebec's southern boundary would now follow the western border of Pennsylvania from Lake Erie to the Ohio River, then down the Ohio to the Mississippi, then up the Mississippi to its source, and then east along the border of the Hudson's Bay Company lands.[3] (In 1670 the company had been granted the entire watershed of Hudson Bay. By the mid-1700s, the borders of the watershed had not yet been found, much less mapped, and so no one knew where the boundary between the company's lands and this newly expanded Quebec might lie. But in 1750 the company had nominated a boundary that would, it believed, enclose most of the watershed—a line along the 49th parallel, from the Rocky Mountains to the border of Quebec.)

Under the Proclamation Line of 1763, Britain had put the lands where settlement was heading under Indian jurisdiction—a condition that might be skirted in practice and perhaps even modified in negotiations with the tribes. But the Quebec Act of 1774 put those lands under the jurisdiction of a sovereign colonial government that was entirely separate. This act was just one of "the long train of abuses and usurpations" that the colonies would cite in 1776—tyrannies that gave them the self-evident right and duty "to throw off such government" and become "free and independent states."

Just as I did with the French and Indian War, I'll skip over the conduct of the Revolutionary War almost entirely and proceed right to the treaty negotiations and their effect on national borders. The one circumstance from the war that you need to know, however, is that near the war's end, Spain allied itself with France on the American side (Spain's primary goal being the return of Gibraltar from the British). Thus, just as had been the case with the French and Indian War, in the negotiations to end the American Revolution, Spain had an interest—only this time while on the winning side.

The story of the treaty negotiations is notoriously convoluted, but it could be boiled down to the following. In 1779 the Continental Congress, in a fit of optimism, sent John Adams to Paris in case the possibility of peace negotiations with Britain arose. His instructions were to negotiate borders only after Britain accepted that the colonies were sovereign states, independent of outside rule. That conceded, Adams was to try for borders that were essentially the Mississippi on the west, the borders of New England as set forth in the Proclamation of 1763, and on the south, the 1763 borders with East and West Florida (not the border of West Florida as pushed north in 1764). The reason for insisting on such a southern border was clear. For the American territory west of the Appalachians to be viable economically, it would need an outlet for transporting its products to the sea, and that outlet was the Mississippi River. With all the borders successfully negotiated by Adams, the United States would control the whole eastern bank of the Mississippi from its source southward—with the sole exception of a narrow strip between the 31st parallel

and the Gulf of Mexico. Surely, it was felt, navigation rights through such an insignificant distance could be worked out.

By 1781 (in June, before the battle of Yorktown) the Continental Congress had lost some of its confidence in victory and began to place its primary hope in the power France might assert, on the field of battle and at the negotiating table. Under pressure from France, Congress expanded its negotiating team, ostensibly to dilute the influence of the superpatriot Adams, and instructed these mediators to make no move in whatever transactions might ensue without the clear assent of the French.

With the American victory at Yorktown in October of 1781, treaty negotiations could proceed in earnest. But the American team got wind of intelligence that France might not join the Americans in insisting on recognition of the colonies' independence prior to negotiations, and that it might accede to a western border for the new nation that didn't reach the Mississippi. Alerted to France's true position in this nest of European intrigue, Adams and his fellow mediators went behind the backs of their French allies and, in defiance of their instructions from Congress, began to negotiate a separate peace with Britain.

Britain saw in these defiant American negotiators a chance to advance its own interests: by acceding to the American demands, it could frustrate the aims France and Spain had hoped to achieve from the negotiations, and thus perhaps undermine their alliance with America. And so Britain agreed both to full American independence and to the 1763 borders. When the Americans presented this draft treaty to their allies in Paris, the French recognized a *fait accompli*

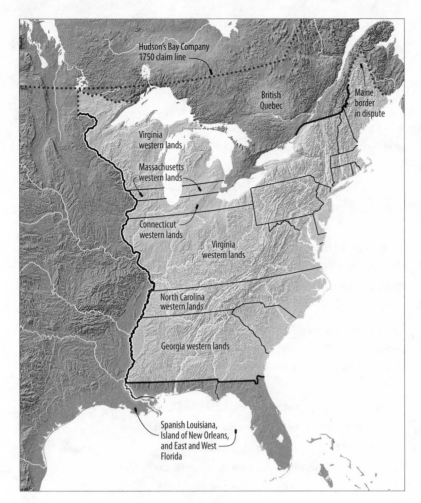

Hudson's Bay Company
1750 claim line

British
Quebec

Maine
border
in dispute

Virginia
western lands

Massachusetts
western lands

Connecticut
western lands

Virginia
western lands

North Carolina
western lands

Georgia western lands

Spanish Louisiana,
Island of New Orleans,
and East and West
Florida

FIG 2.04

when they saw it (not to mention an American phrase of a later century, the "end run"). France leaned on its ally Spain to give up its demand for Gibraltar and settle for the island of Minorca in the Mediterranean, and East and West Florida in America. Spain too saw which way the wind was blowing, and the final treaty was signed in Paris on September 3, 1783.[4]

The thirteen states had gained their independence, as well as sovereignty over a territory— that shown in figure 2.04—they thought was precisely defined. They were right about the sovereignty, but the extent of their territory would be a matter of dispute for a decade.

I've jumped ahead to 1783 to show how the borders of the new nation sketched following the French and Indian War were confirmed by the Revolutionary War. But at the same moment that the Treaty of Paris was being signed—before word had reached America—the Continental Congress established the kernel of what would become a domain of land held by the new nation. The idea of a territory held in common by a national government evolved in the debate over the Articles of Confederation, a debate that began in 1777 and was not completed until 1781. During those four years the idea of a common domain moved from rejection to acceptance and finally, in 1784, to realization.

The debate over the domain concerned whether the states having claims to land beyond the eastern mountains would surrender them to the national government. All of these land claims would have been on the minds of the delegates to the Continental Congress as they convened in Philadelphia in the summer of 1776. Congress did two very important things in July of 1776. One you know about. The other was to stage the first debate on the Articles of Confederation.

On June 12 Congress had appointed a committee of thirteen members—one from each state—to prepare a draft set of articles. Congress had been meeting since October of 1774, making up their rules of governance as they went along; but they knew that if they were to negotiate an alliance with a European power to help them in their struggle with Britain, there would have to be a real government for that power to negotiate with. And should the states eventually emerge victorious, they would certainly want to negotiate peace with Britain on a full nation-to-nation

basis. And so there had to be, at the very least, a government apart from the individual states to which each state had assigned its power to make binding agreements.

That was the absolute minimum government required by the wartime situation, but virtually all the congressional delegates felt that nationhood must consist of more than the mere grant of a power of attorney. The question was, how much more? That was the matter facing the members of the committee drafting the Articles of Confederation.

Even a committee producing only a final draft must begin from an outline, and the committee's first draft was supplied by its driving personality, John Dickinson, a delegate from Pennsylvania. He had thought long about the question of confederation and hoped for a central government with powers more expansive than most delegates were willing to cede. Many of his proposals were rejected by the committee, but three survived into the draft laid before the whole Congress.

In the draft articles, the confederation would have the power to settle disputes between states over shared borders. It would have the power of "limiting the Bounds" of states' claims to western land. And it would have virtually complete power over that western territory, deciding how to sell it off and how to create new states within it. Dickinson included these provisions about land in the hope of resolving some of the disputes that had plagued the colonies from the beginnings of their common history.[5] Present on the delegates' minds in 1776 would have been two particularly bitter disputes that were going on right then.

In 1764 the King's Council had declared that New Hampshire's political jurisdiction extended

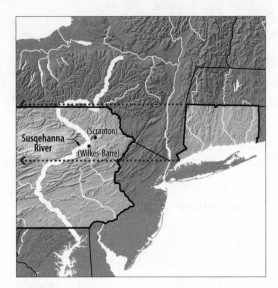

Susqehanna
River
(Scranton)
(Wilkes-Barre)

FIG 2.05

only to the Connecticut River (its present western border). But for decades prior to that ruling, New Hampshire had been laying out townships west of the Connecticut and selling off parcels of land; in 1776 there were still parcels to be sold. Under the council's ruling, New York would have had political control of the land between the Hudson and Connecticut rivers, but would it thereby own all the land New Hampshire had parceled out there? By 1776 the people in the region were organizing to take advantage of the confused situation and declare themselves a new state, separate from both claimants. Their petition for a State of Vermont would arrive at Congress the very next year.[6]

Farther south, the Connecticut colony possessed the happy geographic circumstance of being an east-west rectangle crossed by a comb of south-flowing rivers. Rivers being the primary conduit of settlement in colonial times, Connecticut had been the first colony (apart from tiny Rhode Island) to almost completely populate its allotted territory. With their own land "filled up," its settlers were particularly aggressive in claiming new territory, and nowhere more so than in Pennsylvania's Susquehanna River valley. By the 1770s, Connecticut farmers had moved into the

area. In defiance of the plain intent of the 1681 grant to William Penn, they used the pretext of Connecticut's "sea-to-sea" grant (the dotted lines in figure 2.05) to declare themselves a county of Connecticut.[7]

The drafting committee for the Articles of Confederation knew that conflicts such as these would continue, and probably become worse in the western lands where so many states' claims overlapped. But there was an even more immediate issue before them that involved these western lands.

In addition to Congress itself, all the states were incurring debts to pay for the ongoing war effort, debts that would have to be repaid once peace was achieved. It was obvious that states possessing western lands would be able to use those lands to repay their debts, either by selling land for cash or by granting land in payment for the debt. "Landless" states—specifically Maryland, Delaware, New Jersey, New Hampshire, and Rhode Island—would have no such recourse. (New York and Pennsylvania were a special case: they had big tracts of available land within their recognized borders. Massachusetts did too, in its jurisdiction over Maine; but in the 1770s no one could imagine settling in the rugged Maine woods back of the coast, and so Massachusetts considered itself similarly "landless.") The hope of the landless states was that the debts of all the states would be pooled and then repaid by the national government, in part with proceeds from sales of land in the West. That could happen only if the western lands were held by the confederation, not by the individual states.

The three provisions in the draft articles about land and borders were the committee's attempt to address these problems. But by the time the draft articles were presented to Congress on July 12, 1776, their chief advocate, Dickinson, had surrendered his congressional seat, exhausted by the debate. (He wrote to a friend about that seat: "No youthful Lover ever stript off his Cloathes to step into Bed with his blooming beautiful bride with more delight than I have cast off my Popularity."[8]—that is, his popular election as delegate. Manifestly, the Puritans did not rule in Pennsylvania.)

The delegates debated all the proposals for confederation through August, but could arrive at no consensus (in test votes they had tentatively rejected all the provisions giving the Confederation control of the western lands). At an impasse, they decided to try again the following year. The prospects for a national domain were looking bleak in 1776.

In the year between the first debate on confederation and its resumption in October of 1777, the members of Congress had consulted their constituencies (and each other) enough that they were able to resolve the more contentious issues in a mere seven days of debate. On the crucial question of whether the confederation would decide states' western borders and dispose of the western lands, only the Maryland delegation and a single delegate from New Jersey voted in favor.

With the outstanding issues settled, Congress went on to tighten up the language of the Articles of Confederation, and send the document to a printer so that copies could be carried to all the state legislatures. Once instructed by their legislatures, the delegates would reconvene the following spring for a final vote on ratification.

In December Maryland's legislature debated the articles and charged its delegates in no uncertain terms: refuse to vote for any confederation that left the western lands in the hands of the individual states.[9] And in June, in Philadelphia, the Maryland delegation stood firm, refusing to ratify the articles unless they authorized a national domain in the west. This refusal mattered, because the articles themselves stated that they would become effective only when all thirteen states ratified them. Despite Maryland's rejection, Congress approved a final form of the Articles of Confederation and sent them back to the states for final ratification.

The legislatures of New Jersey and Delaware objected to the absence of a shared western domain but considered confederation to be an overriding concern, and so they adopted the articles with objections. Maryland, however, dug in its heels and again refused to ratify.[10] And thus the articles stood, incompletely adopted for three more years, while Congress continued to operate, with Maryland in attendance. Maryland did eventually sign, and a national domain was eventually established. What follows are some of the events that changed the situation.

One continuing trend that must be kept in mind is that throughout the period between 1777 and 1780, the value of the Continental dollar kept dropping, culminating in Congress's forced devaluation in March of 1780. Many in Congress came gradually to realize that a currency backed up by only a promise to pay could never endure, but if such a currency were backed by a vast fund of sellable land—that would be a dollar people could trust.[11]

"Not worth a Continental"

Do be aware that the "dollars" issued by the Continental Congress were not currency, such as dollars are today. Continental dollars were more nearly promissory notes. If you sold your wheat harvest to an agent of the Continental army, he would pay you in "Continentals," each of which could be redeemed for a set quantity of specie—that is, gold or silver, usually in the form of coins—but only at the conclusion of the war. With these notes in hand, you the wheat farmer could trade them with, say, the blacksmith for new horseshoes for your plowing team. It would naturally be a matter of negotiation between you and the blacksmith over how much your Continentals were worth to him, since no one could know the rate at which the government would actually redeem the notes.

So this redemption rate was an issue of constant concern among the inhabitants of the thirteen colonies. When Congress lowered the rate, everyone who had conducted transactions in Continental dollars worried that they might get only a pittance for their notes at war's end—giving rise to the derisive phrase "not worth a Continental." It was to counter the import of that phrase that Alexander Hamilton later launched the new federal government on a policy of sound credit—from which it has never veered.

In addition to doubts about its notes, Congress was concerned with the mess that had been made of land claims in the west. It wasn't just that the claims of the states overlapped. The states had made grants, some of them huge, to land brokers, and those tracts too sometimes overlapped. To add to the confusion, individual speculators had "bought" land from Indian tribes: apart from the question of who was fooling whom in these transactions, there was the question of which tribes had jurisdiction in the various regions. And, of course, purchases from the Indians had no sanction from any state—if indeed it could be determined which state, or states, the Indian lands fell within.

And there were yet more problems: individual settlers, with no plans for extensive land empires, had begun to "squat" on claims in the West without sanction from any state, Indian tribe, or land company. What status would their modest (and well-meant) claims have in the conflicted world of Indian, state, and speculator grants?

The situation became so bad that in 1779 the Virginia legislature declared that all claims made in its territory prior to the Revolution's start were now void, and that people wanting land would hereafter have to buy it through a state land office. Speculators with claims in Virginia's territory immediately descended upon Congress, protesting Virginia's action. Responding to the lobbying effort, in October of 1779 Congress adopted a resolution (over the objections of Virginia and North Carolina) asking the states to revoke all land sales until the conclusion of the war, and indeed to suspend all land sales made since the Declaration of Independence. The resolution was only a holding action for the advocates of a national domain, but it bought time for other events to unfold.

One development occurred in New York, which was still trying to hold on to the territory between the Hudson and Connecticut rivers where settlers had declared the new state of Vermont. The petition of the Vermonters had been on the table in Congress since 1777, and was yet to be acted on. Late in 1779 the New York legislature had the idea that it might win favor in Congress for its position if it magnanimously surrendered its Indian-treaty claims to land in the west. In early 1780 the legislature adopted such a cession and sent the news to Congress. Congress accepted the cession in October of 1782 but never acted on the Vermont question.[12]

By the spring of 1780, changes were afoot in Virginia. Thomas Jefferson was now its governor, and James Madison was in its delegation to the Continental Congress. Both men believed that achieving confederation was so important that if the states' land claims stood in the way, the states should surrender those claims. With the Virginia delegation so instructed, Congress passed a resolution in September 1780 asking that the states possessing western claims do just that. The way was now open for the creation of a national domain.

Keeping up the momentum, Virginia's legislature passed a resolution on January 2, 1781, ceding to the Continental government all of its claims north of the Ohio River. Naturally, several strings were attached to the package.

Two provisos had to do with Revolutionary War obligations. During 1778 and 1779 the Virginian George Rogers Clark had won a series of victories in what are now Indiana and Illinois. In the

The Erie Triangle

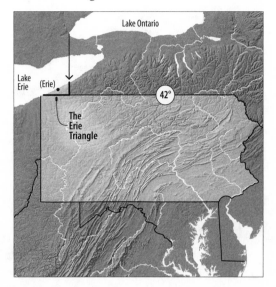

FIG 2.06

No one knew it at the time, but the terms of New York's cession resulted in the "Erie Triangle," that wedge of land at the northwest corner of Pennsylvania (shown in figure 2.06) that gives the state an outlet to Lake Erie. In 1774 surveyors from Pennsylvania and New York had established where their 42nd-parallel border touched the Delaware River, but the line was not run west until 1786 and 1787. Pennsylvania's western border was also not run north to Lake Erie until 1787.[13] Only then, when the two lines had been traced on the ground, was it discovered that these lines met not on land but out in the lake—a convergence which would have given Pennsylvania a laughably tiny shoreline, and no access to the natural harbor formed by Presque Isle (site of the present city of Erie), just north of the 42nd parallel.

Fortunately for Pennsylvania, New York's act of cession had declared its western border (beyond which it surrendered all claims) to be "a meridian line drawn from the most westerly bent of Lake Ontario."[14] That line, it turns out, passes east of the harbor at Presque Isle. But since New York had ceded all land west of that line to Congress, and since by its charter Pennsylvania's territory stopped at the 42nd parallel, the right triangle formed by the Lake Ontario meridian and the 42nd parallel, with the lakeshore as hypotenuse, belonged to neither state but instead to the national government. The federal Congress officially rectified the anomaly in 1792 when it sold the Erie Triangle to Pennsylvania for approximately $150,000.[15]

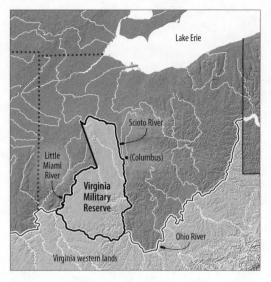

FIG 2.07

course of the campaign, he had liberated from the British several villages populated by French settlers from the time when France had held sway in the region. In gratitude, the settlers had professed allegiance as citizens of Virginia, and Virginia insisted that its land claims be honored. The cession also stipulated that a tract of 150,000 acres at the Falls of the Ohio (across from present-day Louisville) be reserved for Clark and his soldiers as a reward for their service.[16]

Like most states possessing unsettled land, Virginia had paid its soldiers, in part, with promises of free land at war's end. The new state government had set aside a big tract of land for this purpose in what is now western Kentucky, but so many settlers had come into that region, and so many more would doubtless come in the near future, it was feared that once the squatters' claims had been settled, there wouldn't be sufficient land left to honor the bounties granted to its soldiers. If that proved to be the case, Virginia would reserve from its cession a tract of land north of the Ohio between the Little Miami and Scioto Rivers, and use that tract to settle the remainder of the soldiers' claims.[17] ("Scioto," in case you're wondering, is pronounced "Skee-OTT-uh.") That tract, shown in figure 2.07, would become known as the Virginia Military Reserve.

Although Virginia itself had rescinded its grants of large "empires" to land companies, a number of settlers, and groups of settlers, had been granted title to smaller tracts of land under state and even colonial law. Virginia quite naturally asked that those titles be honored when its ceded territory changed hands.

The cession also mandated that any land Virginia granted be used for the common good, and that when it came time to form new states out of the cession, those states would be not less than 100 or more than 150 miles square.[18] This last provision may seem to have come out of nowhere, but it was actually a verbatim repetition of the resolution Congress itself had passed the previous October when it asked the states to surrender their western claims.[19] (We'll return to this provision about the size of states in chapter 4.) Finally, in order to clear the ground of conflicting grants, Virginia's cession demanded that all private purchases of land from Indian tribes be invalidated.[20]

When the Virginia delegation presented its cession to Congress, however, it was rejected. In the debate, the provisions reserving land for soldiers and for honoring the claims of the French villagers proved relatively uncontroversial. The provision for apportioning the territory into states had used Congress's own language. And nobody was going to object to using the ceded land for the common good. What sank the cession were the provisions for honoring Virginia settlers' land claims and—most especially—for invalidating purchases from the Indians.[21] (It seems that several delegates to Congress had invested in Indian lands—including some members and friends of the Maryland delegation and, of all people, Ben Franklin.) These two issues kept Virginia's cession from being accepted until 1783.

But even as Congress was mulling over Virginia's cession early in 1781, the British navy was maneuvering in Chesapeake Bay, threatening Baltimore and Annapolis. Officials from Maryland went to the French ambassador to ask him if France could send help. The ambassador, Chevalier de la Lucerne, diplomatically replied (in tones we can easily imagine) that although France had been helping the colonists, and had even signed a treaty with Congress in 1778, it had been dealing with a body that had not, as yet, been able to constitute itself into a sovereign nation.

The Marylanders got the message, and on February 2, 1781, their delegates in Congress ratified the Articles of Confederation without modification.[22] Ten days later Congress set the official date for the articles to go into effect: March 1, 1781. At last a properly constituted, sovereign nation was brought into being. That nation didn't as yet possess a commonly held national domain, but many could see that it was only a matter of time.

By the fall of 1783, Congress was ignominiously convened in Annapolis, having been forced from Philadelphia and then to Princeton the previous June.[23] In September, Virginia's delegation agreed to a compromise on western land cessions. The state would drop its insistence that its settlers' land titles be honored and relent in its demand that private purchases from Indians be invalidated. (There was some hope that the cession's clause mandating "common good" might give grounds for arguing both points later.)[24]

When the compromise was put to a vote, only New Jersey and Maryland (hoping for still more concessions) refused to endorse it. Under the Articles of Confederation, ratification of the articles had required thirteen votes, but treaties needed only nine votes, and most laws (including land cessions) needed only seven. So the compromise was adopted and sent back to the Virginia legislature. In December the legislature accepted the modified terms and made a conforming offer of cession, which Congress accepted on March 1, 1784.[25]

The new nation thus had the beginnings of its national domain. Virginia had held on to all its land south of the Ohio River: it was plain by then to Virginians that a new state would be created out of their lands west of the Cumberland Gap. None of the states south of Virginia were yet ready to cede land to the nation, and both Massachusetts and Connecticut were, for the moment, holding fast to their western claims. But with New York's surrender of its Indian treaty lands, Virginia's cession meant that the nation as a whole had clear possession of almost all the great triangle of land between the Mississippi and Ohio rivers south of Connecticut's 41st-parallel claim. And although all of this land was still legally in the hands of Indian tribes, Congress presumed that treaties would be negotiated and the land gradually opened not just for squatters already hiding out on unauthorized claims, but for the broad movement of settlement onto land purchased from the national government. And so on the very next day, March 2, 1784, Congress appointed a committee "to devise and report the most eligible means of disposing of such part of the Western lands as may be obtained of the Indians."[26]

Now bear with me as I walk you through a complex chronology. In October of 1783, after agreeing to Virginia's cession, Congress had presumed the Virginia legislature's acceptance, and so appointed a committee to draw up a plan for the governance of the new domain. In February of 1784, the committee was reorganized, with Thomas Jefferson as its chairman. On the very day that Congress accepted Virginia's cession, March 1, Jefferson's committee laid its plan of governance before Congress. It was on the following day that Congress appointed the committee on land disposition, this second committee also chaired by Jefferson. The report of the first committee—on governance—was debated in Congress and approved on April 23, and it has become known as the Ordinance of 1784. The report of the second committee—on land disposition—was submitted to Congress on April 30, debated, and rejected on May 28. Sent back to a new committee, a reworked plan for land disposal was approved on March 4, 1785—a plan that has come to be known as the Ordinance of 1785.

With Virginia's cession, the nation had a shared pool of western land, which would come to be called, formally, the Public Domain. With the Ordinance of 1784, the nation had a plan for managing how the governance of that domain would pass from the national government into the hands of sovereign states to be founded there. With the Ordinance of 1785, the nation had a plan for conveying the land of the Domain out of the possession of the national government and into the hands of individual settlers.

As part of its effort to "aid" the colonies in their revolution, toward the end of the war Spain had moved troops into the British possession of West Florida, and those troops were still in place at the signing of the Treaty of Paris. Spain accepted the treaty's boundaries for its East Florida province, regained from Britain, but disputed the northern boundary of West Florida.[27] (In those days, before anyone had conceived the idea of lying in the sun at the seashore, the East Florida peninsula seemed a worthless swamp, whereas West Florida had a rack of south-flowing rivers giving easy access to a fertile hinterland.) Spain asserted a series of claims, some extending as far north as present-day Tennessee, and more than a decade of negotiations would ensue before Spain agreed to the boundary accepted by the British in 1783: the 31st parallel.

On what basis did the Spaniards make their claim? The most brazen basis was the simple fact of occupation (a tactic Americans themselves would use in Florida thirty years later). There was also a clause inserted into the 1783 treaty to the effect that, if Britain should ever regain West Florida, the northern boundary of that entity would be the line declared in 1764, the latitude where the Yazoo River flows into the Mississippi[28] (figure 2.01).

The issue for the Americans was not so much the location of the border but that border's implications for navigation down the Mississippi to the Gulf of Mexico. The more northerly the border of West Florida, the greater the length of the river where Spain controlled both banks—making it that much easier for that country to close the river to American shipping.

The situation became a crisis when in 1785 Spain did just that. Negotiations ensued, with Spain at one point offering to accept the 31st-parallel boundary if the United States would surrender navigation rights on the Mississippi for twenty-five years. The northern states in the Continental Congress were ready to accept the deal (after all, goods from the west would have to be exported through their ports), but southern delegates, foreseeing their own future interests, defeated the measure.[29]

With the disbanding of the Continental Congress in 1789, matters stood at an impasse until negotiators from the new federal government went to Madrid in 1792. Finally on October 27, 1795, Spain agreed not only to the 31st-parallel border, but guaranteed to American flatboatmen the right both to navigate the length of the Mississippi and to deposit their cargoes in New Orleans for transshipment in oceangoing vessels.[30] With the Treaty of 1795, the United States secured not only the borders won in the Revolution but the economic viability of its western lands.

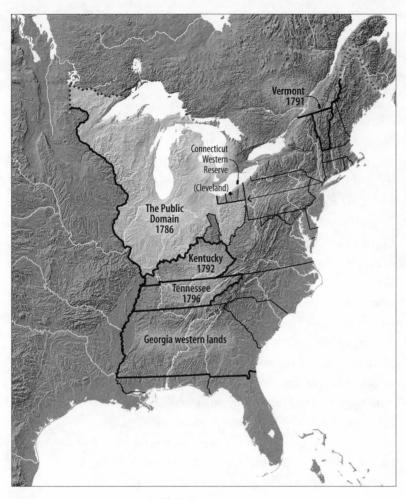

FIG 2.08

Once the idea of a national domain was accepted in 1784—and its manner of future governance set out in 1785—all the states having western land claims eventually fell into line with the new idea.

Massachusetts surrendered its claim in 1785. Connecticut followed suit in 1786, with one proviso: it held on to the 120 miles of its strip immediately west of Pennsylvania, a piece of land "reserved from cession" and thus called ever after the Western Reserve.[31] The state's reasons for keeping the reserve were twofold. The westernmost part was to be used to compensate people in Danbury and other towns whose property had been burned out by British raids during the war (thus the name for the district, the Fire Lands, shown in figure 2.08).[32] Just as important, Connecticut had sold the eastern portion of its reserve lands to a private company the previous year,[33] and that sale (and thus its income) would have become void if the land were to be ceded. Even though it didn't own any of the land, Connecticut maintained governmental jurisdiction over the Western Reserve until the new federal Congress accepted its cession of the tract in April of 1800.[34]

Virginia had held on to its land south of the Ohio and, as anticipated, gave its consent in 1789 to a new state being formed there. In 1791 Congress admitted the new state of Kentucky, with its admission to the Union becoming effective the following year.[35]

Tennessee was likewise formed out of North Carolina's western claims but in a less straightforward fashion. North Carolina ceded its lands west of the mountains to the Continental Congress in 1784, only to rescind the cession the following

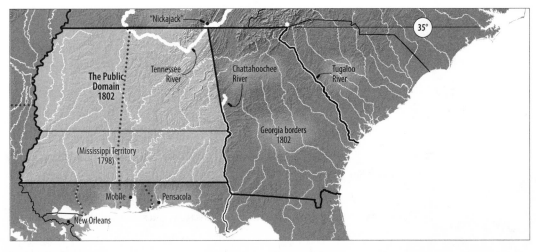

FIG 2.09

year, before Congress had a chance to accept it.[36] In 1790 North Carolina again ceded the territory, this time to the new federal government, which then set up a territorial administration there. The area was called The Territory of the United States South of the Ohio River, because North Carolina's cession was indeed all the land the United States then held south of the Ohio, since Georgia had not yet given up its western claims. Tennessee was admitted to the Union in 1796, the first state to have officially passed from the status of "Territory of the United States" to that of a sovereign state.[37] Thus was set the pattern followed by all subsequent states except California and Texas (exceptions from the ordinary, in this as in so much else). Tennessee, though, had the distinction (which it shares only with Texas) that the federal government did not retain possession of its unsold lands but vested ownership of those lands in the new state government. The lands of Tennessee, like those of Kentucky, were never part of the Public Domain.

The story of the growth of the Public Domain south of Tennessee is a little more complex. Remember that South Carolina claimed a narrow strip of land between the latitude of the source of the Tugaloo north to the 35th parallel—a piece of land we now know not to exist. In August of

1787, South Carolina ceded that strip to the Continental Congress.[38]

The 1795 treaty with Spain definitively set the border of West Florida with the United States at the 31st parallel. So in April of 1798 the federal Congress, following the model used in Tennessee, established the Mississippi Territory in the strip of West Florida (as enlarged in 1764) between the 31st parallel and the latitude of the Yazoo's joining with the Mississippi (see figure 2.09). Congress recognized that Georgia had a claim to this land and so provided for a commission to adjudicate the claims.[39]

Georgia, in the very same year of 1798, adopted a constitution naming state borders all the way to the Mississippi. The southern border was to be the line the United States had negotiated with Spain three years previously. Georgia's northern border, though, had to be made conditional. Remember that by 1798 neither Georgia nor South Carolina knew whether Georgia's northern border (the latitude of the source of the Tugaloo) touched South Carolina's northern border (the 35th parallel). So Georgia's constitution claimed only the land up to the latitude of the Tugaloo's source. Thus, if (as both states believed) the Tugaloo didn't touch the 35th parallel, there would be a thin strip of South Carolina land running west

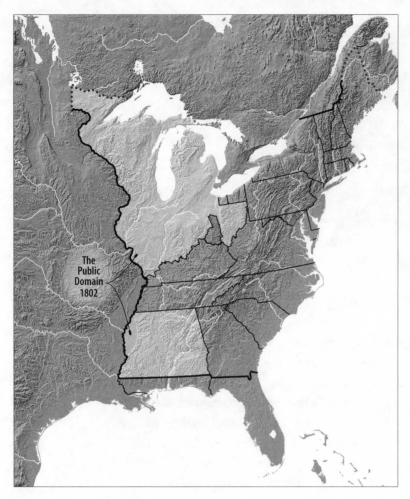

FIG 2.10

would be drawn, up to a little place on the Tennessee River called Nickajack. In return for surrendering its western claims, the United States ceded to Georgia the (nonexistent) strip of South Carolina land at its northern border so that Georgia's boundary would snug up against Tennessee and North Carolina.[41]

With Georgia's 1802 cession, the Public Domain (the pale area in figure 2.10) was finally coterminous with the boundaries of the nation. But there were still two places where the national boundary wasn't quite clear. One was in Maine—a dispute not to be resolved until 1842. The other was in the far northwest of the Ohio Country around the Lake of the Woods in present-day northern Minnesota. The negotiations over the Lake of the Woods were to be the first of many instances where the new nation would gradually expand its boundaries.

That outward push is a story known to every American student: all those Purchases, Cessions, and Treaties which resulted in the shapes on maps that seem so manifest today. Since the narrative is so well known, I'll tell it quickly in the next chapter, stopping only to correct a few commonly held misconceptions about how the borders of the United States—the borders of the shared Public Domain—grew to their present extent.

to the Mississippi along Georgia's northern border, up against North Carolina and the new state of Tennessee.[40]

But by South Carolina's 1787 act of cession, that strip now belonged to the United States. So since both Georgia and the United States had something the other wanted—Georgia, that strip of land; the United States, Georgia's western claims—there was a basis for negotiations to begin. The talks concluded in 1802, when Georgia agreed to its present western border. The boundary line would follow the Chattahoochee River, north from Spanish Florida to Miller's Bend (the place where the river's course makes an abrupt turn to the northeast—thus its westernmost point); from Miller's Bend a straight line

CHAPTER TWO

THREE

The National Domain Expands

1783–1802: THE LAKE OF THE WOODS

The Treaty of 1783 had definitively set the Great Lakes as the border between British America and the United States west of New York: the border would run down the middle of the rivers between lakes, and in the lakes themselves, a line would be drawn that would approximate equidistance from the opposing shores. (Clear enough, but it took until 1908 for the two parties to work out the fine details of which tiny islands fell to which country.)[1]

From Lake Superior to the west, the framers of the treaty presumed that a continuous watercourse ran to the Lake of the Woods, and so they set the next part of the boundary as the center of that stream. (In reality, in that region there is a chain of lakes that don't quite connect; but they come close enough that in 1842 a line could be negotiated, as part of the agreement about the border around Maine.)[2] It is in the Lake of the Woods itself that things get tricky. At the time of the treaty, the Mississippi River had been mapped only a few miles north of the Falls of St. Anthony (present-day Minneapolis and St. Paul), but the negotiators assumed that the river's course took it far north and west of the falls, veering west of the Lake of the Woods, with its source somewhere north of the lake—an imagined course of the river shown in figure 3.01. So the treaty called for the border to be a straight line drawn from the northwesternmost point of the lake straight west until it hit the Mississippi, and then down the Mississippi to the 31st parallel at West Florida.

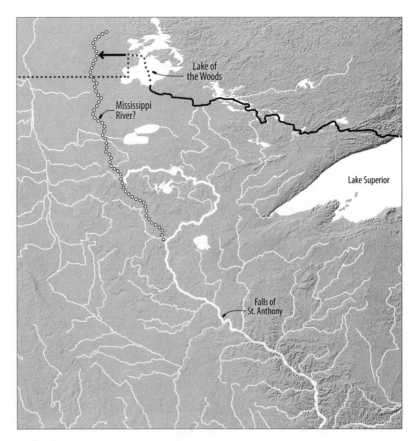

FIG 3.01

Problem is, the generally accepted source of the Mississippi is actually south of the Lake of the Woods. By 1794 rumors of this fact had percolated from the periphery, and so in a treaty signed that year in London, both sides acknowledged that "it is uncertain whether the river Mississippi extends so far northward as to be intersected by a line drawn due west from the Lake of the Woods."[3] They therefore agreed to negotiate further.

By 1802 Secretary of State James Madison could write to his negotiator in London, "It is now well understood that the highest source of the Mississippi is south of the Lake of the Woods." Both sides agreed that the sensible thing would be to draw a line from the northwest corner of the lake to wherever the source of the Mississippi might be. The U.S. Senate quibbled over provisions in the treaty until 1804—by which time the Louisiana Purchase had changed the situation utterly.[4]

THE NATIONAL DOMAIN EXPANDS   57

For the rest of this chapter, we'll be looking at how the borders of the United States expanded outward from the Mississippi to the Pacific. I'm going to be tracing this growth with a series of maps, each recording a moment in that growth; the extent of U.S. territory at that moment will be indicated by a lighter tint enclosed by a heavy line. I talked earlier about the iconic "U.S. Map." In each of the maps to follow, I invite you to contemplate how the pale-gray shape enclosed by the line might have become that classic shape if historical facts had unfolded differently. Some of these possible configurations existed only as proposals in negotiations and so were never present in the minds of Americans; but some of these shapes endured for decades and so were the image which a generation of Americans could have pointed to and said, "That is the shape of my country."

I've placed successive maps in the same position on the pages, allowing you to quickly lift and lower the upper map of any pair to see only the differences between the maps. As you progress through the maps, do try some of these "flip comparisons"—between adjacent moments in time and also over larger intervals. If you lift and lower fast enough, the effect becomes cinematic: the eye is fooled into seeing the two shapes "morph" into each other.

By 1800 Napoleon Bonaparte was exercising influence, if not yet outright governance, over an increasing portion of continental Europe. Spain was induced to show its enthusiasm for the new disposition, and in the Treaty of San Ildefonso, it agreed to return its Louisiana colony to France.[5]

Upon receiving intelligence regarding this transfer in November of 1802, President Thomas Jefferson saw a chance to resolve the question of America's right to navigate the Mississippi River, by gaining possession of the entire eastern bank of the river: he would try to purchase from France both the Island of New Orleans and some portion of West Florida up against the Mississippi, north of New Orleans.[6] Why, though, would Jefferson assume that West Florida was a territory France might cede?

France's original claim to Louisiana rested on the early explorations of La Salle in the seventeenth century. By virtue of having navigated a large portion of the Mississippi, he claimed for France the entire basin drained by the river. France cemented its claim on this "Louisiana" country by seeding it with settlements and trading posts, including sites on the Gulf coast to the east and west of New Orleans. Thus, "Louisiana" might be construed to extend along the crescent of the Gulf coast from West Florida to, perhaps, the Rio Grande.

You will remember that before the treaty negotiations concluding the French and Indian War, France had ceded Louisiana to Spain. The language used in their treaty stipulated that France would surrender "the country known as Louisiana." In the years before the treaty, Spain had come to occupy the Gulf coast, in West Florida and in what would become Texas; but during

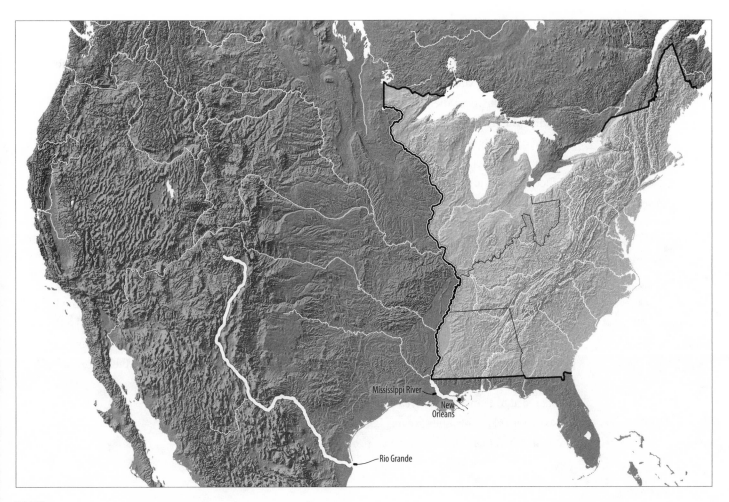

FIG 3.02

those years there had been no formal transfer of claim from France to Spain. It could be asserted that, in inheriting "French Louisiana," Spain was gaining valid title to French lands on the Gulf that it had been "squatting" on.

The Treaty of 1763 gave Britain possession of the eastern bank of the Mississippi down to the 31st parallel, but the treaty left untouched any part of "Louisiana" below that parallel or on the Texas coast. So when the Treaty of San Ildefonso ceded to France "the colony or Province of Loui-siana, with the same extent that it now has in the hands of Spain, and that it had when France pos-sessed it"[7]—this might have meant that France got back not just the Mississippi basin but the Gulf coast as well.

Jefferson hoped to prize away a tiny portion of this vast territory—the Island of New Orleans and some piece of West Florida. What a sur-prise, then, to find Napoleon willing to sell all of Louisiana.

Between 1800 and 1802, Napoleon's tentative American initiative had gone sour. There had been a slave rebellion in what is now Haiti, and of the soldiers dispatched from France to put down the revolt, most died of tropical fevers. Plus, the British threatened to blockade France's Atlantic coast. By entirely absenting himself from America, Napoleon could keep the coming conflict localized to Europe. And if his removal also happened to expand the territory of Britain's recent antagonist, the United States—well, so much the better.

The negotiations were quickly concluded, with the United States agreeing to pay 60 million francs for the territory and to assume any debts France owed the people there—a grand total of just over $2.3 million.[8] The treaty language, though, described the territory being transferred to the United States only as the same as that ceded to France by the Treaty of San Ildefonso—thus, plausibly, the "greater Louisiana" that France had (perhaps) ceded to Spain.

After concluding the negotiations, the United States pressed France to support an expansive definition of "Louisiana." At the time of the transfer, Jefferson wrote to a friend that the Americans had "some pretensions to Rio Norte, or Bravo"[9] (the Rio Grande), and even "better" claims "Eastward to the Rio Perdido, between Mobile & Pensacola, the antient boundary of Louisiana."[10] The French, though, would not be drawn. Napoleon's minister Talleyrand told the American negotiators, delphically: "You have made a noble bargain for yourselves, and I suppose you will make the most of it."[11] The Americans did indeed try. Following Jefferson's interpretation, in 1803 the United States claimed that the Louisiana Purchase included the land south of the 31st parallel, from the Mississippi to the Perdido. Spain never accepted the claim, and there matters stood until 1819.

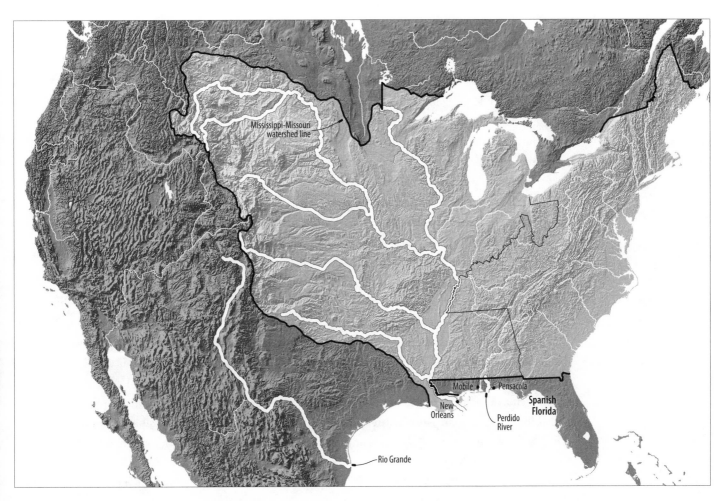

Mississippi-Missouri
watershed line

New
Orleans

Mobile   Pensacola

Perdido
River

**Spanish
Florida**

Rio Grande

FIG 3.03

What was undisputed about Louisiana, no matter how defined, was that it encompassed the whole western watershed of the Mississippi River. One of Jefferson's motives in sending Lewis and Clark to Oregon in 1804 was to see how far north the tributaries of the Missouri River (a tributary, of course, of the Mississippi) might extend.[12] Lewis, on the trip back from the Pacific in July of 1806, pinned his hopes on one Missouri tributary, the Marias River. After heading upriver toward the northwest for some days, he found his path along the river turning southward. And so at a spot he named Camp Disappointment, he wrote, "I now have lost all hope of the waters of this river ever extending to N latitude 50 degrees."[13]

Why, though, this concern about reaching the 50th parallel? Remember that although the eastern parts of present-day Canada were administered much as the thirteen colonies had been, most of western British North America was under the control of the Hudson's Bay Company. And in 1750 the company had described its territory west of the Great Lakes as all the land north of the 49th parallel[14] (as we saw in figure 2.03).

The British government never explicitly approved this claim, but in the negotiations around the 1713 Treaty of Utrecht, which settled the War of Spanish Succession, the British negotiators proposed the 49th parallel as a dividing line between French Louisiana and the Hudson's Bay Company lands. France countered with a line a degree or more further north, but no agreement could be reached.[15]

So, even though it had never been made official, the idea of a 49th-parallel border to British North America was in the air when negotiators sat down to work out Britain's border with the new Louisiana Purchase in 1807. The American commissioners proposed the parallel as the boundary, and the British readily agreed.[16] The agreement was doubtless helped along by two happy facts. First, the map of Lewis and Clark's expedition had not yet been compiled (and would not be until 1810,[17] with publication to follow in 1814[18]), and so it was still possible to believe that the watershed of the Missouri might extend well north of the 49th parallel. (Lewis was unable to explore his second candidate for "northernmost tributary," the Milk River; its watershed does indeed extend beyond the 49th parallel.) Second, it was well known by all that the Red River extended well south of the parallel. The waters of the Red flow north into Lake Winnipeg and then into Hudson Bay, making its watershed plainly not part of Louisiana. Splitting the difference at the 49th parallel let both sides feel they had made a roughly equivalent land swap.

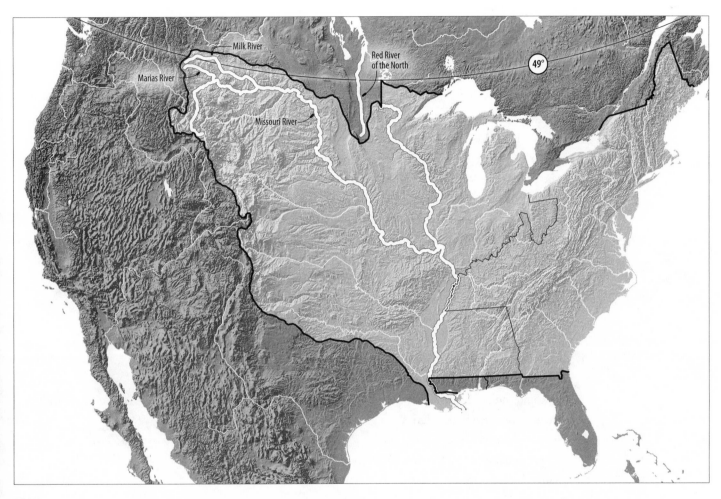

Marias River
Milk River
Red River of the North
Missouri River
49°

FIG 3.04

One problem remained, though: where on the land did the 49th parallel actually fall? Specifically, where did it fall with respect to that border point in contention since 1783, the northwest corner of the Lake of the Woods? The negotiators decided to cover all possibilities. They agreed to draw a line "due north or south (as the case may require) from the most northwestern point of the Lake of the Woods until it shall intersect the forty-ninth parallel of north latitude."[19] From that point the border would extend due west on the parallel to the crest of the Rocky Mountains—or, as we now like to say, to the Continental Divide, the line west of which all waters flow to the Pacific, and east of which all flow to the Gulf of Mexico: the very definition of the Mississippi watershed.

Unfortunately, the 1807 agreement was not ratified, nor was a similar one in 1814. Only in 1818 could the governments of both countries convince themselves to sign the agreement reached eleven years previously.

The negotiators of 1818 were unable, though, to come to an agreement about the territory *beyond* the Continental Divide, the watershed into the Pacific of the Columbia River, already becoming known as the Oregon Country. With no way to resolve the issue, the parties agreed that the territory beyond the Rockies would be "free and open for the term of ten years"[20] to citizens of both countries. As the 1818 agreement was about to expire, in 1827 the United States and Britain agreed to extend joint occupation indefinitely,[21] and there matters stood in the Oregon Country until the 1840s.

The Northwest Angle

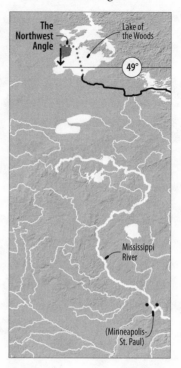

FIG 3.05

The northwesternmost point of the Lake of the Woods was found by British surveyors in 1824, and the location—which they marked with a pile of logs 12 feet high and 7 feet square—was accepted by the United States. When the 49th parallel was actually marked on the ground in 1872, the surveyors confirmed that the parallel ran south of the Lake of the Woods. And so, true to the provisions of the treaty signed in 1818, the surveyors appointed to mark the border between the United States and (by now) Canada found the pile of logs at the northwesternmost point of the lake and ran a line due south to the 49th parallel. As it turns out, this north-south line cuts across a peninsula jutting eastward, so that east of the line there is a little piece of the United States bounded on the west by Canada and on the other three sides by water.

And thus quite legally (but probably against good sense), the famous Northwest Angle of the United States was produced—the northernmost point of "the Lower 48," and a testament to what can happen when people at the centers of power project borders onto an unknown periphery.

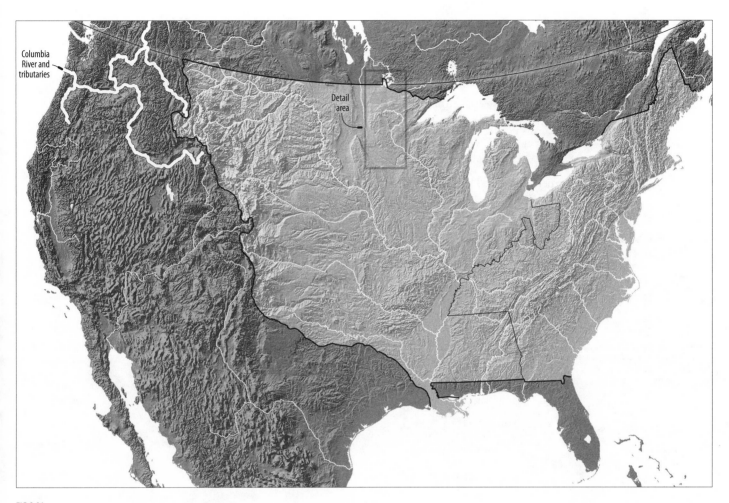

Columbia
River and
tributaries

Detail
area

FIG 3.06

In the years after Americans took possession of the Island of New Orleans in their new Louisiana territory, the border to the north of the "island," with West Florida, remained in dispute. A Spanish government was in place, but Americans were moving in to settle. In 1810 the settlers in the area north of New Orleans and just west of the Pearl River revolted against the Spanish government and asked to be admitted to the United States. The American territorial government in New Orleans sent troops to their aid, skirmishes with the Spanish authorities resulted, and the Americans took control, annexing the area into the new state of Louisiana when it was declared in 1812.[22]

The new state also claimed an enlarged western border—not just to the watershed line of the Mississippi, but westward to the Sabine (pronounced "sa-BEAN") River (on the "Jeffersonian" pretext that French "Louisiana" had extended down the Gulf coast to the Rio Grande). Louisiana's declared border ran along the Sabine from its mouth on the Gulf of Mexico up to its intersection with the 32nd parallel, and then north from that point to the 33rd parallel.[23] (These are the present borders of the state of Louisiana.)

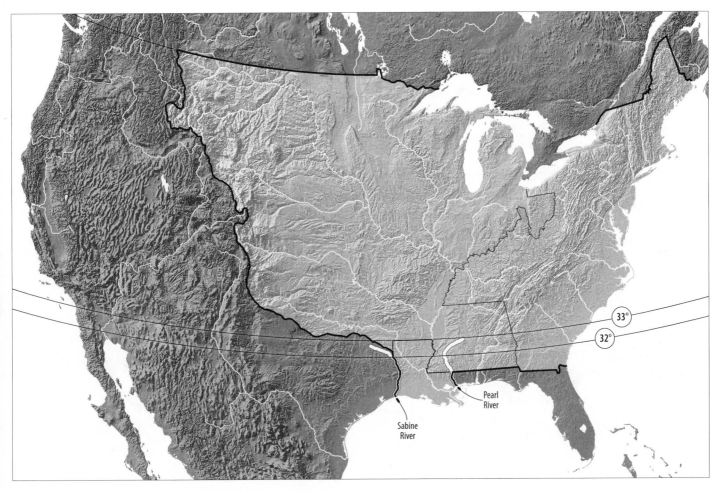

Pearl
River

Sabine
River

33°

32°

FIG 3.07

Further provoking Spain, the United States in 1812 also annexed the area just east of the Pearl River, eastward along the Gulf coast to the Perdido River (the area comprising the two back-to-back "boot heels" of Mississippi and Alabama), declaring it to be under the governance of the Territory of Mississippi.[24] Then in 1818 General Andrew Jackson, interpreting ambiguous orders to suit his own aims, led a force of soldiers into what is now the Florida Panhandle, deposed the Spanish governor, and raised the American flag at Pensacola, east of the Perdido.[25]

Jackson's invasion had the effect of galvanizing the negotiations then going on between Secretary of State John Quincy Adams and the Spanish minister to the United States, Luis de Onís. The bargain they struck was this: Spain would cede both Floridas to the United States, which would in turn pay off $5 million of the claims American settlers there had against Spain. The United States would abandon its "Jeffersonian" claim to the Texas coast, and a definitive line would be drawn to mark the boundary between the United States and Spain's possessions in North America.[26]

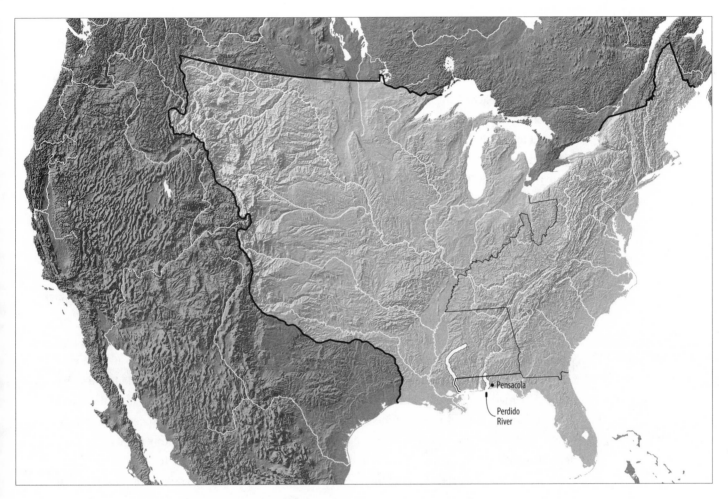

FIG 3.08

The northern border of Spain's possessions would start at the declared Louisiana state border: the Sabine River from its mouth up to the 32nd parallel. From there the line would proceed north until it reached the Red River (the Red River of the South, not the one in Minnesota). It would follow the Red River westward to the 100th meridian, then go north on that meridian to the Arkansas River. Then it would proceed west up the Arkansas "to its source in latitude 42 north" and then west on that latitude "to the South Sea."[27] This meant that Spain was surrendering all land east of the Adams-Onís Treaty line to the United States, and that it was giving up any claim to lands that fell west of the Continental Divide and north of the 49th parallel.

The negotiators, as before and after, were working with a then-current map (in this case, one printed in 1818 in Philadelphia); but maps of the time made little distinction between geographic features located by longitude and latitude, features noted by explorers but not placed accurately with surveying instruments, and features merely conjectured. Among the mountain men of the time—the beaver trappers who were then the only white people yet to visit the Rockies extensively—there was a legend about a plateau, or perhaps a high meadow, in the mountains somewhere southwest of present-day Denver, out of which all the great rivers of the West sprang. From this one place (they conjectured) flowed the Colorado to the southwest, the Rio Grande to the south, the Arkansas to the east—perhaps even branches of the Columbia and Missouri to the northwest and northeast.[28] It was misapprehensions like these that negotiators were constantly forced to deal with as they drew boundaries over unexplored territory. When John C. Frémont finally pinned down the location of the Arkansas' source in 1844,[29] it was found to be some 200 miles south of the 42nd parallel. Luckily the drafters of the Adams-Onís Treaty did with the Arkansas River what had been done the previous year with the Lake of the Woods: they agreed that if the Arkansas' source was found to be south or north of the 42nd parallel, a line would be drawn from the source to the parallel, north or south as needed.

When all the geographic features named in the Adams-Onís Treaty were finally located within the net of longitude and latitude, the shape of the resultant border could be correctly drawn on maps. The familiar stair-step shape of the Adams-Onís line has been reproduced on maps for well over a century now, and our eyes have become so accustomed to that distinctive shape that we simply assume it to be the actual shape of the boundary negotiated in the treaty. But it's important to remember that at the time of the treaty—and indeed for thirty years thereafter—not one person could have visualized that stair step as the shape of the boundary. That border outlines a shape that seems familiar, even manifest, to us; but it is a shape that would have looked unfamiliar, perhaps even strange, to the people of the time.

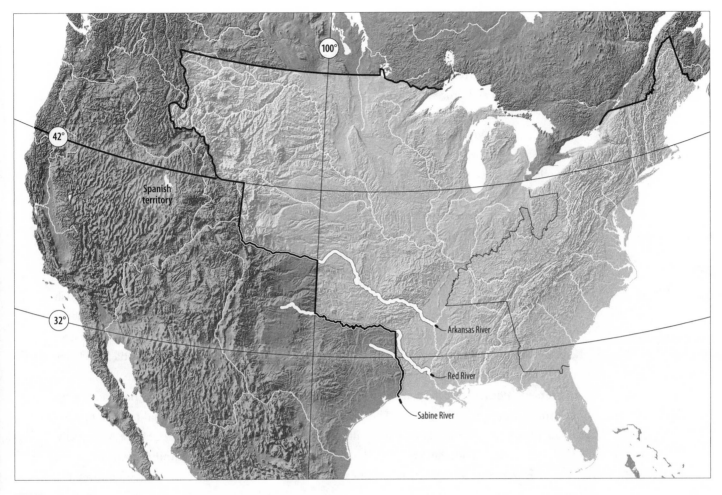

FIG 3.09

The Treaty of 1783 had said that the border between the United States and the British possessions to the northeast would start at the mouth of the St. Croix River (Maine's present eastern border) and proceed up the river to its source. Unfortunately, contemporary maps of the region showed the name *St. Croix* applied to three different rivers. The United States naturally claimed the easternmost (the one that bears the name today), while the British claimed the middle one of the three. A joint commission was authorized in 1794, by 1796 they had identified the river to be the boundary, and by 1798 surveyors had mapped it to its source.[30] Beyond that point the two sides agreed to disagree about their border. That unresolved situation still prevailed when Maine was divided from Massachusetts and became a state in 1820.

In 1829 the two sides, unable to reach an agreement, decided to call in an independent arbitrator. They chose the king of The Netherlands (a surprising choice to us today, but this was neither the first nor the last time a "neutral" monarch would be brought in to adjudicate a border). Here was his dilemma:

The Treaty of 1783 had been fairly clear about this portion of the U.S. border. From the head of the St. Croix, a line was to be drawn due north to the ridgeline that separates the watershed of the St. Lawrence River from that of the Bay of Fundy on the Atlantic side. Now, watershed lines are fairly easy to find on the ground. You traverse the territory, and where you cross a stream, you note whether it is flowing (in this case) north to the St. Lawrence or south to the Atlantic. You follow each stream up to its source, locate that source point by latitude and longitude, and

plot that location on a map as, say, a dot with an arrow pointing toward the stream's eventual outlet. When you have done that for every consequential stream in the region, your map has two parallel strings of dots, their arrows pointing in opposite directions. Plot a line between the opposing dots, and you will roughly mark the ridge of land between the two watersheds.

So mapping the border ought to have been a simple task: plot the boundary line of the St. Lawrence watershed, run the line north from the St. Croix's source to where the two lines intersect, *et voilà* (as they'd say in Quebec): everything south of those two lines would be in the United States, everything north would be British territory. But Britain, dismayed to learn that the watershed line veered so far northward to the St. Lawrence as to pinch off Nova Scotia from lands to the west, held out for a better deal.

In 1831 the king of The Netherlands delineated a compromise line that is close to what would become the border of Maine. Essentially, he stopped the line running north from the source of the St. Croix at the St. John River, followed a branch of that river to its source, then projected a line due west until it intersected the watershed line of the St. Lawrence.[31] The U.S. and British negotiators accepted the compromise. Maine, however, balked at the agreement, and so the U.S. Senate, acceding to the wishes of a new state, refused to ratify the Dutch king's compromise.

There matters stood until 1842 when Britain, hoping for a final disposition of the entire U.S.-British border from the Atlantic to the Pacific, sent a special mission to America, headed by Lord Ashburton. The negotiations that ensued

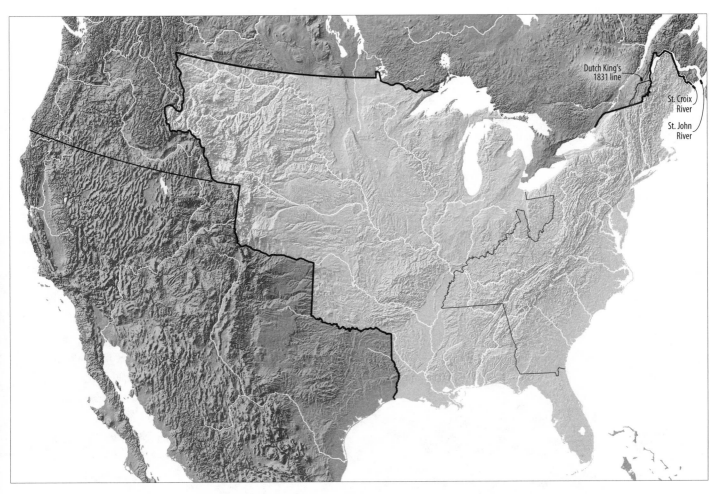

FIG 3.10

produced a Maine borderline that takes several
hundred words to describe and is punctuated
with those "if it should be discovered that . . ."
clauses so often found in treaties drawn prior
to a survey. Maine was induced to accept the
treaty worked out between Lord Ashburton and
Daniel Webster, and the agreed-upon line was
finally surveyed between 1843 and 1847.[32] Once
the surveyors had done their work, it was found
that Maine had gotten about one thousand fewer
square miles than it would have under the line
set by the king of The Netherlands.

When the only white people in the Oregon Country were fur trappers and traders, joint occupancy by Britain and the United States (and thus no real government) was a tenable situation, and was probably preferred by the trappers. But once American settlers began to farm the Willamette Valley in the early 1840s, governance became a real issue. Farmers need title to their land, and only a government can issue, record, and enforce land titles. But which government—the U.S. or the British?

Not only were settlers in Oregon anxious for a decision, the issue had figured prominently in the 1844 presidential campaign. The Democratic candidate, James K. Polk, had secured the nomination by unabashedly advocating expansion. Going further than his rival Democrats and much further than the opposing Whigs, he favored both admitting the recently formed Republic of Texas to the Union and taking possession of all of Oregon—the second position sloganeered into the phrase "Fifty-four Forty or Fight!"[33] The meaning of the slogan was that the United States would take all the territory west of the Continental Divide, north from the 42nd-parallel Adams-Onís line all the way up to a line at latitude 54° 40'. But why that particular—and peculiarly precise—latitude?

By navigating the Columbia River and being "successor" to the lands Spain had surrendered north of the 42nd parallel—lands that might have been construed to extend up to Russian-claimed Alaska—the United States had some basis for a claim into the Far North. And so it approached Russia to see whether the two countries could recede their overlapping claims back to a common border. In that part of Pacific coast, there are only a few navigable water passages into the interior. One of these is a river channel now called the Portland Canal, whose mouth is found, from the sea, by rounding the southern tip of Prince of Wales Island. That tip—at 54° 40' north latitude—is where Russia and the United States agreed in 1824 to split their claims.[34] A line at that latitude would be projected east to the Continental Divide; above that line would be Russian territory, and below the line, American. Such a line would give Russia an inlet, via the Portland Canal, to its interior lands.

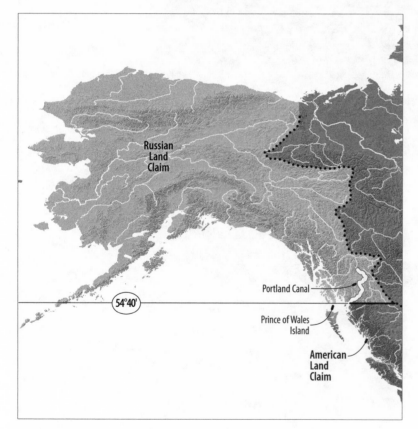

Russian Land Claim

54°40'

Portland Canal

Prince of Wales Island

American Land Claim

FIG 3.11

Columbia
River

Willamette River
and Valley

Continental
Divide

54°40'

FIG 3.12

But Britain had a claim to that same territory, and so the following year, 1825, Britain and Russia negotiated a boundary between their claims. Britain adopted the same 54° 40' latitude as the Americans but drove a harder bargain: Russia's lands would reach that far south, but instead of extending into the Rockies, they would be confined to a narrow strip paralleling the coast—meaning that the Portland Canal would be a water route inland to British, not Russian, territory. The narrow strip of Russian land would skirt the coast up to the northwest until it intersected the 141st meridian. The border would then run due north, up the meridian, until it hit the Arctic Ocean (the present boundary between

Alaska and Canada's Yukon). Britain would own everything east of the line, Russia everything west of it.

So after 1825 all the claimants except the United States and Britain had backed away from the territory west of the Rockies between the 42nd parallel and latitude 54° 40'. For the United States to demand all of this Oregon Country was to demand 54° 40'—or fight.

During the on-and-off negotiations over Oregon, up to those in 1842 with Lord Ashburnham, the British had consistently offered the Columbia River as a border. The Americans had countered with a series of lines farther north, including a line at the 49th parallel. Part of the British insistence had arisen because their primary shipping point for furs was Fort Vancouver (now Vancouver, Washington) on the northern bank of the Columbia (opposite present-day Portland). By the 1840s, beaver in what is now Washington State had been pretty well trapped out, and the British had moved their base of operations northward from Fort Vancouver to Fort Victoria at the southern tip of Vancouver Island[35] (not to be confused with the city of Vancouver, British Columbia, which came later—a source, in Canada, of the kind of confusion Americans endure over Washington, DC, and the state of the same name).

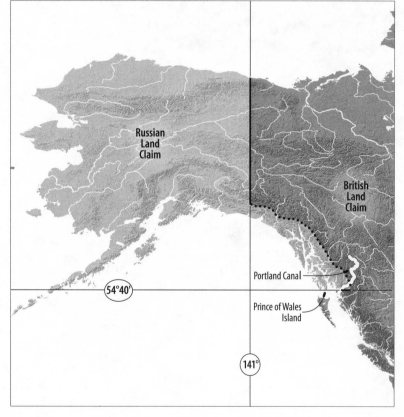

Russian Land Claim

British Land Claim

Portland Canal

Prince of Wales Island

54°40'

141°

FIG 3.13

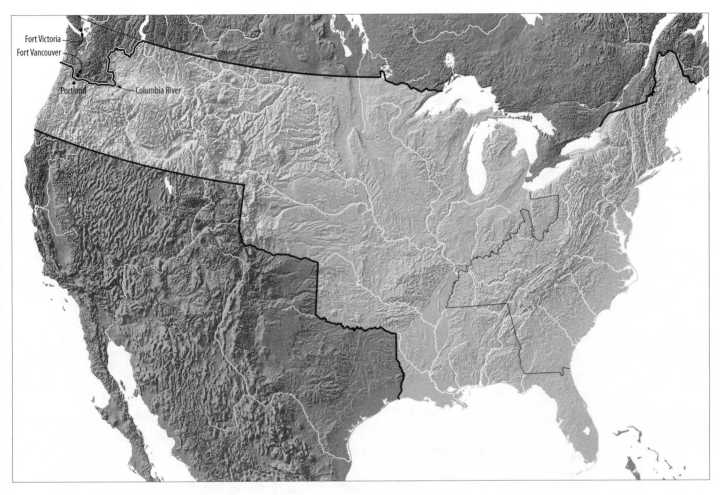

FIG 3.14

As the joint-occupation era wore on, Britain found its interests receding northward. Finally, in 1849 the British accepted the long-standing American proposal to extend the 49th-parallel border westward to the Pacific. The one sticking point might have been that Vancouver Island extends below the 49th parallel (potentially placing Fort Victoria within American territory), but the United States readily acceded to running the line south of the island, and so a deal was struck.[36] With the 1849 Buchanan-Parkenham Treaty, the United States achieved what the colonial grants had promised—land stretching from sea to sea.

Port Roberts

True to the established pattern, the negotiators "at the center" left one crucial ambiguity in the Buchanan-Parkenham Treaty, which resulted in one borderline absurdity.

When the border veered south from the 49th parallel to skirt around Fort Victoria, the treaty language dictated that it was to go "through the middle of the channel which separates the continent from Vancouver Island." Unfortunately, the strait between the mainland and Vancouver Island is strewn with smaller islands: there are any number of potential channels through that strait.

Naturally, the United States nominated the westernmost channel, the Haro Strait, as where the border should go, since that would throw many of the islands into its territory. Britain chose the easternmost, the Rosario Strait, for the same reason. Neither side would budge, and the matter stayed unresolved until 1872, when the negotiators decided to put the issue to an impartial arbitrator—this time, the emperor of Germany. The delegates packed their charts and traveled to Berlin to acquaint the emperor with the intricacies of Puget Sound geography; he ruled that the American case had more merit, and so the border was charted westward around the islands.[37]

That was the ambiguity in the treaty. The borderline absurdity is Port Roberts, a tiny spit of land that, as surveyors later found, projects south from Canada just barely across the 49th-parallel line. Like the Northwest Angle in Minnesota, this is a little piece of the United States that can be reached on land only by going around through Canada.

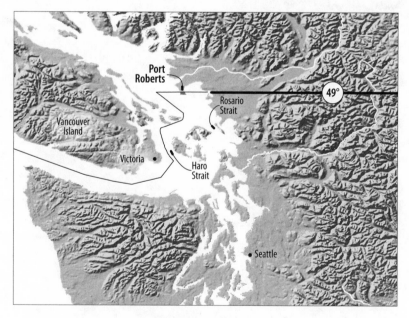

FIG 3.15

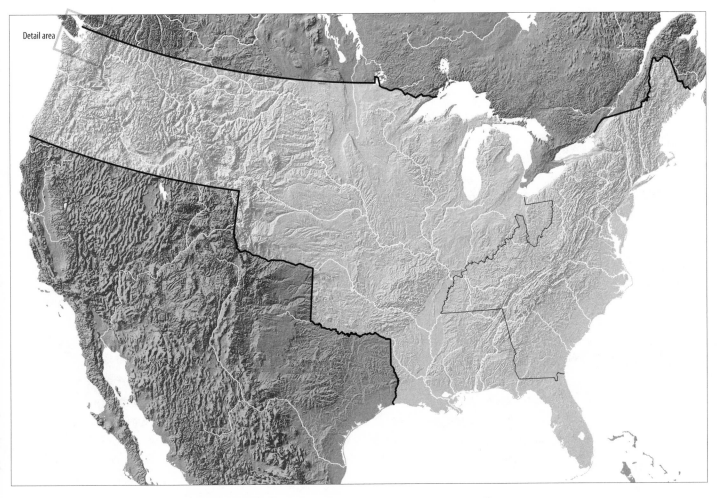

FIG 3.16

The founding of Texas and its struggle for independence have become the stuff of American and Hollywood legend. Can any of us think of the Alamo without hearing the voice of John Wayne—even if we never saw the movie? No one can hope to correct all our mythologies about Texas (and who'd be so mean-spirited as to try?), so let me focus on verifying only those details that affect the borders of Texas and thus those of the United States.

Upon independence from Spain in 1821, Mexico faced a dire geographic problem: it had a territory that extended up to the Adams-Onís line of 1819, yet its center of population and national identity was a thousand miles to the south, around Mexico City. An obvious lesson of the whole colonial era was that a central government could assure control over an outlying territory only by settling it. Yet Mexican farmers could not be induced to settle up against the frontier with the United States. So in that part of Mexico that is now East Texas (where pressure from American incursion was greatest), Mexico hoped to co-opt American settlement through the *empresario* system. Under the system, an American entrepreneur could get a large land grant if he could bring in people to settle on his grant. The catch: every settler had to profess allegiance to Mexican jurisdiction. Several Americans took up the offer to become *empresarios* of new colonies in Texas—Stephen Austin is the most famous— and by 1830 thousands of Americans had settled in the region.

At first the settlers enjoyed a kind of autonomy through neglect much like their colonial predecessors had experienced, before the 1760s, under English rule. But once the Texas colonists began to assert themselves, the Mexicans, like the British before them, attempted to tighten control over their unruly "citizens." The Texans (not for the last time seeing their situation as parallel to that of the colonies in 1776) eventually revolted, defeating the Mexican general and president, Antonio López de Santa Anna, on April 21, 1836,[38] at what is now a shrine among Texans, the little town of San Jacinto. Taking him captive, the Texans forced Santa Anna to sign a treaty; and on the basis of that treaty, Texas set forth a set of borders in its constitution when it declared itself a republic on December 19 of that year.

The northern border of the new Republic of Texas would be "the boundary line as defined by the treaty between Spain and the United States"—the 1819 Adams-Onís line. Its boundary to the south and west would be the Rio Grande, up from its mouth "to its source, thence due north to the forty-second degree of north latitude."[39]

The shape of the nation thus described certainly looks anomalous to our eyes—that weird "tail" stretching hundreds of miles to the north. But of course no Texan was envisioning this shape when the borders were set out. As we saw earlier, the shape of the U.S. border at the Adams-Onís line was not then known, since the source of the Arkansas had not yet been found; and the source of the Rio Grande had likewise not been accurately mapped.

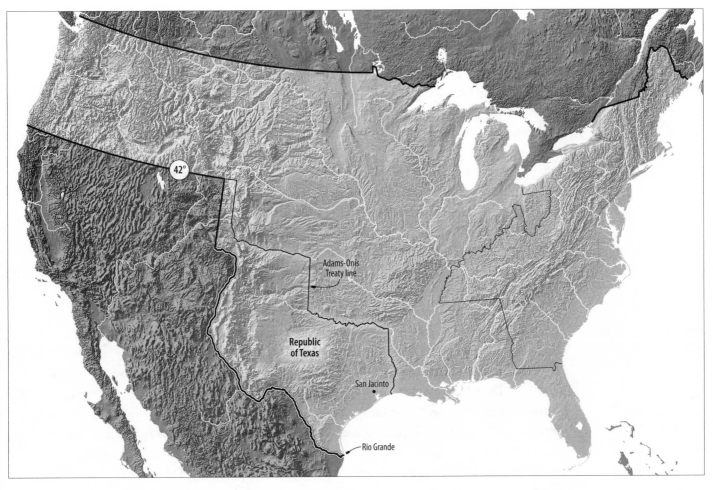

FIG 3.17

The border the Texans declared at the Rio Grande was anomalous not just geographically but politically. Unlike other expansive border claims of the previous centuries, this claim was not projected from a center of power into a "void" whose characteristics were unknown to those staking the claim. Everyone at the time knew that the Rio Grande (in present-day New Mexico) was the thread that linked, and irrigated, the centuries-old mixed Spanish-Indian settlements centered on Santa Fe. A border at the river would split that area down the middle, half in Mexico, half in Texas—and put Santa Fe itself inside the new republic.

The Santa Fe region had even been given the name New Mexico by the Spanish, then Mexican, authorities. "Texas," by common usage, extended only from Louisiana to about the Nueces River, some miles south of the area settled by the *empresarios*. Though the borders of both territories were indistinct, Santa Fe was indisputably not part of any Texas anyone then imagined.

So the claim of Texas was not just expansive but provocative. But as long as the claim remained only on paper, in the Texas Constitution, Mexico felt no compelling need to respond forcefully to the provocation. It was partly fear of a conflict with Mexico that caused Congress to reject Texas's appeal for annexation to the Union in 1836; the greater obstacle was the North's opposition to such a huge expansion of the area of the United States being opened to slavery.[40] And so Texas remained an independent nation with disputed borders for nearly a decade.

By the 1840s Texas was making overtures of an alliance with Britain. To head off such a threat, John Tyler hoped to cap the final year of his presidency with a treaty annexing Texas to the Union. But when the Senate took up the treaty, in April of 1844, the twin issues of potential Mexican conflict and slavery doomed it to defeat.

In early 1845 Tyler achieved his goal by other means. (Be aware that until 1936, presidents were inaugurated in March, not January, and that the Congress elected two years earlier—in this case, in 1842—might remain in session until the inaugural.) Knowing he could never get the two-thirds majority in the Senate required for a treaty of annexation, he asked instead for a joint congressional resolution of annexation, which would require only a simple majority of both houses. The new strategy got the bill passed— just barely; and on March 1, three days before the end of his presidency, Tyler was able to sign the resolution admitting Texas to the Union.[41]

Thus, contrary to popular Texas legend, the state was not joined to the United States by "a treaty between two sovereign nations." But its status as a republic did have one huge effect on the expansion of the national domain. Unique among states west of the Mississippi, Texas retained title to its lands, and so they never became part of the Public Domain—a telling circumstance we'll revisit at the conclusion of this book.

Mexico, upon hearing of the annexation, did what any self-respecting nation would do: it broke off diplomatic relations. The brand-new Polk administration responded by ordering U.S. troops across the Nueces River into Corpus Christi and then, in March of 1846, all the way to the Rio Grande—to which the Mexicans responded with a full-scale mobilization for war.

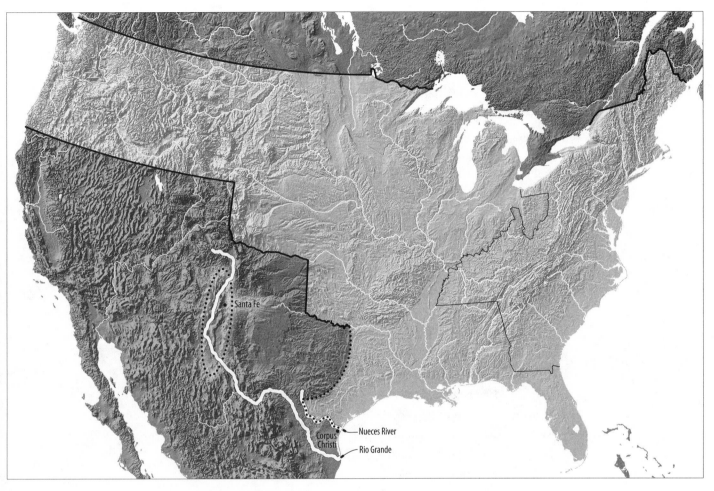

FIG 3.18

As I did with American wars of the eighteenth century, I'm going to skip over the battles of the Mexican War and jump right into the treaty negotiations that concluded the war and set new boundaries. But to understand those talks, you need to know a little about the borders that existed in the region before the war.

A key to the negotiations lies in the definition of the boundaries of the Spanish province, later the Mexican state, of New Mexico. Spanish America did not have the tradition of describing boundaries with the legalistic (if often ambiguous) precision that prevailed in "Anglo" America; but all participants knew that New Mexico as a cultural entity comprised the eastern and western watersheds of the Rio Grande north of the Paso del Norte. This *paso* ("pass") was the place where the Rio Grande breaks through a north-south chain of mountains and provides an alluvial bench wide enough for a town site: the town of Paso on the southern bank of the river (the present Mexican city of Ciudad Juarez).

These, then, were the "facts on the ground" in 1845: a coherent cultural region on both sides of the upper Rio Grande, a distinct cultural region in the area north of the Nueces up to Louisiana, and a lower Rio Grande valley largely uninhabited but crossed down its middle by a river that seemed—to border-obsessed Americans—like a manifest dividing line.

So when in the wake of the Texas annexation the new President Polk initiated talks with Mexico over a new border, he looked at his maps and saw not cultural regions but a long unbroken line—the Rio Grande—as a natural border for the United States. Polk family members, with access to the president's private diaries, tell us that he had hopes for the Rio Grande not as a border but as a border *marker*. He envisioned a latitude line being struck west from the mouth of the river all the way to the Pacific, and then south to encompass the whole of the Baja California peninsula.[42] (Now, *there's* a possible "Map of the U.S." to contemplate!)

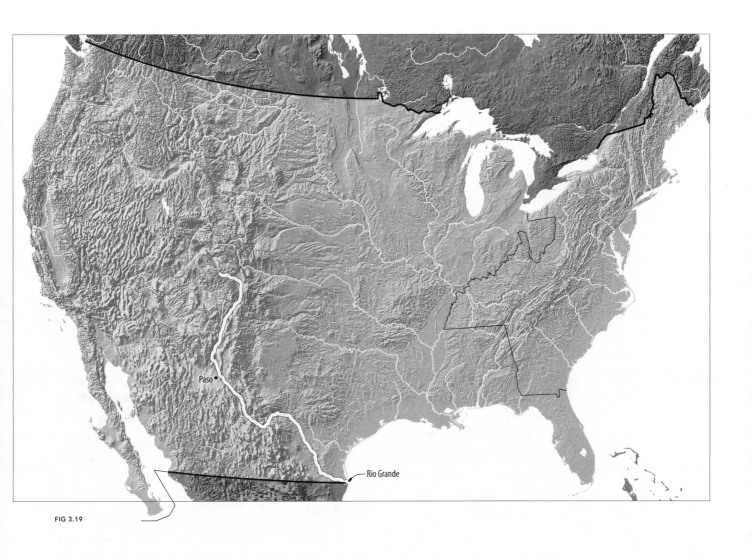

FIG 3.19

Paso •

Rio Grande

In November of 1845, before the U.S. incursion, Polk had sent his envoy John Slidell to Mexico City to negotiate a common border at the Rio Grande. Slidell was also authorized to try for three even more expansive borders: all of New Mexico; New Mexico plus California south to San Francisco Bay; and all of New Mexico plus California south to Monterey Bay. For each of these cessions, the United States would pay Mexico sums ranging from $5 million to $25 million.

To Mexico—which might well have granted the United States a border at the Nueces—the cession of indisputably "Mexican" territory would have been an assault on a national dignity barely two decades old. Far from entering into negotiations with the American delegation, Mexico did not even accord it recognition.

The situation was decidedly different in the spring of 1847, after American forces had invaded Vera Cruz on the Gulf coast and penetrated deep into Mexico. President Polk sent another envoy, Nicholas Trist, to negotiate. This time the American delegation was received, but what it had to offer had changed. With the successes following Vera Cruz, Polk had instructed Trist to obtain a Rio Grande border for Texas, plus all of New Mexico and Upper California, and if possible, Lower California as well.

Mexico did not appoint its peace negotiators until late August. Then the delegations traded several border schemes back and forth, with the Mexicans offering parts of Upper California but holding fast to New Mexico and the Nueces border. But while the delegates talked, American troops approached Mexico City, and by September 14 they held it. Negotiations were broken off, not to resume until December.

In the December talks, the Mexicans proposed a border: up the Rio Grande to a point 2 leagues north of Paso, then west on a latitude line to the crest of the Sierra Mimbres (the watershed line of the Rio Grande and natural "geographic" border of New Mexico); then north along that crest to the Gila River (pronounced, as you probably know, "HEE-la"), and west on the Gila to its confluence with the Colorado. From that point the line would go due west to the Pacific—unless that line cut through San Diego, in which case the boundary of Mexican territory would shift to a latitude line 2 leagues north of San Diego's harbor.[43]

(A marine league is 3 nautical miles. Since a nautical mile is 1 minute, or $1/60$ of a degree, of the earth's 360° circumference, a league is $1/20$ of a degree—or in landlubber's terms, about 3.25 statute miles. It is worth contemplating the course of the development northwest Mexico might have taken if it had had the port of San Diego at its upper corner—a sort of Mexican Seattle.)

Negotiations ensued and finally, on February 2, 1848, Trist and the Mexican delegates signed a draft treaty that named as the U.S.-Mexican border a line similar to the one offered by the Mexicans the previous December. The prime exception was that San Diego would go to the Americans: from the point where the Gila intersected the Colorado, a straight line would be run to a point on the Pacific 1 marine league *south* of the southernmost bend of San Diego Bay. For all of this the United States would grant Mexico $15 million and pay off any claims American settlers had against the Mexican government.

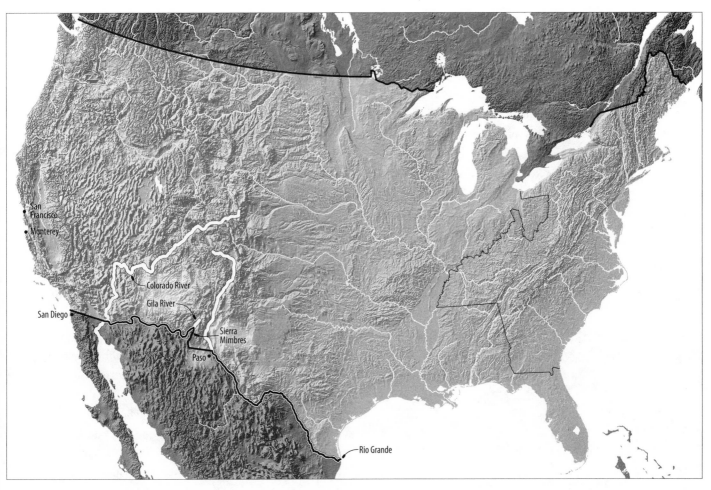

FIG 3.20

When Trist returned to Washington, President Polk summarily fired him for not having obtained more of Mexico, but submitted the treaty to the Senate anyway—where the Treaty of Guadalupe Hidalgo passed handily on March 10, 1848.[44] As drawn, the treaty boundary had four segments. Three of them were quite precise: the courses of the Rio Grande and the Gila River, and the straight line from the Gila to a point 1 league south of San Diego Bay. But the fourth segment—between the Rio Grande and the Gila—was described only as the southern and western boundaries of the Mexican state of New Mexico as drawn on a "Map of the United Mexican States" published in 1847 by J. Disturnell of New York. It was this map that Mexican and American commissioners had in hand when they gathered at the Rio Grande in 1850 to mark the new boundary on the ground.[45]

As measured with the graphic scale on the map, the southern border of New Mexico started on the Rio Grande about 7 miles north of Paso and extended about 3° to the west before it turned north. One solution for the commissioners would have been to trace the border in just that way: measure 7 miles north from Paso and then project the line 3° west, then north to the Gila River.

But the commissioners wanted to tie the new border into the global net of longitudes and latitudes, and so they took out their surveying instruments to determine the precise positions of the town and the river. The map was crossed by latitude and longitude lines, and when they drew the river and the town at their correct latitude and longitude, the "corrected" Rio Grande was 2° west of the river on the map, and the "corrected" Paso was some 30 miles farther south than the town on the map. What to do now?

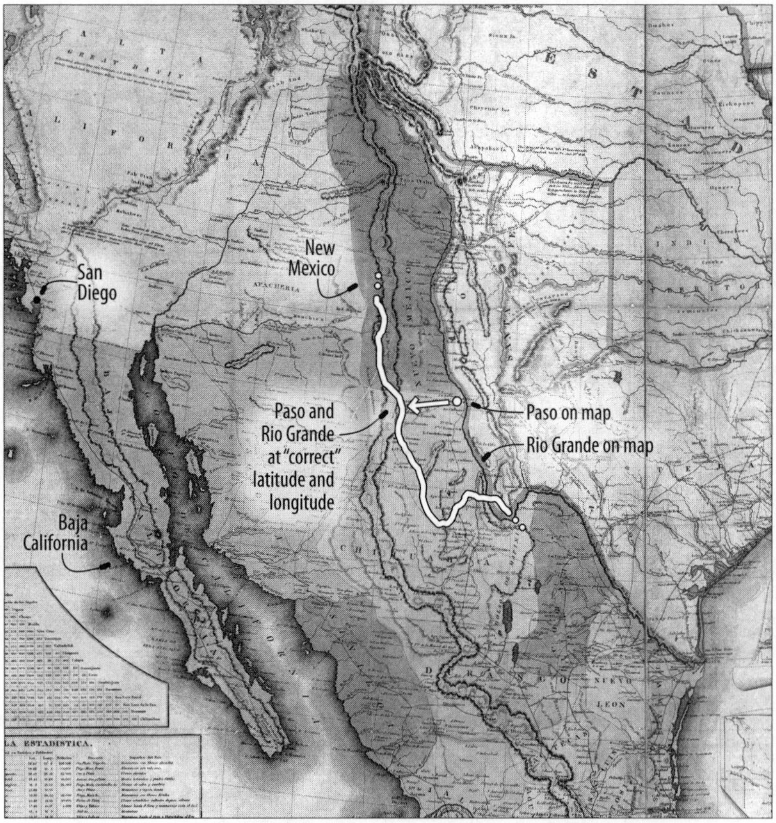

The Mexicans proposed to leave the border in the same position in which it had been printed on the map, but redraw the town and the river in their "corrected" latitude-longitude positions. The border could then be projected by measuring the resulting distances on that modified map—meaning that the east-west line would start 30-some miles upriver from Paso and extend about 1° to the west before turning north.

The Americans, seeing that such a solution diminished their territory, proposed a compromise: use the upriver latitude line, but extend it west by the full 3-degree length the line had possessed on the map. The Mexicans accepted the idea, and the party moved up the river to the agreed-upon starting point and began surveying the line westward.

After they had run about half of the line, the chief American surveyor, who had been detained, finally arrived on the scene. Seeing what had been done, he strongly protested. The New Mexico border on the map had been located with reference to the river and the town, not with reference to longitude and latitude: if the river and the town are moved, the border must be moved along with them.

Agreement being impossible, the two proposals were forwarded to Washington, where in 1852 Congress sided with the surveyor by refusing to appropriate any more money to complete the compromise line. The question was finally rendered moot by the negotiations over the Gadsden Purchase in 1853.[46]

FIG 3.21   Library of Congress, Geography and Map Division.

CHAPTER THREE

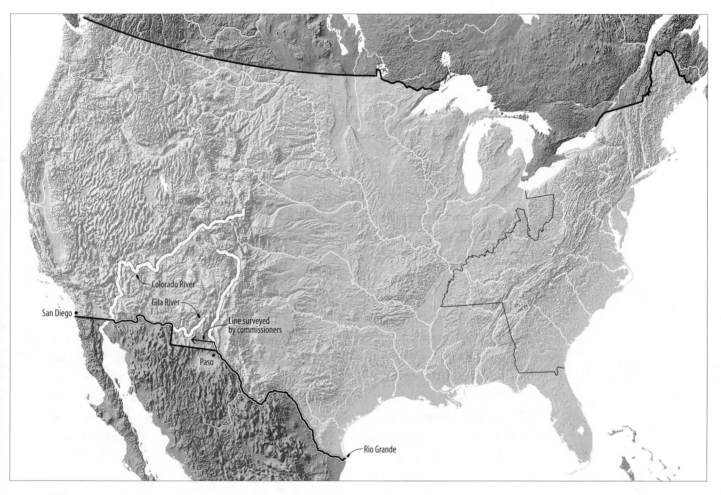

San Diego

Colorado River

Gila River

Line surveyed
by commissioners

Paso

Rio Grande

FIG 3.22

The second theater of the Mexican War was in Upper California, and to get there, General Stephen Kearney (pronounced "Carney," according to family tradition) moved his troops out from St. Louis along the Santa Fe Trail in the summer of 1846. On August 18 Kearney captured Santa Fe and began preparations for the second leg of his mission, the trek to San Diego. At Santa Fe he split his forces: he would lead a horse troop quickly over the mountains, and the heavier supply wagons would try to find a flatter route around the mountains.

The wagon train worked its way south and west but could find no mountain pass until it met a band of Mexican traders, who described a route suitable for wagons: around the mountains and then west to the town of Tucson. From there the party could make its way back up to the Gila.[47] The Americans followed the advice, and you can experience parts of their route on Interstate 10 through New Mexico and Arizona.

That route was suitable for horse-drawn wagons, but by the 1850s Americans were thinking about routes for transcontinental railroads, which require shallower gradients than wagons do. Accordingly, the War Department sent topographical engineers from the U.S. Army into the area to survey possible routes. With the survey completed in 1853, President Pierce sent James Gadsden, a senator and promoter of a transcontinental railroad through New Mexico, to negotiate the purchase of the land through which the potential routes passed.

The army's survey maps formed the basis for Pierce's instructions to Gadsden, which took the form of six possible borderlines. The positions of these lines are complex, but their import is plain:

the United States would make one more try for Baja California, and would see if it could get the southern half of the Rio Grande watershed; but above all else, it would get all the possible railroad routes around Tucson.

Mexico was not about to give up Baja California, or its access to the Rio Grande, so the real negotiations were over the portion of the border between the Rio Grande and the Gulf of California. Mexico was amenable to selling land in the middle of this region; the problem was with the eastern and western ends of Gadsden's proposed line.

The Mexicans were particularly concerned that their national borders provide an overland route across the Colorado River to the peninsula of Baja California: they would not accede to any border south of the Colorado's outlet to the Gulf of California. So the best Gadsden could get from the Mexicans was a border consisting of two straight lines. The first line would begin at the point where latitude 31° 47' 30" crossed the Rio Grande (almost precisely the point demanded by the chief U.S. surveyor in 1850) and go straight to the intersection point of the 31st parallel and the 111th meridian. The second line would run from there straight to a point on the Colorado River 2 marine leagues north of the northernmost bend of the Gulf of California.[48]

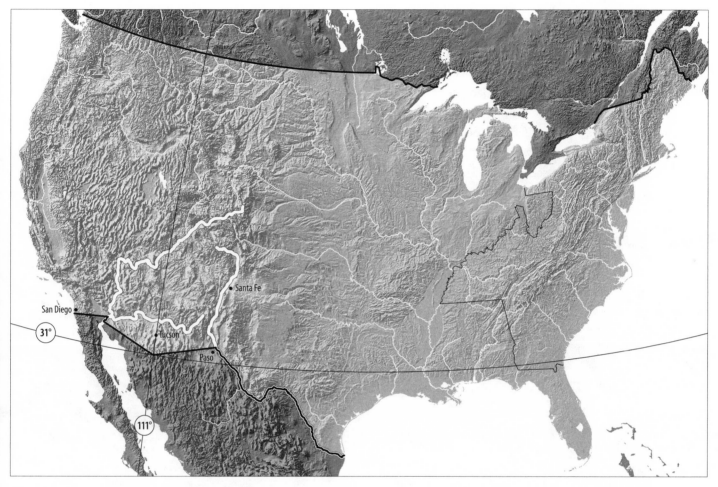

FIG 3.23

Gadsden and the Mexicans signed the treaty on December 30, 1853, but in the Senate debate that followed, virtually every aspect of the line was changed. As amended, the line would start at the point where latitude 31° 47' crossed the Rio Grande (a shift northward from the negotiated point of some 3000 feet) and then follow that parallel west for 100 miles. There the line would turn south to latitude 31° 20' and then run west on that parallel until it intersected the 111th meridian. From there a straight line would angle up to a point on the Colorado River 20 miles below its junction with the Gila—and from there, up the Colorado to the previously established California border that headed south of San Diego Bay.[49] For this sliver of desert the United States would pay Mexico $10 million in gold.[50]

The Senate approved the amended treaty on April 17, 1854, and the Mexican government followed suit later in the year. The Senate seems to have outsmarted itself, though, since its amendments meant that Mexico would actually give up less land than it had originally agreed to.[51]

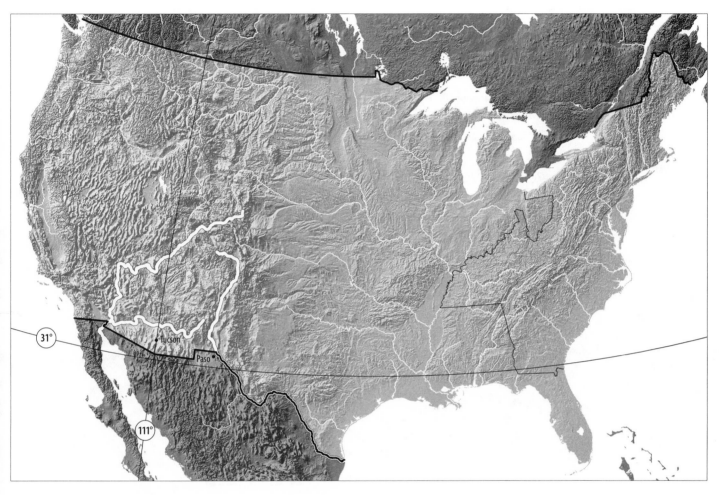

FIG 3.24

Here is one more "U.S. Map" to contemplate: the shape of the territory over which Congress would eventually have the authority to sell land directly to settlers and to establish new states. You'll note the huge chunk taken out of this territory, the state of Texas, which, of all the western states, was the only one to retain its unsold land after statehood.

Because this vast region fell under the control of the federal government, this special world-within-a-nation gradually acquired the name by which we have come to identify it—the Public Domain.

To remind you of where we are going with this story, in part 2 we will see how this domain was apportioned into territories and then states. In part 3 we will see how each of the states in the Public Domain was apportioned into the rectangles of the Public Land Survey System. But before we do that, I must acquaint you with the final addition to the Public Domain, Alaska.

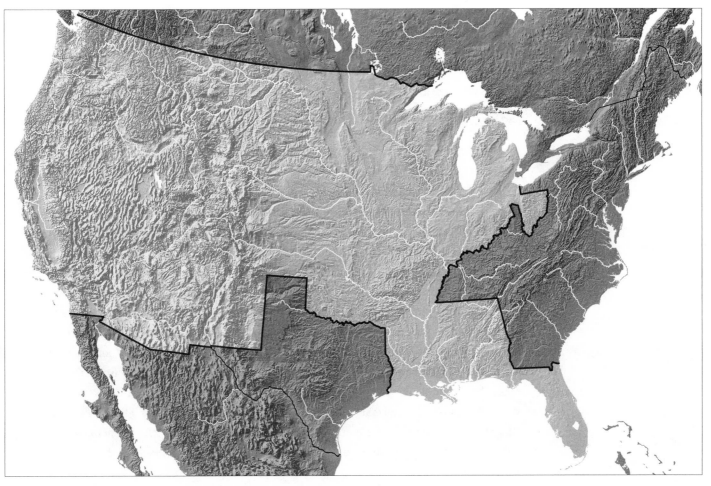

FIG 3.25

It is an index of their perceived relative values that while the United States paid Mexico $10 million for the Gadsden strip, it paid Russia $7.2 million for all of Alaska. By 1867 Russia had concluded that it was not worth its while to expend money on naval forces to defend its North American territories, and it did not want them to fall into the hands of Britain. So when Secretary of State William Seward approached the Russian ambassador in Washington with a proposal to purchase the Alaskan territory, a deal was quickly struck, on March 30 of that year.[52]

The agreement explicitly stated that the border of the purchase would be the boundary Russia had negotiated with Britain in 1825 (the agreement even requoted the earlier language). Utterly clear was Alaska's long north-south border at the 141st meridian.[53] Much less clear was the border of the panhandle stretching down the Pacific coast to the southeast.

The treaty with Britain had stated that the panhandle boundary would start at the southern tip of Prince of Wales Island, run east to the center of the Portland Canal, and then follow the center of that channel up to the latitude of the 56th parallel. From that point the border would "follow the summit of the mountains situated parallel to the coast" up to the 141st meridian. As was typical of treaties for unsurveyed territory, there was an exception: if at any point on the coast there was found to be no peak within 10 marine leagues of the coast (30 nautical miles, or about 34 statute miles), then the border would be a line traced parallel to the coast, 10 leagues inland.[54]

Unfortunately, the geography of the coastal area makes this border description uninterpretable. Mountains come right down to the ocean, but in a tumbled fashion with no dominant range of peaks "parallel to the coast." Threaded between these peaks are a series of narrow fjords which penetrate inland.

When a region is lightly inhabited, such ambiguities can be finessed through local accommodations; and this was the case for decades in Russian, then American, Alaska. But when in the 1890s the Klondike (in Canada just northeast of the Alaska Panhandle) boomed into a gold-mining region, it became imperative to determine where the border lay between the United States and Canada.

After some inconclusive negotiations, a commission was formed in 1903 to adjudicate the border. There were six delegates: three American, two Canadian, and one British, who proved to be the swing vote. Two possible borders were under consideration, both equally untenable.

The Canadians proposed a border like the one in figure 3.26, a line connecting the peaks of the first mountains to rise from the Pacific oceanfront. The American position was that since no range of mountains truly ran "parallel to the coast," the peaks should be ignored and the border drawn as a line 10 leagues inland from the general run of the coastline. The "coastline" was interpreted as the shorelines of the fjords that penetrated inland—resulting in a border something like that shown in figure 3.27. Canada's proposal would have given Alaska a territory that was cut every few miles by deep fjords, with no possibility for traverse by land—a territory that in some places would extend inland only a few hundred yards. The American proposal would have produced a scalloped border in which, from the head of every fjord, a semicircle of 10-league radius would bulge into Canada.[55]

But the compromise that was reached split the difference and garnered the approval of the American delegates and the British commissioner. By a careful perusal of the maps, it was possible to identify a series of mountain peaks (in no sense a "chain") that fell roughly midway between the two proposed borders. Fortuitously, a line connecting these peaks fell just inland of the heads of most of the fjords, giving the Americans the possibility of overland passage.

The commission majority could thus contend that it had adhered to the peak-to-peak "letter" of the 1825 treaty, even if the treaty's "spirit"—its geographic intent—had been unclear. But the real reason for the border's eventual acceptance by all sides was that it seemed to give the Americans what they needed for a viable territory—and not one square mile more.

With last-minute local accommodations, this was the boundary that was finally marked on the ground, with the survey being completed in 1914. Where the designated peaks were accessible, brass or concrete markers were set into their summits. These monuments were placed so as to be "intervisible": from any one monument on the line, a surveyor could see both the "previous" marker and the "next" one. Thus, even though the boundary would never be marked on the ground, if ever a question arose about the jurisdiction under which a human intervention fell, a determination could be made by going to the nearest mountaintop marker and, from there, sighting on the next marker in the chain. With the sights thus "looking down the line," the surveyor could simply tilt his scope downward and see whether the disputed point was inland of the line (Canada) or seaward of it (Alaska).

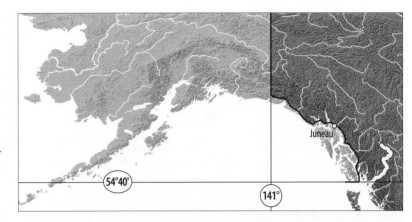

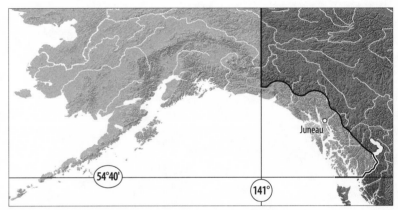

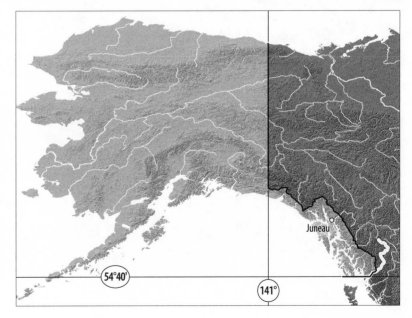

FIG 3.26–28

PART 2

Apportioning the Domain into States

OVERVIEW

If you look in the U.S. Constitution for how to amend the document, you'll find a detailed procedure of precise steps, each to be enacted in sequence, each with its mandated percentage of adoption. But if you look for how to accomplish the potentially more consequential act of expanding the nation through the creation of new states, the Constitution is curiously laconic: Article IV, Section 3 says merely,

New states may be admitted by Congress into the Union; but no new state shall be formed or erected within the jurisdiction of any other state, nor any state be formed by the junction of two or more states, or parts of states, without the consent of the legislatures of the states concerned, as well as of the Congress.

Once invested with that momentous power, Congress never felt constrained to hold itself to a consistent, Constitution-like procedure for state formation. Rather, an "ideal" model evolved, which exerted a countervailing influence on the expedient desires of the moment.

The ideal that evolved in the nineteenth century was that as settlement in the wilderness coalesced into a coherent pattern, Congress would draw a boundary around that area of coherence, declare it a "Territory," and appoint a provisional government. When the territory's population had grown to a sufficient level, and economic and cultural patterns had emerged, a constitutional convention would be called (on the model of America's own founding convention), the delegates would draft a constitution, and—with that proof of graduation from wilderness to self-government—apply to Congress for admission as a new state, fully coequal with all the states previously established.

That was the model, but it was an ideal obeyed only fitfully. In the chronicle of state formation, certain states skipped territorial status altogether (Vermont, Kentucky, California, and, effectively, Tennessee and Louisiana). One state was promulgated despite a minuscule population (Nevada), and another state with a population both substantial and organized was denied admission for decades (Utah). Some territories had borders drawn not to encompass settlement patterns but to split them (Idaho, Washington, and Montana). And one state had borders with no stronger cohering principle than to draw together all that had been left by previous congressional actions (Oklahoma).

The wonder of this ad-hoc process is that, to their inhabitants, the borders of the fifty states feel neither conditional nor arbitrary but natural, right—manifest.

FOUR

A Method of Forming New States Emerges

QUESTIONS FACING THE CONTINENTAL CONGRESS

As the idea of a national domain arose in the 1780s, so too did the idea that new states would be created out of that land. With the coming of independence and the severing of ties with Britain, it came to seem untenable that certain states (Massachusetts, Connecticut, Virginia, North Carolina, Georgia) should grow to become "superstates," with vast western hinterlands, merely because of the vagaries of grants from the throne of England.

And as the idea arose that the states thus unjustifiably blessed would eventually surrender those claims to a central government, it came to seem equally untenable that a central government would itself forever administer the whole of the West surrendered by the states. Americans had thrown off the idea of an empire conducted from a distant center: to reerect that pattern by making all of the West a "colony" of the central government—such a prospect was simply unimaginable.

The assumption by the early 1780s was that the territory of the West would be apportioned into new states, but what sorts of states? Would they be "lesser" states, forever inferior to the original thirteen? The experience of the Continental Congress, where all the colonies were fully equivalent—tiny Rhode island and Delaware each having the same single vote as vast Virginia and populous Massachusetts—had taught Americans the idea that "statehood" was an all-or-nothing concept: there were no categories of statehood.

So by the mid-1780s there was a developing consensus that the West would be apportioned into new states and those states added to the national confederation as equal partners. There was not a consensus, however, on two crucial questions:

- *How would the boundaries of the new states be decided?*
- *How would governments be established in those new states?*

There was of course a model, right at hand, for how to bound and govern new states: that system of royal grants by which many of the colonies had themselves been founded. We saw in chapter 1 how the kings of England (and later the Board of Trade) would make grants of bounded tracts to a syndicate or a powerful individual who would be responsible for governing the colony. But the resulting colonies had found that this system had not worked well for them. For one reason or another, most of the proprietorships or corporations had been dissolved by the time of the Revolution: by then nine of the colonies had their governors appointed directly from London; only Maryland and Pennsylvania remained nominally in the hands of their original proprietors, the Calverts and the Penns; and only the Connecticut and Rhode Island corporations were still functioning.[1]

Nonetheless, this land-grant model had a certain familiarity to recommended it, and it might have been adopted as a way to apportion and govern the West. Certainly, the new states that had thrown off the Crown and hereditary nobility were not about to anoint American noble families on the model of the Penn and Calvert proprietorships. Placing new states in hands of joint-stock corporations might have been a possibility, but the experience of the colonists argued otherwise.

For decades leading up to the Revolution and for several decades after, Americans repeatedly experienced the spectacle of syndicates whose purpose was not the orderly settlement of the wilderness but speculation at the expense of settlers; they would grab hold of the lands that seemed most desirable and then force later buyers to pay inflated prices for their farmsteads. Starting about midcentury, as Americans began to accumulate serious wealth, investors began pooling their money in land companies. With cash in hand, the speculators would approach any entity that had a claim on western land, however tenuous—individual colonies, Indian tribes, the British, French, and Spanish governments—and pay them some nominal sum for "title" to a vast acreage deep in the wilderness.

As the decades of the eighteenth century wore on, and the prospect of colonial settlement beyond the Appalachians became more certain, the number of these syndicates increased, and the efforts of their land jobbers became more frenzied. We saw earlier, in the discussion of Virginia's land cession, how the claims of these companies often had only the most tenuous basis in law, indeed held titles from entities whose land claims sometimes overlapped. You will recall that it was, in part, Virginia's insistence that all such claims be abrogated that delayed the Continental Congress's accepting Virginia's cession of its western lands.

The story of one of these companies can stand as an index of the growing antipathy in colonial, then confederation, America to the idea of trusting the apportionment of the West to corporations.

"VANDALIA" WARNS AGAINST NEW STATES FROM SYNDICATES

Although Vandalia could have become the four-teenth American colony, its genesis lay within one of the many land companies trying to corner the market on western lands in the last decades of the 1700s. What distinguishes this particular syndicate is the illustriousness and near success of its incorporators.

Centered in Philadelphia, the company managed to attract the attention and funds of the region's most eminent lawyers and bankers, chief among them the banker Samuel Wharton and William Franklin, beloved but illegitimate son of Benjamin and, during the time of the company, royal governor of New Jersey. (William was later a Loyalist who fled the colonies for Britain.)

The plan was to buy a tract of some 2.4 million acres adjoining Pennsylvania's southern border in what is now West Virginia (the tract marked 1769 in figure 4.01), and agents of the company approached the Board of Trade in London with that proposal. But on December 20, 1769, the board indicated that it would look favorably on an application for a bigger grant, one large enough for the creation of a new proprietary colony.

A week later the Philadelphians had reorganized themselves as the Grand Ohio Company and presented the board with a proposal for a colony encompassing (in present-day terms) almost all of West Virginia and the portion of Kentucky that wedges in between Virginia and West Virginia (the two areas marked 1770). This would be a colony of some 20 million acres, all of it carved out of Virginia's original grant (and all of it, at least under the Proclamation of 1763, closed to settlement by whites). The Board approved the company's application and sent it on to the Lord of the Treasury (as we have seen, effectively the

FIG 4.01

prime minister at this time), who on January 20, 1770, also approved.

All these approvals, however, came from a government in real turmoil. During the first decade of his reign (1760–70), George III went through a whole series of first ministers. William Pitt the Elder had been forced to resign in 1768 due to what we now might call bipolar disorder, to be replaced by the Duke of Grafton. It was under Grafton's ministry that the Philadelphians had their success, but his approval of their colony was one of his last acts before being forced from his post and replaced by Lord North, who would hold the office until 1782.

So with more pressing matters at hand, the charter for the new colony was shelved until August of 1772, when the Board of Trade again approved the application. The board now called the colony Pittsylvania in honor of William Pitt the Elder, still active in politics despite his setbacks in that arena (and an advocate in Parliament for colonial interests).

In February of 1773, the board authorized a new charter, with even larger borders (the two areas marked 1773 in figure 4.01) and a new name—Vandalia—in honor of George's wife Queen Charlotte, "as her Majesty is descended from the Vandals" (the word then carrying its original Germanic-tribe sense). But that was as far as the charter went. As so often happens with deals like this one, momentum spun down, opposition mounted, and the idea of doing nothing came to seem the best course of action. It would not help that in 1773 Lord North's tax plans for the colonies were being thwarted, especially his tea tax, culminating in the Boston Tea Party on December 16. And it couldn't have helped the case for the new colony that the representative to the court for both bothersome Pennsylvania and rebellious Massachusetts was Ben Franklin, father of one of the Vandalia investors. Franklin, in and out of London since 1764 as spokesman for various colonies and groups, had seen his welcome wear thin as the British government took an increasingly hard line on colonial matters. When he returned to America in 1776, the Vandalia matter was just one of many initiatives that ended with his departure from England.[2]

By September of 1783, George Washington had these thoughts about settlement in the West:

To suffer all the wide extended Country to be overrun with Land Jobbers, Speculators, and Monopolisers or even with scattered settlers is, in my opinion, inconsistent with that wisdom and policy which our true interest dictates, or what an enlightened People ought to adopt, and, besides, is pregnant of disputes both with the Savages, and among ourselves, the evils of which are easier to be conceived then described; and for what? But to aggrandize a few avaricious Men to the prejudice of many, and the embarrassment of Government.[3]

So much for the idea of letting land companies settle the West—and this from the most respected man in the new nation. The words came in a letter written in reply to a request from James Duane, chairman of the Congress's committee on Indian affairs. (The Continental Congress was then convened in Princeton, conveniently just a few miles from Washington's Rocky Hill, New Jersey, headquarters.) Washington's letter is prescient, in almost every particular, of eventual American policy toward settlement, state boundaries, and even relations with the Indians. (Unlike most members of Congress, Washington had actually been in the West in his military campaigns before and during the French and Indian War, and so knew its geography and inhabitants.)

The best policy, he advised, would be to "people the country progressively"—that is, to not allow people to settle anywhere they liked but to compress new settlement into areas adjacent to regions already populated. The proper method for accomplishing this would be to negotiate a succession of boundaries between Indian lands and lands open to settlement, taking care "neither to yield nor grasp at too much," making it "a felony to survey or settle beyond the line, and let the frontier garrison enforce the order."[4] By this means, settlements would be compact enough that the new residents could form effective governments; there would be no strife between Indians and settlers; and speculators would be thwarted from staking out lands far into the wilderness and holding them until settlement reached their holdings, when a premium price could be extracted.

Washington's vision of gradually receding "negotiated" boundaries for Indian lands of course came to pass. And his suggestion for holding settlement back from Indian lands became national policy—though seldom enforced with the vigor he had imagined.

Having replied to Duane's inquiry about Indian matters, Washington asked his indulgence to talk also about the formation of new states, one certain consideration for state boundaries being what borders might be negotiated with the Indians. For the western border of a first new state, Washington proposed projecting a line that would begin at the point where the Great Miami River joins the Ohio, go north to Fort Wayne on the Maumee River, and then curve around and include the fort at Detroit.

If Detroit could not be included, "a more compact and better shaped district for a State" would be to stop the western border at the Maumee River and then erect a second state north of the Maumee, to encompass the peninsula between Lakes Huron and Michigan.[5] In this, plotted in figure 4.02, Washington very nearly predicted the borders of Ohio and the Lower Peninsula of Michigan.

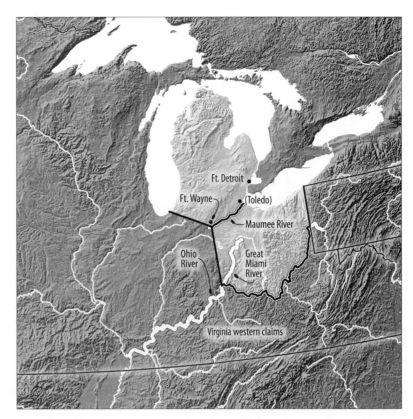

FIG 4.02

Duane must have appreciated Washington's thoughts on new states, but the problem his committee faced was the possibility of war with the Indians over unauthorized settlers crossing the Ohio River into Native territory. The committee had originally been charged only with drafting laws that would enable Congress to regulate trade with, and perhaps settlement among, the Indians. But with Washington's counsel, Duane's committee voiced to Congress a fear that such laws would be insufficient to prevent strife: only the erection of a functioning government in the region of settlement and trade might secure the peace.

Congress agreed and on October 15, 1783, explicitly resolved "to erect a district of the western territory into a distinct government," and to form a committee to propose a plan

for connecting with the Union by a temporary government, the purchasers and inhabitants of the said district, until their number and circumstances shall entitle them to form a permanent constitution for themselves, and as citizens of a free, sovereign and independent State, to be admitted to a representation in the Union; provided always, that such constitution shall not be incompatible with the republican principles, which are the basis of the constitutions of the respective states of the Union.[6]

Congress thus committed itself to the idea that though it would be the body to form new "districts" that would become states, the inhabitants of those districts would perform the primal, government-forming act of writing their own constitutions. (If you wonder at all about this emphasis on the new states adopting "republican" constitutions, remember that beginning in 1776, all thirteen newly declared states had had to write new constitutions, several of them quite unlike the Constitution adopted by the nation in 1788 but nonetheless "trial runs" for that document.

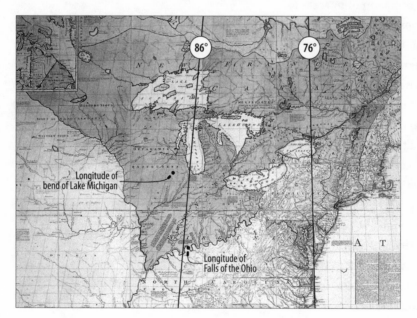

86°

76°

Longitude of bend of Lake Michigan

Longitude of Falls of the Ohio

FIG 4.03

A line of longitude through the Falls of the Ohio actually passes east of Lake Michigan. But if you look at this map, used in the 1783 Treaty of Paris negotiations (compiled by John Mitchell of London in 1755) and circulated widely afterward, you can see how Jefferson might have presumed that a line from the falls would "dead-end" on Lake Michigan's south shore. The maps of Jefferson's plan for states presume that to be the case.

Library of Congress, Geography and Map Division.

The issue of constitution writing was thus very much on the minds of the delegates sent to Congress by the states.)

The committee on governance beyond the Ohio was appointed that same day, October 15, but despite the felt urgency of events, Congress didn't take up the matter until the following April. In fairness, Congress did have a lot on its plate during those intervening months:

- *Later that fall it moved its seat of operations from Princeton to Annapolis.*
- *In December it was presented with Virginia's second offer to cede its western lands, in which the new "districts" suggested by Washington would be located.*
- *And on January 14, 1784, it debated and ratified the Treaty of Paris, finally bringing the Revolution to a close and establishing the United States as an independent nation.*

Little wonder, then, that the Indian Trade and Settlement Committee went through changes, emerging in February of 1784 with a wholly new membership: Jeremiah Chase of Maryland, David Hall of Rhode Island, and Thomas Jefferson, now a delegate from Virginia, as chairman.[7] As we saw in chapter 2, on March 1, 1784, the very day that Congress accepted Virginia's cession of western lands, Jefferson's committee tabled the report on western governance it had prepared the previous month. Then Congress debated the proposal, changed some features, and passed the Ordinance of 1784 on April 23. For a sense of what changed in the debate, the provisions of the committee's March 1 proposal follow.

The committee had worked under the presumption that all former colonies would follow Virginia's lead and cede their western claims. Taking that entire unbounded West as their legitimate focus, the committee proposed that the entirety of the West be apportioned into new states, with their borders set by this formula:

- *Begin at the 31st parallel and draw lines of latitude every two degrees to the north up through the 45th parallel.*
- *Then, from the Falls of the Ohio (present-day Louisville), draw a meridian line north and south throughout the territory.*
- *Then, from the confluence of the Ohio and Kanawha (it's pronounced "kuh-NAW") rivers, draw a second meridian north and south. This grid of lines would set the borders of the new states. (This grid is drawn in figure 4.04 on the following page—with the caveat outlined at left in figure 4.03.)*

The only exceptions would be that all of the Michigan peninsula north of the 43rd parallel would be one state, and the state east of the Kanawha meridian would extend past the 41st parallel all the way to Lake Erie. In addition, the new states would all become members of the confederation, each by a three-step process:

- *When the free male citizens of one of these territories felt themselves ready (or when Congress mandated), they would convene and adopt, as a temporary form of government, the constitution of any of the original thirteen states.*
- *When the district's population reached 20,000 free inhabitants, they would reconvene to write a permanent constitution.*

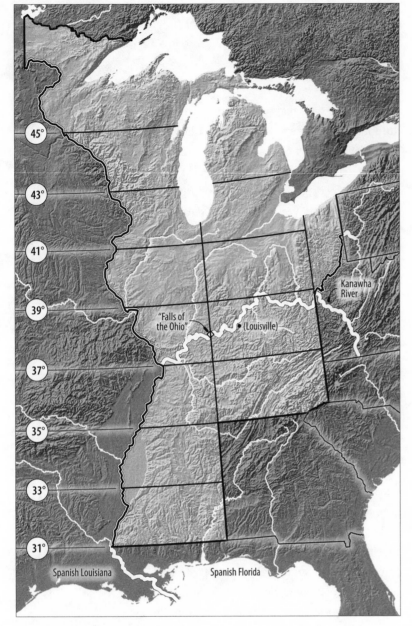

FIG 4.04

The maps of Jefferson's plan are based on a sketch in his papers, which shows the new states arrayed as they are here and Lake Michigan bent eastward as on the Mitchell map. I've drawn these maps with the lake in its actual shape, moving the longitude line to meet the lake at its base.

- *Finally, when its population matched that of the least populous of the thirteen original states, the territory would be admitted to the confederation "on an equal footing" with the original states.*[8]

There would be, of course, some "fine print" in the admissions process:

- *New states would be admitted only upon the consent of two-thirds of the states in the confederation, the number required to reach that fraction increasing as the number of states increased.*
- *New states would "forever remain part of this confederacy of United States of America."*
- *New states would be subject to the laws of the national government to the extent the original states were.*
- *New states would be responsible for paying a part of the national debt, apportioned according to a rule applied equally to all states.*
- *The governments of all new states would be "in republican forms," and no citizens of them would be allowed to hold any hereditary title.*
- *After 1800, slavery would be prohibited in any of the new states.*[9]

This is a truly remarkable and audacious document. Congress had asked for advice on how to control trade with the Indians and set up a temporary government west of Pittsburgh to hold off an imminent Native uprising, and it got instead a complete plan for apportioning the entire West and bringing it into the confederacy!

Where did these stipulations come from? Scholars will forever adduce the provenance of one or another provision from newly discovered or reinterpreted correspondence. Perhaps all we can safely say is that parts of the plan came from ideas in the air at the time, that some resulted from haggling among the members of the committee, while others could only have sprung full-blown from the brow of Thomas Jefferson.

Some of the provisions can be partly explained by events of the time. In light of subsequent American history, the prohibition of secession might seem to foreshadow threats to the Union over slavery, but that was not the issue facing Congress in 1784. The fear then was that states beyond the Appalachians, especially in western Kentucky and Tennessee, might feel so distant from the Atlantic coast that they would form their own independent confederation. Worse, they might be tempted to join their destiny with the Spanish territories just across the Mississippi. (Spain, and later Mexico, would hold out such possibilities well into the nineteenth century. The town of New Madrid, at the boot-heel of Missouri, memorializes a Spanish offer accepted by Americans in 1789.)[10]

The provision outlawing slavery after 1800 seems equally remarkable to us, but that too comes from our knowledge of events that the Continental Congress couldn't then foresee. Toward the end of the eighteenth century, many Americans—even in the South—imagined that slavery might gradually wither away. Tobacco lands were becoming exhausted, and for most of the crops grown inland, slave labor was uneconomical. We know (as Congress could not) that what would make slavery so very profitable, and cause the South to be so vociferous in its defense, would be the perfection of the cotton gin in the 1790s. In colonial America, cotton was what we might today call a specialty crop (the nearest anal-ogy today might be the growing of flax for producing linen): having to pull the tenacious seeds out of cotton bolls by hand limited the quantity that could be sent to mills. Once the seeds could be extracted quickly by machine, vast crops could be processed for the market. Before the cotton gin, the fertile lands west of Georgia seemed to offer land best suited for crop farmers, not plantation owners. A corn or wheat farmer would need only his family, not gangs of slaves, to get the crop in. Or so many Americans expected, and some hoped.

As for the report's other provisions about governance—equal assumption of debt, disavowal of hereditary titles, "training-wheel" adoption of an existing constitution, population requirements—they seem exactly what a people who had just defeated a monarchy and erected thirteen republics would mandate. What seems to our eyes strange and unaccountable is the plan for apportioning the West into small rectangular states. Where did that idea come from?

Scholars agree that this plan for state formation has Jefferson's fingerprints all over it, but he didn't invent the system out of whole cloth. It was a synthesis of habits that Americans had evolved during the colonial period, some mandates from Congress itself, Jefferson's vision for an agrarian America, and some quirky Jeffersonian predilections for what humans should rationally accede to.

It's important to note what's *not* surprising about this plan for forming states—those parts that seem to be not "Jeffersonisms" but natural outgrowths of thoughts in the air at the time. We should not be surprised, for example, by the rectangularity of the states in the plan. We have seen how most of the colonial grants were described, in part or whole, by lines of latitude. That so few lines of longitude were cited in the grants flows naturally from how many of the grants were "from sea to sea" and thus had no need for north-south borders to the interior.

We also should not be surprised by the particular latitudes the plan cites. Its starting point, the 31st parallel next to Spanish Florida, was in 1784 the most definitive border named in the just-signed Treaty of Paris, apart from the water boundaries cited. (Remember that at this point Spain had not yet dug in its heels on the West Florida boundary question.) And the 35th parallel was the agreed-upon southern border of North Carolina's territory. At some point soon, that line would need to be surveyed. Why not put that inevitable effort to good use? And there was the further circumstance that Connecticut, which was then implacably opposed to ceding its lands, had a claim bounded on the south by the 41st parallel. Might not some use be made of that fact?

And finally, we should not be surprised that the north-south borders are not precise lines of longitude, but projections from identifiable landmarks. As an amateur of surveying, Jefferson would have known how difficult it is to determine the longitude of a place, and so his choice to specify longitude by landmark rather than by cited degree number reflects commonsense surveying practice.

No, what surprises us about Jefferson's plan, especially its image when drawn on a map, is that it is so completely unlike the state borders that actually came to pass. Apart from their strangeness, the borders seem to have that quality of

rationalism gone blockheaded that so beguiles us in Jefferson's mountaintop home, Monticello: his bed in a niche between two rooms, the entry-hall clock whose cannonball counterweights descend through a hole in the floor, the dumbwaiter for wine hidden in a mantelpiece. This rack of states appears to be just the kind of artifact you would get if, in a parlor game (or in your study on a mountaintop), you allowed an abstract principle to play itself out unconstrained by real-world contingencies.

The plan certainly contains elements of a thought experiment, but it is perhaps more accurate to say that Jefferson took mandates that Congress had enunciated earlier without too much thought, and then ran with them far beyond where the government would have gone, because doing so turned out to accord so well with his political agenda. Follow this reasoning.

Recall that when Congress first began urging states to surrender their western claims in October of 1780, it passed a resolution to the effect that new states carved out of the cessions be not less than 100 or more than 150 miles square. Virginia, in both its 1781 and 1783 offers of cession, cited this provision again as one of the conditions under which it would cede its claims. (Jefferson, of course, had a hand in both cessions: he urged the 1781 act as governor, and carried the 1783 cession to Congress as a delegate.)

Add to this one of Jefferson's lifelong fascinations, the rationalizing of weights and measures, which also extended to currency (we can thank Jefferson for the fact that the dollar contains 100 cents). For example, he proposed, both informally in notes to himself[11] and formally to Congress,[12] that the standard for distance measurements in the United States be that length of a pendulum which will oscillate once per second. That length, "by Sir I. Newton,"[13] is 39.2 inches—this a decade before revolutionary France set its meter at 39.4 inches. (The French meter would be based on the supposed circumference of the earth, specifically one ten-millionth of the longitude line, from equator to pole, passing through Paris.)

Jefferson too had toyed with the idea of basing the primary unit of length on the planet's circumference, but he seized upon the 360 degrees into which that circumference is conventionally divided. Specifically, he looked to the minute of circumference (one-sixtieth of a degree, you'll recall) by which the nautical mile is defined, that 6,076-foot distance (Jefferson in 1784 presumed it was 6,086 feet)[14] we encountered in the negotiations over the U.S.-Mexican border. We can easily see the good sense of using the nautical mile at sea—sail north one degree, and we traverse 60 nautical miles—but Jefferson was convinced that it would make equally good sense to measure distances on land by the same standard of length. Here's how this idea of a reformed mile, which Jefferson called a *geographical mile*, connects to his plan for new states.

Recall that in addition to chairing the committee apportioning the West into governable states, Jefferson also chaired the committee on apportioning the West into salable parcels. Could the two plans (he must have asked himself) be brought into conformance? They could, if the parceling was done not in traditional statute miles but in his new geographical miles.

If a state spanned 2 degrees of latitude, north to south, then there would be by definition precisely 120 geographical miles between its borders.

That being so, if the land of the West were to be apportioned into parcels one geographical mile square (which is what Jefferson would propose in his land-parceling report), then the boundaries of the parcels would align precisely with the latitudinal boundaries of the new states. Each state would consist of exactly 120 strips, north to south, of one-mile-square parcels.

And there was yet one more argument for the "2-degree-tall" plan for states. Remember the additional provision, from Congress and from Virginia's land-cession offers, for states 100 to 150 miles square. That provision asked for states between 10,000 and 22,500 square miles in area. A state 2 degrees from north to south is about 138 statute miles "tall" (close to midway between the 100- to 150-mile stipulation); but to keep within the implied 10,000- to 22,500-square-mile area, a state that tall could be no more than about 163 miles wide. Even with the imperfect cartographic knowledge of 1784, Jefferson would have known that two ranks of such states—about 325 statute miles from east to west—would not reach from the Appalachians to the Mississippi.

But if the new states' areas were to be gauged in *geographical* miles, a 22,500-square-mile state that was 120 miles tall would be 187.5 geographical miles wide (in statute terms, 215 miles); and twice that length (430 statute miles) *will* extend from the Appalachians to the Mississippi.

In early 1784 Congress convened in Annapolis, and in the course of one day, March 1, the delegates adopted Virginia's offer of cession. Then Jefferson presented his committee's proposed ordinance on forming new states. The delegates debated the ordinance from late March to late April, and their amendments were recorded directly onto the sheets of the committee's report. As was often the case when Congress met, all thirteen states were not represented: the number of states present via their delegates' attendance during this debate varied from nine to eleven. As mandated in the Articles of Confederation, each state had one vote, which was determined by polling whatever delegates from that state were present that day. A simple majority determined a state's vote; if its delegates split evenly over a question, that state was recorded as not voting.

This voting procedure is essentially why, in the adopted ordinance, there is no mention of slavery. Toward the conclusion of Congress's discussions, on April 19, delegates from North and South Carolina introduced a motion to consider the provision outlawing slavery in states entering the Union after 1800. On that day twenty-three delegates were in attendance, representing nine states. When the vote was taken, sixteen delegates voted to retain the prohibition, while seven voted to remove it. But the sixteen delegates represented only three states, whereas the seven represented six states; and so, having failed to gain the support of seven of the thirteen states, the provision was deleted.[15]

Two other changes Congress made to the ordinance, though, have remained in force to this day, and are very much a live issue for people in what is now the West. First, a provision was added that none of the new states "shall interfere with the primary disposal of the soil by the United States in Congress assembled."[16] A little background here: implicit in the evolved idea of the Public Domain was the principle that the U.S. government would have title to the land not yet formed into states. Also implicit was the idea that the U.S. government would sell land directly to private owners. It would seem to follow that when a territory attained statehood—became "sovereign"—that state would assume title to all the land within its borders that had not yet passed into private ownership: after statehood it would be the *new state* that would sell land to private owners. This provision of the ordinance overturned that assumption: even after statehood, any unsold Public Domain land within a state's borders would remain the property of the U.S. government. It is this provision, and its continuation in later laws, that have led to the federal government retaining ownership of huge tracts in the Far West.

Moreover, not only would the national government retain ownership of unsold land, but "no tax shall be imposed on lands [that are] the property of the United States."[17] The principle stated in this second change to the ordinance has also endured, through reenactment, down to the present day. Combined with federal ownership, this prohibition of taxation has had major and continuing effects on states in the Far West.

Congress's other changes to the ordinance were largely cosmetic. The delegates dropped the clause banning hereditary titles, not because they endorsed titles, but because they thought an ordinance on new states was an inappropriate place to mention the issue.[18] And they refined the language describing the new states' boundaries, omitting Jefferson's invented names (for which we may thank them) and beginning the sequence of latitudes at the 45th parallel going south, rather than from the 31st parallel heading north. They also pushed the border of one new state southward, from the 39th parallel down to the Ohio River. (The amendments to the plan are shown in figure 4.05.)

With all that accomplished, Congress passed the revised Ordinance of 1784 on April 23 by a vote of ten states to one. Only South Carolina dissented.[19]

Jefferson must have been pleased overall with the ordinance in its final form (although he may have been too busy just then to contemplate it: he would submit his committee's report on parceling the public lands just two days later, on April 25). The antislavery clause had been deleted (he voted to retain it), but the plan for transition to statehood and the scheme for state boundaries had been heartily endorsed. Congress had been offered the more modest plan Washington proposed—a single state with compacted settlement behind negotiated and fortified boundaries, with the implication of further states similarly formed—but had chosen Jefferson's more visionary scheme instead.

Yet just three years later, Congress would reject both Jefferson's mechanism for forming state governments and his map of state boundaries, and would adopt a plan very much like the one Washington had proposed. What had changed? Events had intervened, of course, but what undid Jefferson's plan was more likely Congress's gradual realization of what his plan presumed and what it implied—its hidden agenda, if you will.

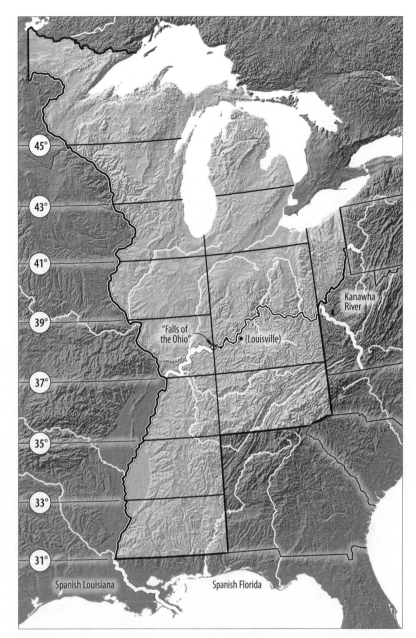

FIG 4.05

George Washington and James Monroe were quick to spot this agenda. Washington, in March of 1785, worried that in laying a grid of states across the entire West, Congress would give people implicit license to settle anywhere they chose, resulting in sparse settlements scattered thinly across the whole territory. Scattered settlements, he feared, would be difficult to defend against Indian attacks, and difficult to govern effectively. Monroe's fear, penned to Jefferson himself in January of 1786, was that the new states would eventually outnumber the old ones in Congress:[20] if fully implemented, Jefferson's plan would produce something like fourteen new states.

But one has only to invoke the resonant term *yeoman farmer* to see how, to a mind like Jefferson's, these were not faults to be avoided but advantages to be sought. What could be better for the nation than to have small bands of farmers forming their own governments, insulated by miles of wilderness from the pernicious, urbanizing East? Who could be more qualified to form new states? How much better for the nation to have its national affairs influenced by the votes of men with self-sufficient, agrarian habits? This, of course, caricatures Jefferson's thought, but when the prospect of only some of these outcomes was pointed out to Congress, the delegates became convinced of the unwisdom of Jefferson's plan, and the need to adopt another.

James Monroe had taken the lead in unveiling the plan's implications. In 1785 he took a trip through the Northwest (one of the first congressional fact-finding trips on record) and returned "with a conviction of the impolicy of our measures respecting it."[21] To his eyes the regions south of Lakes Michigan and Erie were swampy, and the lands along the Illinois and Mississippi rivers were barren (like most people at that time, Monroe thought that the presence of trees was the only trustable gauge of a soil's fertility: soil covered only in grasses must therefore be useless). Jefferson's plan would, he feared, result in two states almost wholly swampy, and two others utterly barren. Might it not be better—for the new states themselves—to lay out bigger states so that, though each would be burdened with poor lands, each would also have good lands to compensate?[22] (Bigger states in the West would also mean *fewer* states in the West, a situation whose advantages were only too apparent to delegates from the East Coast.)

In 1786 Monroe was in Congress as a delegate from Virginia, and early that year he moved that the state boundaries of the 1784 Ordinance be reconsidered. A committee was formed, and its report—submitted on March 24—recommended repealing the part of the ordinance that prescribed boundaries for the new states.

But questions were being raised, not just about the ordinance's plan for boundaries, but about its plan for organizing governments in the new states. When delegates to Congress looked closely at the idea of dividing the West into many small states, they had not liked what they saw. Similarly, when they truly examined the idea of letting frontiersmen set up their own state governments, that idea also came to seem unwise. But whereas they could only imagine the long-term consequences of following the small-state plan, by the end of 1785 an actual instance amply demonstrated what sorts of troubles could result if frontiersmen were encouraged to set up their own state governments.

From the time when settlement first crossed the Appalachians, people in the backcountry had been proposing, and even establishing, governmental entities on their own, in defiance of the central authorities on the coast. The settlers acted out of intentions that ran the whole gamut of political motivation. Some thought themselves ignored or misunderstood by the coastal authorities and merely wanted governments that were more effective and attuned to their needs. Other movements were driven by a backwoods desire to escape government control entirely. And inevitably, some were the brainchildren of individuals or small "cabals" whose motives were mere aggrandizement—personal, political, or economic. Of these many campaigns, the proposed state of Franklin is perhaps the most noted. It is also, in its confused mix of motivations, typical of such movements and can stand (as did Vandalia) as an index for all of them.

Recall that on June 2, 1784, North Carolina ceded its territory west of the ridge of the Smoky Mountains, only to repeal the cession on November 20, at the very next sitting of the legislature. But in the months between, people from the Holston River area (just west of the cession line in the northeastern corner of present-day Tennessee), suddenly without a government, met and on August 23 resolved to enclose their region in a new state; its described borders are plotted in figure 4.06 on the following page. Word of this development soon reached one Arthur Campbell, and with him begins the confused tale of the aborted state of Franklin.

Campbell was the military commandant of Washington County, which Virginia had formed in 1777 as a government in what is now the triangle in Virginia's far southwest. Since 1782 he had been agitating to take the county out of Virginia (under his governance), and after the Holston convention he worked up a petition to Congress from "the people of Washington County" proposing a new state that would include both the Holston area and his own county. He based the boundaries of his proposed state on provisions in the Ordinance of 1784, but with important modifications.

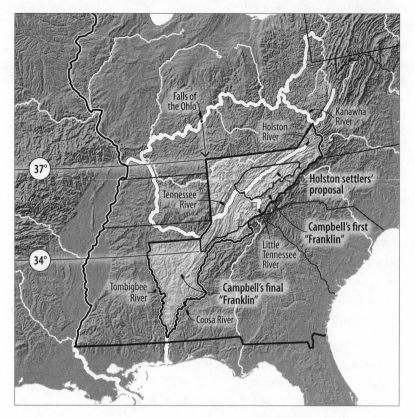

FIG 4.06

The state's northern and western borders would be the 37th parallel and the meridian of the Falls of the Ohio, as specified in the 1784 Ordinance. On the east it would not follow North Carolina's cession line at the Smoky Mountains but instead the eastern watershed line of the Mississippi. Under one reading of Campbell's intent, this line follows the ridge of the Cumberland chain—the Mississippi-Atlantic watershed divide—then meanders between the heads of rivers flowing to the Mississippi and those flowing directly to the Gulf of Mexico. The new state's southern boundary would not be the 35th parallel specified in the ordinance, but the 34th, 70 miles farther south. And finally, in its northeastern corner, the new state would extend to the Kanawha River south of the Greenbrier River, to encompass Campbell's home jurisdiction. Last but not least, Campbell gave a name to his proposed state, one he was sure would find favor in Congress: the new state of Franklin.

When the Holston settlers reconvened late in the year, Campbell showed up with his own, more expansive petition. Despite some confusion, the group was able to adopt a constitution. Following the plan in the Ordinance of 1784, it was chosen from among those of the thirteen original states: the one most familiar to them, North Carolina's. Following a somewhat loose interpretation of the ordinance, the group also resolved to reconvene the following November to modify that document into a permanent constitution; and in the call for that third convention, it declared the borders of the new state to be those as set forth in the ordinance (that is, presumably, the eastern state between the 35th and 37th parallels). The Holston settlers next chose a governor from among their number, and a delegate who would carry their petition to the Continental Congress. But which petition would he carry—Campbell's or the one from the Holston settlers?

After the turn of the year, the delegate traveled to New York City (where Congress had moved), carrying with him both petitions. Congress took no action on the petitions, but that stopped neither Campbell nor the Holstonites.

Since the boundaries specified in the call for the third convention included at least a part of Washington County, when the meeting convened in November, Campbell was there, carrying a radical new constitution as well as yet another boundary for the state—this one extending into Alabama almost to Florida. During the convention, one of the delegates, the Reverend Samuel Houston (cousin of the future Texas hero), spoke strongly for Campbell's constitution, but the Holston settlers preferred working from the North Carolina Constitution, and they won the day.[23]

The Franklin movement eventually ran out of steam, just as did virtually all the attempts by settlers to form new states. But the story serves to illustrate a lesson Congress would come to learn: when government is allowed to arise spontaneously, it can easily fall under the sway of petty demagogues. It would be going too far to say that the Franklin episode "convinced" Congress of the inadvisability of Jefferson's plan for forming settlements—many factors and considerations were at play—but when the delegates convened to amend Jefferson's plan for state borders, they threw out his plan for "bottom-up" governance as well.

Not just second thoughts about Jefferson's plan but other events turned Congress's attention to the question of establishing states. On April 19, 1785, the delegates accepted Massachusetts's cession of its western lands, and on May 26, 1786, they would accept Connecticut's cession, with its Western Reserve exceptions.[24] None of the states south of Virginia seemed disposed to follow suit just then, so Congress had a real incentive to focus its attention not on the whole of the West, as Jefferson had done, but on only that part over which it had (or soon would have) undisputed control, what was then called the Northwest.

One issue to clear up was the condition Virginia had attached to its cession, that its ceded territory must be divided into states 100 to 150 miles square. In July of 1786, a motion was brought asking Virginia to release Congress from this condition, proposing instead that the area northwest of the Ohio River be divided into "not less than two nor more than five" new states— modified in the final resolution to "not more than five or less than three," as compared with the nine states in Jefferson's plan.[25]

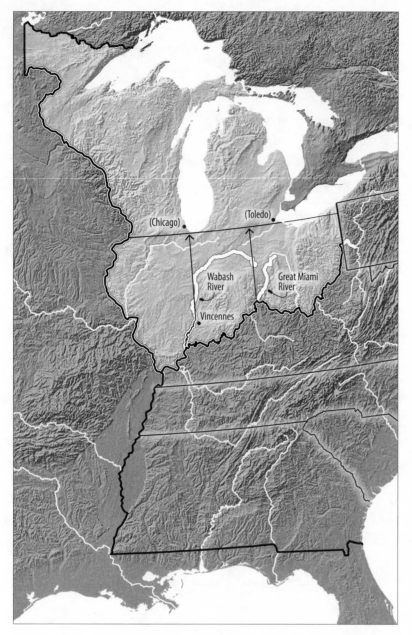

FIG 4.07

states, using lines easy to locate on the ground. Congress saw good sense in Grayson's plan, incorporated it into its request to Virginia, and passed the resolution on July 7, 1786.

James Monroe, meanwhile, had been at work as chairman of a committee on governance for the Northwest. He submitted his report in May of 1786, but before it could be taken up, his one-year term as delegate from Virginia expired, so he was not present for the substantive debate on his plan.

During the debate, Monroe's report was greatly expanded, and in July of 1787 Grayson's plan for state boundaries was folded into it, with the exception (shown in figure 4.07) of moving the starting point of the Wabash meridian up the Wabash to the town of Vincennes.[26] Finally, on July 13, 1787, the plan was passed into law, known ever after as the Northwest Ordinance of 1787.

It was a measure radically different from Jefferson's 1784 Ordinance, and not just in its state borders. Significantly, the ordinance outlawed slavery in the Northwest, and guaranteed religious freedom as well. The change that most affects our story, though, is the mechanism it set out for establishing governments in the new states.

Under the Northwest Ordinance, the settlers in a congressionally bounded state-to-be would not set up their own government; they would be given one, right from the start, in the form of a governor, a state secretary, and judges, all appointed by Congress. When the number of free males reached 5,000, they could elect an assembly and submit a list of residents, from whose number Congress would choose a governing council (for those times when the assembly was

During the debate, Virginia delegate William Grayson proposed state boundaries that would accomplish this division: project a parallel of latitude east and west so as "to touch the most southern part of Lake Michigan," and then run two meridians north to it, one from the mouth of the Wabash River where it joined the Ohio, the other from the similar mouth of the Great Miami. Such a plan would neatly carve the Northwest into five not sitting). Congress, though, would retain the power to appoint whomever it chose as governor, as well as the right to veto legislation passed by the assembly.

Finally, when the population reached 60,000, the residents of the proposed state could write a constitution and apply to Congress for admission to the Union. If accepted, the new state would be a full and equal partner of the original thirteen—with the exception that Congress would retain title to any lands not yet sold, and those lands would be exempt from taxation by the new state.[27]

To historians the Northwest Ordinance was one of the signal achievements of the Continental Congress, because it set the template for all subsequent state formation. This is not to imply that the ordinance was "reused" to form all later states: there were variations from the mechanisms it presented, and some significant departures as well, throughout the settlement period. But from the ordinance came principles and practices that Congress and the public would come to see as the proper way to form new states.

First, there would never again be an attempt at a comprehensive, binding map that apportioned all unsettled territory into future states. Rather, the determination of state borders would be delayed until more facts were in. The process that evolved, especially after the Civil War, was to divide the West into vast territories, but the borders of those territories would be only for administrative convenience; they were not meant as precursors of future state borders. Only when a region seemed to be coalescing into a foreseeably coherent settlement would Congress draw a border around that protostate (and call it, con-fusingly, a territory) and authorize its settlers to begin in earnest the progression of government-forming steps toward statehood.

And it would most definitely be Congress that drew those borders. After the Northwest Ordinance, no local group would be allowed to declare borders of a state (although settlers of Ohio, Indiana, and Illinois were able to modify the borders set out for them in the Northwest Ordinance). In fact, toward the end of the state-forming period, Congress increasingly overrode local desires for borders. After the Northwest Ordinance, governance beyond the established states would be very much a "top-down" affair, directed by Congress through the increasingly powerful territorial governors it appointed.

But another provision of the Ordinance of 1787 would become an ironclad practice in state formation: the use of longitude and latitude lines, and river courses, for state borders. The only exception to this practice in the whole of the Public Domain would be the watershed border between Idaho and Montana.

That last fact points up an interesting fact about American states: almost none have what could be termed "organic" borders, borders that enclose a geographic, economic, or cultural entity.

Watersheds are perhaps the paradigm of a geographic entity: draw a line around a watershed, and every drop of water that falls within that border will stay within that border, either soaking into the ground or moving across it in streams to a single point of outlet. In human terms: place a raft on any stream in the watershed; if left to drift, it will converge with every other raft similarly launched.

When water transport was the only sure way to move goods to market, both downstream with the current and upstream against it, a watershed was not just a geographic entity but an economic and cultural whole as well. When the settlers of the Holston River area set up their first small "state," their border comprised essentially the part of the Holston watershed below the Virginia line. The "Franklin" that Arthur Campbell carried to the December 1784 convention merely completed the rough bounding of the watershed of the upper Tennessee River.

It's a wholly inadvertent happenstance of Jefferson's plan, but a curious number of his proposed states encompass regions that would come to have a kind of geographic and cultural coherence. Let's indulge in a little alternative-history fantasizing in which the United States developed as it has, except only within the area of the Treaty of Paris boundary, and with the West apportioned into the states of Jefferson's plan—a "U.S. Map" depicted in figure 4.08.

- *Sylvania is a state entirely of woods and lakes—impoverished once, but later to become rich, first from iron deposits and then from vacationers.*
- *Michigania is indisputably the Dairy State.*
- *St. Paul and Milwaukee, though, have to deal with straddling a state border (like that town of actual American history, Kansas City).*
- *Assenisipia is the hinterland of Chicago.*
- *Metropotamia is the Land of Detroit, with the resulting state of . . .*
- *Cherronesus being almost wholly agricultural, and a paradise for hunters and fishermen.*
- *With Illinoia in place, "upstate-downstate" differences with Chicago vanish.*
- *Both Cleveland and Cincinnati get states of their own to dominate.*
- *Nashville no longer has to contend with the very different culture of Memphis: it and the Capital of the Delta each get a state to dominate.*
- *And Georgia becomes not only the biggest state in the Union, but one of the richest, with steel and cotton, peaches and railroads.*

This alternative fantasy of small, culturally coherent states (and one powerhouse exception) could never have come to pass, and perhaps that's not such a bad thing. Look, by way of contrast, at the three states that actually did come directly out of Monroe's initiative: Ohio, Indiana, and Illinois. Anyone who has lived or traveled through these states knows how different their northern and southern halves are, with divergent economies, conflicting politics, even different accents. The states enclosed in those Northwest Ordinance boundaries are decidedly "inorganic."

This fact would have displeased Jefferson. Where there is conflict to manage, there must be intrusive government; only in an organic entity, where people have a near unanimity of interest, can there be a "government that governs least." But the split economies and cultures of these three states have helped to ensure their long-term stability, just as Monroe might have predicted.

But these states, as they have turned out, would have pleased no one so much as a third Founding Father, James Madison. All during the debate on the federal Constitution, it was his great contention that the surest way to avoid the tyranny of a majority would be to throw divergent interests together into an "extensive republic." One effect of forming expanded states in the West was to produce just such a condition, not just in each of the states formed under the Northwest Ordinance, but (as its principles were reapplied) in nearly all the states apportioned out of the National Domain. Look at the geography of almost any state in the West, and you'll find that it has at least two landscapes. Think of Oregon and Washington (forest and desert); or Montana, Wyoming, and Colorado (mountains and plains); or all the states in the column from the Dakotas down to Oklahoma (enough rain for corn or wheat in the east, only enough for grazing toward the west). And even where a state's boundaries happen to enclose relatively unified geographies (Nevada, Iowa), the resulting states were big enough that cultural differences found sufficient distance to take root.

Almost none of this pervasive yoking of opposites happened by design. In all of the Public Domain, only two principles for state shaping were consistently applied: make 'em rectangular and make 'em big. Those two precepts, when imposed with indifference to geography, would

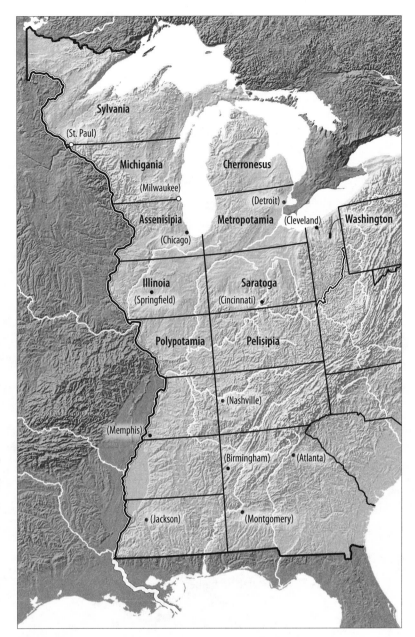

FIG 4.08

almost inevitably result in states containing "factions" (Madison's term) that would be forever contending.

This outcome would perhaps not surprise the drafters of the Northwest Ordinance. In actuality the outcome only confirms what they were beginning to experience in their own states as settlement moved west: the growing divergence of interests between tidewater and hill country.

These colonial variations happened because the original English grants were formed on the same principles Congress would use: the royal grants were as big as practicable, rectangular where possible, and made without too much regard to geography (which was, in any case, largely unknown then). The inadvertent result in the larger colonies was a happy, "Madisonian" balancing of interests.

Before we move on to the chronology of how the states in the Public Domain were apportioned, let me broach two thoughts. In the story of the formation of the original colonies, we saw one colony's borders that were particularly "inadvertent," as its eventual shape was formed almost entirely by the more determinate boundaries of the colonies around it. The state that resulted has, of all the thirteen original states, the greatest number of identifiable geographic and cultural regions. It's why New York has been able to boast, with some justification, that it is the Empire State.

And there is a counterexample to this rule of states being best formed by benevolent inadvertence (or, perhaps, inadvertent benevolence). I said that after the Northwest Ordinance, residents of a region were never permitted to set the borders of their state. That is not quite true. The border between Virginia and West Virginia is almost entirely the result of county votes on the issue of secession, taken when Virginia chose to join the Confederacy early in 1861. With a few exceptions, the counties voting yes on secession stayed in Virginia, while those voting no coalesced into West Virginia. To the extent that a position on secession can be said to mark a divide between cultures, West Virginia's eastern border encloses behind itself a unitary cultural whole. To that extent, it is the closest we come in the United States to an "organic" state borderline.

The maternal side of my parentage comes from Virginia, but the paternal side hails from West Virginia; and I've never encountered people more devoted to, and proud of, their native state than my father's folks. But I think even my West Virginia forebears would at least entertain the thought that determining your own borders so as to enclose an organic whole may not be the best way to form a state, at least not in this Union. Better perhaps to have the task done by an outside party, by abstract principles inadvertently applied by the powers at the center, who are *just ignorant enough* of your parochial concerns.

FIVE

The Evolution of the Territories and States

In the political realm, there have been debates in recent years about whether one institution or another "looks like America," but "what America looks like"—in a cartographic sense—has been a settled matter for over a century. America looks like the image in figure 5.01 on the next page. This particular configuration of forty-eight shapes has constituted "the U.S. Map" since the Dakota Territory was split into a stack of two states in 1889. Since the 1959 admission of Alaska and Hawaii, the iconic image has been modified somewhat: by common consent we seem most often to show those two states to ourselves as islands somewhere south of San Diego, with Hawaii somewhat too large, and Alaska vastly too small. But though we mean no disrespect to our newest states, most of us find our eyes pulled back ineluctably to the curious animal that stands on two pointy feet and holds up a tiny head, its body carved up into forty-eight interlocked pieces.

But in the century between the Revolution and the division of the Dakotas, an American asked to imagine a map of the nation would have conjured a wholly different image. As the national domain expanded across the continent and was divided into territories and states, "the U.S. Map" changed continually. If you take every single change into account, there were over thirty different "maps of the U.S." Some of the configurations lasted for only a matter of days, but a few were in effect long enough, and were disseminated widely enough, that for significant numbers of people, those maps were "what America looks like."

I've reconstructed twenty-eight of those "U.S. Maps." Bear in mind that for some of the early renderings, the shapes drawn by cartographers and imagined by the people of the time would have been different from the shapes we can construct with our present-day geographic knowledge. And note as well that some of the shapes on the maps—those of "unorganized" territories, for example—have boundaries that were never intended as definitive borders at all; those entities merely take on an aura of "intentionality" when drawn on a map. Nonetheless, do let yourself contemplate some of the shapes on the maps as "might have beens." Imagine Idaho and Montana divided, as they might have been, along the Continental Divide. Take a look at Missouri as it was originally constituted—and recognize that for seventeen years, Americans saw that shape, so strange to us, as "the map of Missouri."

"Fun Facts"

Much of the data for the following maps come from a book compiled by Edward M. Douglas and published in 1932 under the auspices of the Department of the Interior, with a title that says it all: *Boundaries, Areas, Geographic Centers and Altitudes of the United States and the Several States, With a Brief Record of Important Changes in Their Territory and Government.* At the end of a book filled with facts and elegantly hand-drawn maps, Douglas adds an appendix filled with the kind of statistics that once appeared on the paper placemats of truck-stop restaurants. He tells us that the two points farthest apart in what is now the contiguous forty-eight states are Cape Flattery in Washington State and "a point on the Florida coast south of Miami, 2,835 miles apart."[1]

And have you ever wondered where the geographic center of "the Lower 48" is? You'd find that point by cutting the precise shape of the contiguous states out of a sheet of material with an absolutely uniform thickness and weight, and then finding where the shape would balance on the point of a pin. It turns out that the balance point is midway between the eastern and western borders of Kansas (which seems only fitting), just a few miles south of the Nebraska line. If you're ever on Interstates 80 or 70, veer off toward the little town of Lebanon, near the intersection of U.S. Routes 281 and 36.

You will recall that in its Northwest Ordinance of 1787, the Continental Congress specified the borders of three new states in the territory north of the Ohio River. The first state was to span from Pennsylvania to a meridian at the mouth of the Great Miami River; the second would extend to a meridian drawn north from the town of Vincennes; and the third would stretch to the Mississippi. All three states would be bounded on the north by a parallel of latitude drawn across the southernmost point of Lake Michigan, with Congress reserving for itself the right to form states north of that line. The new federal Congress carried forward the provisions of the Northwest Ordinance in 1789, and there matters stood in "the Northwest" for over a decade.

In "the Southwest," Georgia maintained a claim on lands all the way to the Mississippi, but Spain still had a legalistic hold on the part of its West Florida colony between the 31st parallel and the mouth of the Yazoo River. With its 1795 treaty with Spain, the United States took control of that strip of land, and on April 7, 1798, declared it to be the new, federally governed Territory of Mississippi.[2]

Cape
Flattery

Geographic Center of
the Contiguous States

"point on
Florida coast"

FIG 5.01

FIG 5.02

The author, in younger days, at the geographic center of the contiguous states, near Lebanon, Kansas. Some years ago, local civic organizations took notice of their fortuitous proximity to this cartographic fact, bought land for a small park, and erected this handsome monument at the park's center.

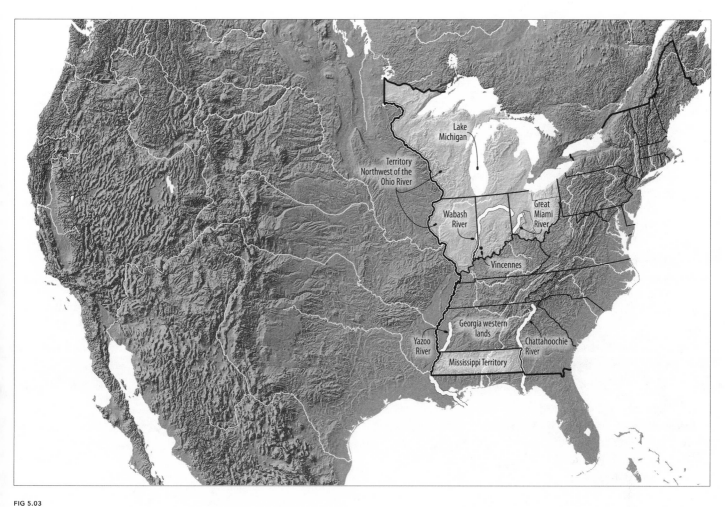

FIG 5.03

Georgia naturally contested the establishment of the Mississippi Territory, but eventually ceded its western lands in 1802. In 1804 Congress expanded the Mississippi Territory all the way up to Tennessee.[3]

On May 7, 1800, Congress modified the Northwest Ordinance, setting a precedent it would follow (with exceptions) for all the subsequent states. Rather than declaring boundaries for protostates, as had been done in 1787, provisional territories—larger than any potential states—would be set up, with the question of state boundaries to be settled later. In chapter 7 you will read about the Greenville Treaty of 1795, under which Indian tribes ceded much of what is now Ohio; one of the borderlines of that cession ran from the mouth of the Kentucky River northward to Fort Recovery. With its act of 1800, Congress adopted that line as the eastern border of a new Indiana Territory, extending the line from the fort due north to the British possessions in present-day Canada.[4]

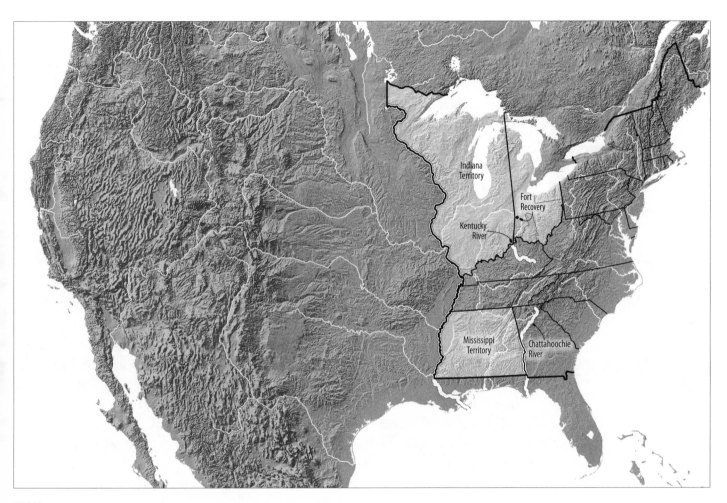

FIG 5.04

By 1802 there were sufficient settlers in what is now Ohio for Congress to begin the process by which it would erect the first truly new state (in the cases of Kentucky, Tennessee, and Vermont, Congress had essentially acceded to developments already well advanced). Congress had set a precedent in establishing the vast Indiana Territory, and it set another precedent in the procedures it mandated for Ohio's statehood. On April 30, 1802, it adopted an *enabling act*, setting forth the boundaries of the new state and authorizing the citizens within those bounds to gather together and adopt a state constitution under which they would govern themselves. The Ohioans adopted a constitution on November 29, and Congress approved the charter the following February 19, authorizing statehood to take effect on the constitution's first anniversary, November 29 of 1803.[5]

There was a problem, however. In its enabling act, Congress had gone back to the borders specified for the "easternmost state" in the 1787 Northwest Ordinance—on the west, a meridian line from the mouth of the Great Miami River, and on the north, "an east and west line, drawn through the southerly extreme of lake Michigan."[6] The

new Ohio Constitution cited the same boundaries but inserted a proviso, to the effect that if an east-west line from the foot of Lake Michigan should intersect Lake Erie below the mouth of the Maumee River, then,

with the assent of the Congress of the United States, the northern boundary of this State shall be . . . a direct line running from the southern extremity of Lake Michigan to the most northerly cape of the Miami [now Maumee] Bay, after intersecting the due north line from the mouth of the Great Miami River.[7]

Clearly the intent of Ohio's founders was that the natural harbor at the mouth of the Maumee River would go to the new state, not to the still-unnamed territory to the north. Congress, however, took no explicit action on this crucial proviso in Ohio's constitution, and there matters stood until 1817.

By that time, Indian tribes had ceded the lands around the Maumee River, and it became common knowledge that an east-west line through the southerly bend of Lake Michigan would cut Ohio off from Maumee Bay. So the surveyor

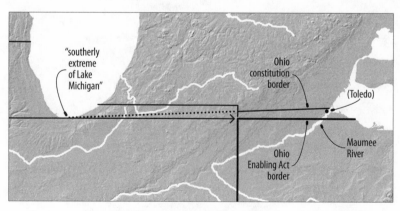

FIG 5.05

CHAPTER FIVE

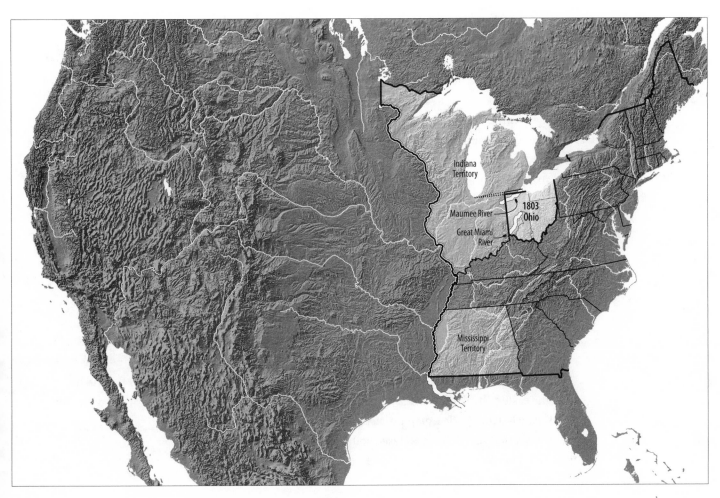

FIG 5.06

William Harris was sent out to run the angled line from the foot of the lake to "the most northerly cape" of the bay. By now, though, the Territory of Michigan had been established, and the governor of the territory not only protested to Congress but in 1818 sent his own surveyors out to run the east-west line. Neither Ohio nor Michigan would accept the other's line, and matters had so deteriorated by 1835 that armed troops patrolled the border region. Finally, in June of 1836, Congress settled the matter in the enabling legislation for Michigan's own statehood, declaring for the angled line and giving subsequent generations the city of Toledo, Ohio, rather than Toledo, Michigan.[8]

Once the Louisiana Purchase was approved, Congress set about to provide governance for its vast new domain. In 1804 it chose the 33rd parallel as a dividing line: all of the purchase north of the line would be called the District of Louisiana (changed in 1805 to the Territory of Louisiana), and everything south would be the Territory of Orleans.[9] You will recall that President Jefferson had an expansive view of what he had purchased, and the declared borders of the Orleans Territory reflect this. It would extend to the Mississippi watershed line on the west, but on the east it would include all the land east of the Mississippi lying south of the 31st parallel, all the way to the Perdido River. The Spanish authorities naturally contested this: their view was that their West Florida province comprised all the land east of the Mississippi below the 31st parallel except the Island of New Orleans (which, you'll remember, was the eastern bank of the Mississippi below Lake Ponchartrain). The dispute stood until 1819, when the United States annexed both East and West Florida.

To the north, Indian cessions around Fort Detroit convinced Congress of the necessity of erecting a government for the region, and so on June 30, 1805, it carved a new Michigan Territory out of the previously established Indiana Territory. Its southern boundary was to be the east-west line drawn through the foot of Lake Michigan that Ohio had disputed, with the western boundary being "a line drawn from the said southerly bend through the middle of said lake to its northern extremity, and thence due north to the northern boundary of the United States."[10]

Clearly the intention here was to give the new territory jurisdiction over what present-day Michiganders call the Lower Peninsula, but because of the configuration of Lake Michigan, a strict interpretation of Congress's words would add in a portion of the Upper Peninsula as well. In the following maps, watch as the Michigan Territory comes to claim more and more of the Upper Peninsula.

CHAPTER FIVE

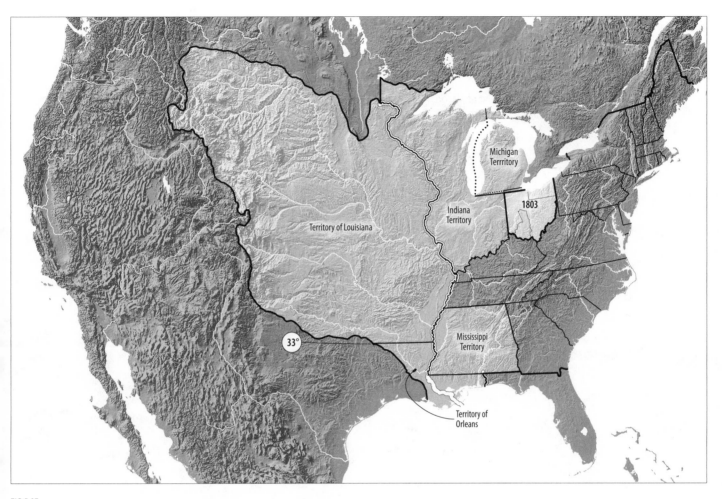

Territory of Louisiana

Michigan Terrritory

Indiana Territory

1803

Mississippi Territory

33°

Territory of Orleans

FIG 5.07

On May 1, 1809, the Indiana Territory was diminished yet again, with the establishment of a new Illinois Territory. The new entity would extend from the Mississippi eastward to "a direct line drawn from the said Wabash river and Post Vincennes, due north to the territorial line between the United States and Canada."[11] Indiana Territory thus reverted to the borders of the "middle state" described by the Northwest Ordinance 'way back in 1787—and Michigan Territory acquired a little more of the Upper Peninsula.

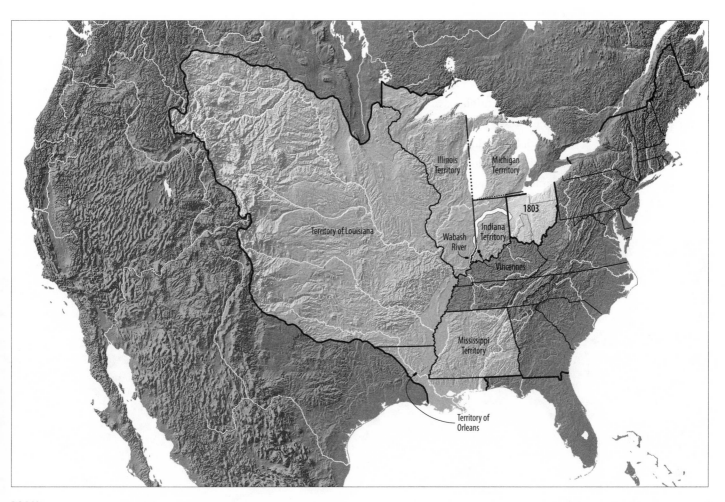

FIG 5.08

As we saw in chapter 3, in 1812 Congress took the provocative step of declaring expansive borders for a new state of Louisiana—borders that pushed into Spanish territory on both the east and the west. Curiously, Congress brought this about in two steps. On April 8, 1812, it set forth a western border for the new state that went up the Sabine River to the 32nd parallel, north on a meridian to the 33rd parallel, and east on that parallel to the Mississippi. Then six days later, on April 14, it added to the brand-new state the area south of the 31st parallel and west of the Pearl River—giving Louisiana the borders it has today. Compounding the offense to Spain, Congress also took the area between the Pearl and Perdido rivers and officially added it to the Mississippi Territory.[12]

Having assigned the name *Louisiana* to the new state, Congress needed a new name for the remaining lands of the Louisiana Purchase. It chose the mellifluous name of the great river that stretched so far to the northwest, establishing the Missouri Territory on June 4, 1812.[13]

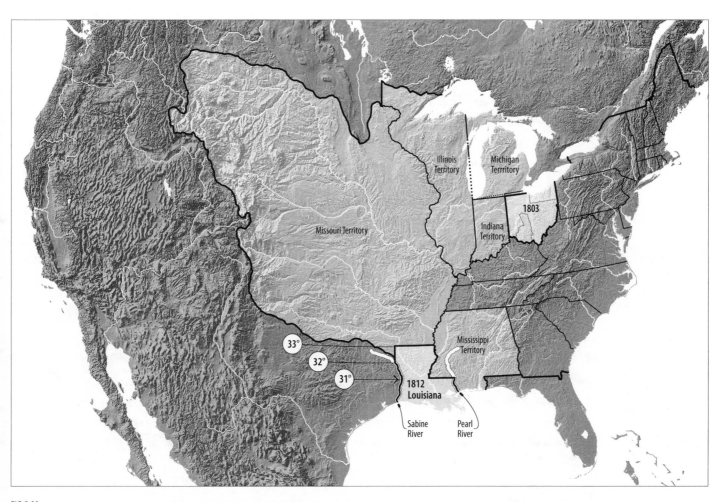

FIG 5.09

By 1816 Congress determined that Indiana Territory was ready to become a state, and on April 19 it passed an enabling act to that effect. In the act it wisely corrected two more of those borderline absurdities that keep cropping up in any history of America's borders.

The first had to do with the meridian drawn north from the town of Vincennes on the Wabash River. By 1816 everyone knew that above Vincennes the course of the Wabash describes a shallow arc to the west before veering back east farther upstream. A line straight north from Vincennes would slice across that arc, placing a thin sliver of land into Illinois Territory and cutting Indiana off from the eastern bank of the Wabash. So Congress instead declared that Indiana's western border would run up the middle of the Wabash "to a point where a due north line drawn from the town of Vincennes would last touch the northwestern shore of said river." In other words, the border would follow the river northward until the river crossed back over the meridian from Vincennes.

The second correction had to do with the east-west borderline drawn tangent to the southern-most bend of Lake Michigan. Congress finally recognized how foolish it would be if this new state sat right next to a vast inland sea and yet had a shoreline on that sea that could be measured in feet. So the northern border of Indiana was declared to be "an east and west line drawn through a point ten miles north of the southern extremity of lake Michigan." Both of these borders took effect when Indiana's statehood became official on December 11, 1816.[14]

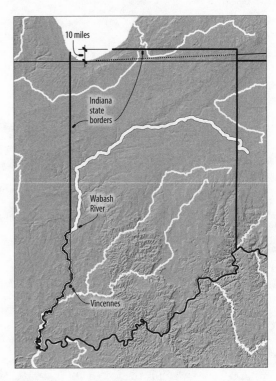

FIG 5.10

Just after the turn of the year, Congress took up the matter of a new state in the Mississippi Territory. By this time, Americans had been settling in the western half of the territory for some time, both along the Gulf coast (establishing "facts on the ground" in Spanish West Florida) and in the lands along the Mississippi north of the Louisiana line. In March of 1817, Congress adopted a pair of measures dividing the Mississippi Territory into a new state of Mississippi and a new Territory of Alabama. Their common border would start from the north, where the Tennessee River crossed the 35th parallel, then follow the Tennessee south to where Bear Creek flowed into it (giving the new state an outlet onto the Tennessee River). From there a straight line would be drawn to the northwest corner of the previously established Washington County in Mississippi; and from that corner, the boundary would run due south to the Gulf of Mexico. Statehood took effect on December 10, 1817;[15] and ever since, the Alabama-Mississippi border has had that slight "back bend." Now you know why.

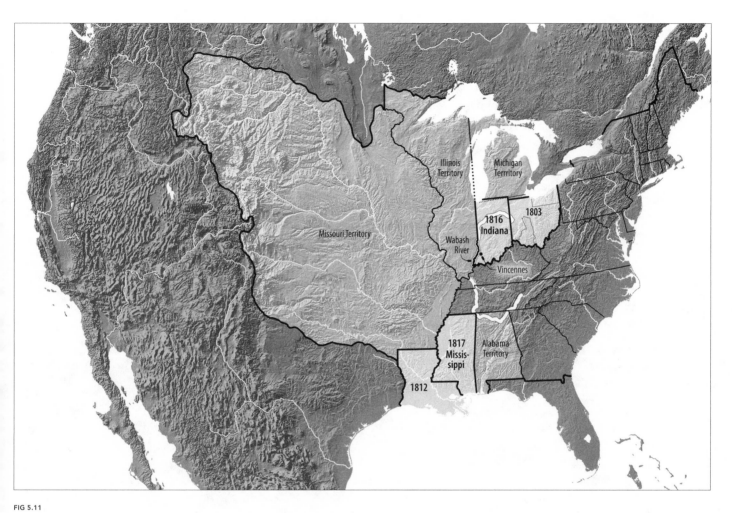

FIG 5.11

Having modified the Northwest Ordinance boundaries for Indiana, Congress did so again for the new state of Illinois. The northern border was declared to be the parallel of latitude at 42° 30'. This latitude gave Illinois not only the nascent port at the Chicago River but also the lead mines around Galena in the northwestern corner of the state. (Galena is the common name for the mineral lead sulfide, the primary ore from which lead is refined.) Illinois' statehood became effective on December 3, 1818, at which time the remainder of the former Illinois Territory was added to an expanded Michigan Territory.[16]

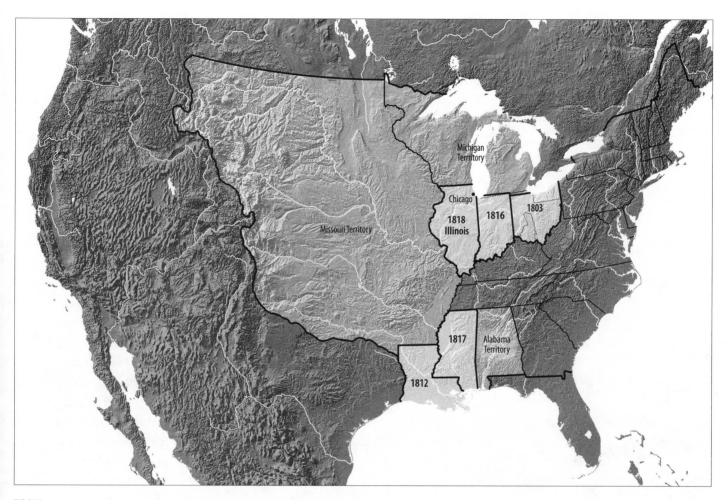

FIG 5.12

On December 14, 1819, the state of Alabama came into existence, with the same boundaries as Alabama Territory.[17] Earlier in the year the acquisition of Spanish Florida had been completed, but Congress chose, for the moment, not to erect a territorial government there. This situation occurred several times during the territorial period (sometimes inadvertently), and to it historians have given the name *Unorganized*.

As part of the same negotiations, you'll remember, the border of the Louisiana Purchase was redrawn at the Adams-Onís Line, reshaping the Missouri Territory. On March 2, 1819, Congress carved a new territory out of Missouri and chose for it the name *Arkansaw*.[18] The territory's name would later be changed to the more familiar spelling, but its northern border would remain, and would become the basis for the Missouri Compromise the following year.

You'll recall from chapter 1 that the royal grant to the Lords Proprietors of Carolina in 1665 set the border between Virginia and the future North Carolina at latitude 36° 30'. In 1779 and 1780, the two colonies dispatched a joint commission of surveyors to mark the line on the ground. The surveyors knew that they would have to stop their work at the Tennessee River, for beyond lay Indian country. As they drove the line through the woods, it veered gradually northward, so that by the time the team reached the shore of the Tennessee, it was about 3° north of 36° 30'. Nonetheless, when Virginia and North Carolina surrendered their western lands, this became the border between Kentucky and Tennessee.

In 1819 a treaty was signed with the Cherokee that extinguished Indian claims in the area west of the Tennessee River, and so it became possible to complete the survey of the Kentucky-Tennessee border. The Kentucky legislature instructed Robert Alexander and Luke Munsell to proceed up the Mississippi River to the point where it crossed 36° 30' of north latitude, and then run a line due east to the Tennessee River. This they did, the line crossing over a great oxbow in the Mississippi's course, cutting off a tiny piece of Kentucky from the rest of the state.[19]

Regardless of the consequences, the position of latitude 36° 30' had been marked with great accuracy on the banks of the Mississippi, and Congress seized on that fortuitous circumstance. The northern border of the new Arkansaw Territory would be latitude 36° 30'—with one exception.

You may have heard of the great New Madrid earthquakes of 1811 and 1812; you may even know that the town's name is pronounced "New

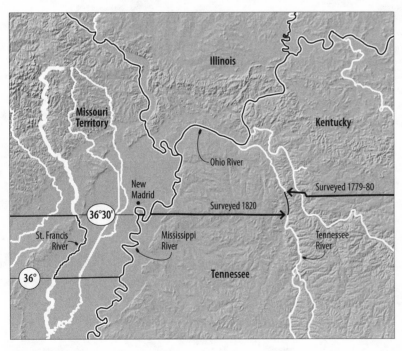

FIG 5.13

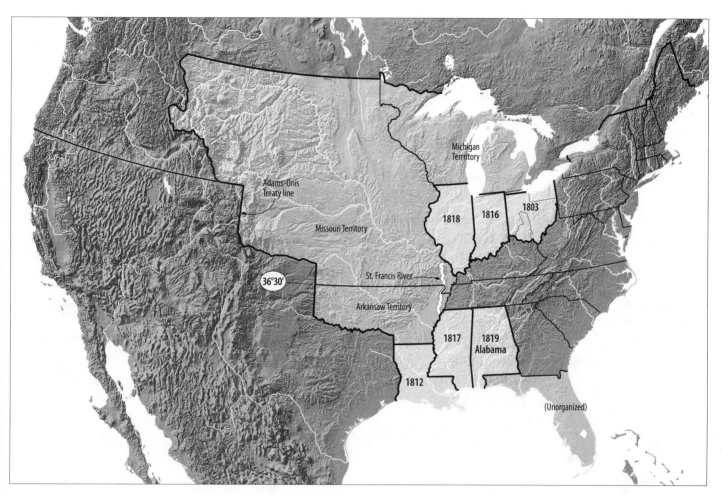

FIG 5.14

MAD-rid"; but you might wonder why American settlers in what is now Missouri would commemorate the capital of Spain. The reason is that they had been drawn to the region at the turn of the nineteenth century by offers of land from the then-Spanish government of Louisiana.[20] Their farms stretched up and down the Mississippi from the New Madrid settlement, and they asked not to have their region split in two by the 36° 30' parallel. Thus, a "boot heel" was created for them:

the border of Akansaw Territory would start at the 100th-meridian line set by the Adams-Onís Treaty and go east on latitude 36° 30'. At the St. Francis River, though, the border would follow that river's course southward to the 36th parallel, and from there run east to the Mississippi. The new Territory of Arkansaw came into existence on Independence Day of 1819.

All through 1819 and into 1820 Congress was debating the real purpose behind organizing Arkansaw Territory: the admission to the Union of the state of Missouri to provide a government for the thousands who had settled there. Many of the settlers had come from Virginia (some, after first settling in Kentucky), and they carried the tradition of slavery with them. The Missouri Compromise resolved the debate over slavery in that its provisions included Missouri's admission as a slave-holding state, but to maintain a 12-to-12 balance of slave and free states in the Senate, Maine would be split off from Massachusetts as a free state—with the further proviso that north of latitude 36° 30', slavery would be "forever prohibited."[21]

The enabling act for Maine was passed on March 3, 1820; the Missouri act, three days later. Maine was admitted to the Union effective March 15, 1820; Missouri's admission was delayed until August 10, 1821.

Missouri's western border was to be a line of longitude struck through the confluence of the Kansas and Missouri rivers (present-day Kansas City). The northern border would then be a "parallel of latitude which passes through the rapids of the river Des Moines, making the said line correspond with the Indian boundary line."

By 1822, Congress had determined that it was time for Florida to have a true government, and so on March 30, the Territory of Florida came into existence. Two years later Congress shrank the Arkansaw Territory (and renamed it Arkansas), moving its border to a line of longitude 40 miles west of the southwestern corner of the new state of Missouri. West of Arkansas Territory would now be Indian Territory,[22] a foreshadowing of the policy, soon to be adopted, of "removing" all eastern Indians to reservations in the West.

CHAPTER FIVE

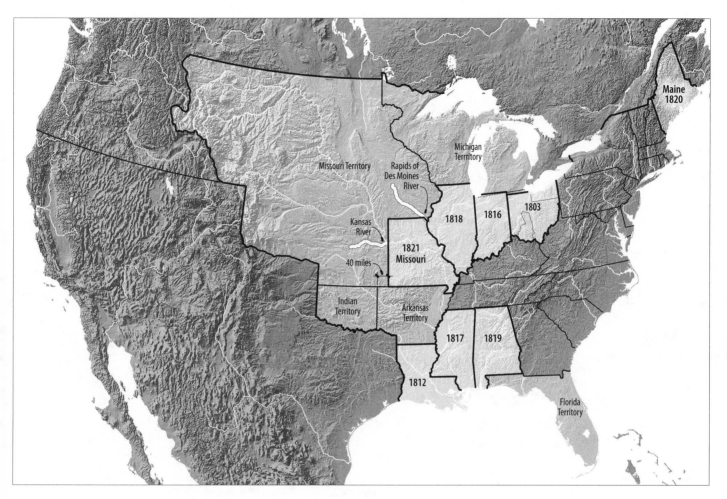

FIG 5.15

Arkansas Territory was diminished once again in 1828, this time by two treaties with Indian tribes. On May 6 a treaty with the Cherokees defined the western border of Arkansas Territory as

commencing on Red river, at the point where the Eastern Choctaw line strikes said river, and run due north with said line to the river Arkansas; thence in a direct line to the South West corner of Missouri.[23]

The Eastern Choctaw Line had been set by a treaty with the Choctaw Nation on January 20, 1825. The line was to begin "one hundred paces east of Fort Smith, and running thence due south to the Red River."[24] In 1836 Arkansas would become a state within these territorial borders; the angled-back shape of its western border was due entirely to the location of a frontier fortress.

Indian Territory was extended in 1834 with the expansion of the Michigan Territory. On June 28, Congress chose to extend Michigan's jurisdiction to the Missouri River. The border would run northward along the Missouri to its junction with the White Earth River, which would be the extreme western border of Michigan Territory.[25] (The White Earth extends almost to the 49th parallel, so we can draw Congress's intent by projecting a line northward from its source point.) Beyond the Missouri would be a territory reserved, for all time, for Indians.

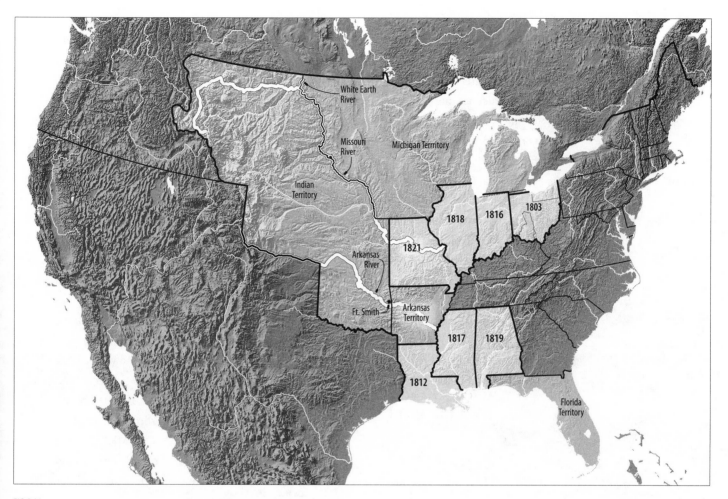

FIG 5.16

In 1836 Missouri amended the provisions in its constitution that describe its borders, extending the state's northwestern corner to the Missouri River. Congress agreed to the change in June, and the new border became official on March 28, 1837.[26] While Congress considered Missouri's application, it approved statehood for Arkansas, which took effect on June 15, 1836.

On that same day Congress passed the enabling act for Michigan's statehood. The legislation called for the new state's southern border to run along the boundary with Indiana, but then to jog south and run along the angled-upward boundary that Ohio had requested in its constitution—if, that is, the people of Michigan gave their approval at a constitutional convention. At first the Michiganders refused, but in December of 1836 they relented,[27] and that is why Michigan has had that zigzagged southern border ever since.

Settling the state's northern border was also difficult. No "natural" boundary occurs where what is now called the Upper Peninsula can be said to diverge from the main body of the land, but there was some hope that two watercourses—one flowing into Lake Superior, the other into Lake Michigan—could be located whose source points would almost touch. That became the northern border under which Michigan was admitted to the Union on March 28, 1837. It was later determined, however, that more than 50 miles separated the rivers, so in 1846 the legislators improvised. Essentially, the border would begin at the mouth of the Montreal River on Lake Superior, then follow the river up its course. From the river, Congress called for the border to jump in a straight line to the western end of "the Lake of the Desert" (now called Lac Vieux Desert), go to the lake's eastern end, then jump to the nearest headwater of the Menominee River (at Brule Lake). From there the border would follow the river to Lake Michigan, but once at the lake it would veer northward around the peninsula that now contains beautiful Door County, Wisconsin.

Michigan's admission set in motion a reorganization of the old Michigan Territory. On June 7, 1836, Congress ratified Missouri's "bite" out of the territory's southwestern corner. On June 15 it passed the Michigan enabling act, and on July 3 it created a new Wisconsin Territory out of the resulting remnant of Michigan Territory.[28]

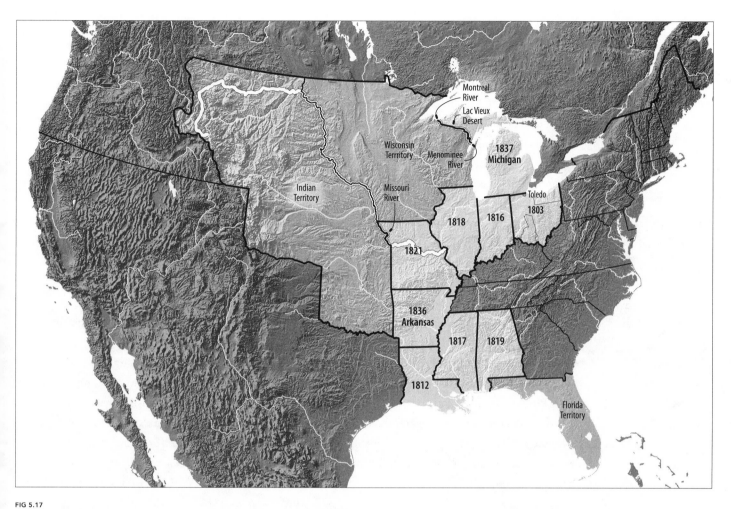

FIG 5.17

You will recall that in 1842, the issue of Maine's northern border was finally resolved when the state accepted the treaty worked out between Lord Ashburton and Daniel Webster. But that event falls in the middle of the story this section tells. Let me jump back—and forward—a few years.

In 1838 Congress reorganized the Wisconsin Territory, deciding that all its lands west of the Mississippi would be put into a new Iowa Territory. By 1845 Congress felt the southeastern part of the new territory to be ready for statehood, and on March 3 it passed enabling legislation, with the special proviso that the proposed border would have to be approved by a referendum of the people in the area.

The southern border of the new state of Iowa would naturally be at the northern boundary of Missouri, with its eastern border at the Mississippi River: that much was undisputed. At issue was the matter of the state's northern and western borders. The enabling act called for Iowa's northern border to be a line of latitude through the mouth of the "Mankato or Blue-Earth river" at the Mississippi (the river is now called the Whitewater, and its mouth is at about 44° 10' north latitude). The western border would be a meridian line "seventeen degrees and thirty minutes west of the meridian of Washington city." (This is approximately longitude 94° 30' west of Greenwich; I'll explain about the Washington Meridian in a moment.) When put to the vote, the people of Iowa rejected these boundaries, throwing the question back into the lap of Congress.

Iowa's 1845 border is the first instance of Congress specifying boundaries with reference to a meridian at Washington, DC. From colonial times it had been the practice in America to calculate longitudes with reference to the meridian at the Greenwich Observatory, but in 1842 Congress authorized the construction of an equivalent American observatory, to be built west of the White House on a promontory above the banks of the Potomac. A bit of national chauvinism entered into this decision, but some practicality as well. One of the main functions of an observatory then was the signaling of precise moments of time—solar noon, for example—to which other, more portable clocks could be synchronized. Having a "master clock" on this side of the Atlantic made perfect sense.

The observatory was up and running by 1844, ready for Congress to ask that its calculations be used to set Iowa's longitudinal border. In 1850 Congress went further, mandating that even though the Greenwich Meridian would continue to be used for "nautical purposes," within the United States the Washington Meridian would be the one used for "astronomic purposes" (meaning, especially, measurements on land). The new meridian was defined with precision: it would be the line of longitude passing through the center of the dome of the new observatory. Unfortunately for us in this story, precision at the Washington observatory leads to imprecision when talking about American borders. At the time, the Washington Meridian was calculated to be 77° 03' 02.3" west of Greenwich—meaning that, to get approximate "Greenwich longitudes," 77° 3' would be added to "Washington longitudes." Thus, Iowa's proposed western border at 17° 30' W (Washington) falls at about 94° 33' W (Greenwich). (To make things simpler for you, in the maps to follow I'll be using

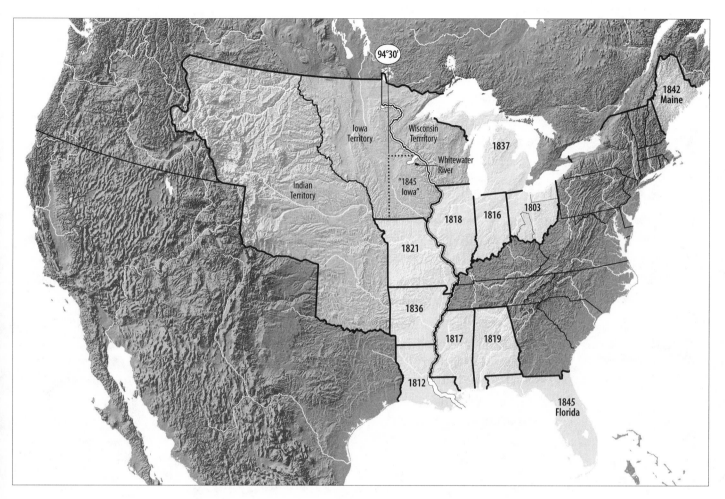

FIG 5.18

Greenwich longitudes, making the conversion by adding a rounded 77 degrees to the Washington longitudes.)

With some exceptions, Congress used the Washington Meridian to specify longitudinal boundaries through the 1880s, by which time virtually all of the present state borders had been specified. In 1884 an international conference was held, setting the Greenwich Meridian as the world standard; afterward, when Congress spoke of longitude, it did so in Greenwich terms. None-

theless, the 1850 mandate to use the Washington Meridian was not repealed until 1912.[29]

Congress had a busy day on March 3, 1845: not only did it pass the enabling legislation for Iowa, it elevated the Florida Territory to statehood.[30]

As I mentioned earlier, 1845 marked the beginning of a three-year period in which the Public Domain would expand all the way to the Pacific. The joint congressional resolution admitting Texas to the Union was passed on March 1, 1845, with formal statehood coming on December 29. On June 15, 1846, the Buchanan-Parkenham Treaty was signed, bringing the Oregon Country under sole American jurisdiction.

A few months later, on August 4, Congress passed a modified enabling act for Iowa, with new borders. The new state would now extend all the way to the Missouri River, its western border following the river up to its confluence with the Big Sioux River. The border would then run up the Big Sioux to latitude 43° 30', and head east on that latitude to the Mississippi. Iowa accepted these borders, and statehood was declared on December 28, 1846. With Iowa's statehood, the rest of the old Iowa Territory was left "unorganized" and would remain so until 1849.[31]

Also in 1846, Congress passed an enabling act for a new state of Wisconsin, specifying the borders the state has today. On December 16 of that year, a state convention adopted a constitution that specified those borders, but the delegates asked if Congress might accept a slightly different border on the northwestern corner (curiously, one that would diminish the area of the state). Under the request, Illinois and Michigan would bound the state on the south and east, and the Mississippi would be the southwestern border. The border to the northwest, though, would start at the mouth of the St. Louis River on Lake Superior, follow the river up to its first rapids, then head directly south, through Lake St. Croix, to Lake Pepin, a prominent widening in the course of the Mississippi.

Congress accepted these borders, amending its enabling act on March 3, 1847. On April 5 the people of Wisconsin rejected the proposed constitution. There were no exit polls in 1847, but apparently the proposed borders were not the determining issue in the referendum.[32]

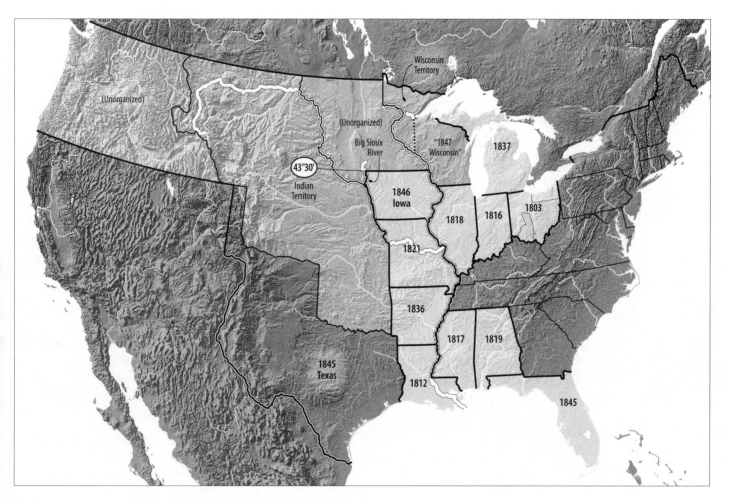

FIG 5.19

Wisconsin Territory

(Unorganized)

(Unorganized)

"1847 Wisconsin"

1837

Big Sioux River

43°30'

Indian Territory

1846 Iowa

1818

1816

1803

1821

1836

1817

1819

1845 Texas

1812

1845

On February 1, 1848, Wisconsin adopted a new constitution, again accepting the borders set out in the original enabling act of 1846 (the present borders). Once more, though, the convention asked Congress to accept a different border on the northwestern corner—but one that would *expand* the area of the state. The proposed border would run up the St. Louis River to its rapids, but would then run in a straight line to the mouth of the Rum River on the Mississippi. If Congress had accepted this change, we might now speak the Twin Cities of Minneapolis, Minnesota, and St. Paul, Wisconsin.

Congress, though, declined to act on Wisconsin's proposal, and so the borders of the new state reverted to those set out in the original enabling act. Under that act, Wisconsin would be bounded by Michigan, Illinois, and the Mississippi on the north, south, and southwest, respectively. The northwestern border would run up the St. Louis River to its rapids, but then run due south to the St. Croix River, and follow the course of the St. Croix to the Mississippi. On May 29, 1848, Wisconsin was admitted to the Union with these borders. One consequence was that the leftovers of Wisconsin Territory were now added to the part of territory left "unorganized" by Iowa's statehood.[33]

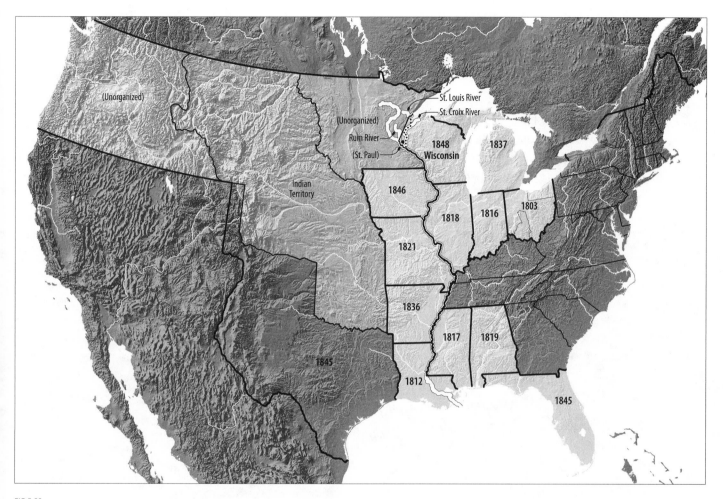

FIG 5.20

Congress soon rectified the "unorganized" situation northwest of Iowa and Wisconsin. On March 3, 1849, it declared the lands east of the Missouri and the White Earth rivers to be a new Minnesota Territory.[34]

On July 4 of the previous year, the Treaty of Guadalupe Hidalgo came into effect, giving the United States possession of its continentwide sweep. A month later, on August 14, Congress declared all the land west of the Continental Divide between latitudes 42° and 49° to be a new Oregon Territory.[35]

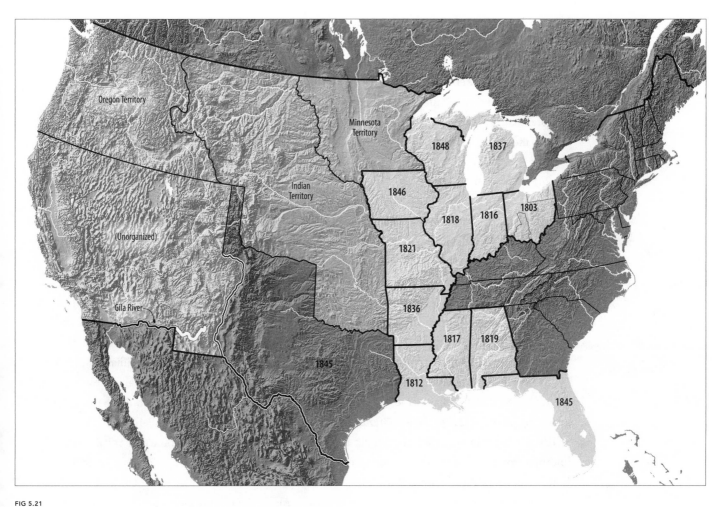

FIG 5.21

In 1850 Texas was persuaded of the wisdom of selling its far western reaches to the national government. Congress would pay the state $10 million—curiously enough, not in cash but in stocks bearing 5 percent interest for 14 years. As part of the agreement, Texas was confirmed in its right to the public lands within its new borders. It is those borders that give this state the most recognized shape of all the states of the Union.

The borders up against Louisiana and Arkansas, and along the Red River, were set in the Adams-Onís Treaty. New was the border that touched the 100th meridian. Texas was a slaveholding state, and since, according to the Missouri Compromise, slavery was prohibited (except in Missouri) north of latitude 36° 30', the territory of Texas had to stop against that line. So the new border of the state ran along latitude 36° 30' to its intersection with the 103rd meridian, then south on that meridian to the 32nd parallel, and west along that line to the Rio Grande.

Texas, though, immediately presented a novel interpretation of one of its borders. We have seen how, in the naming of rivers, sometimes one branch of a river will bear the name of the river from its source to its mouth (as does the Mississippi), while other rivers begin to bear their name only upon the confluence of two or more smaller rivers (the Ohio River is called that only after the confluence of the Allegheny and Monongahela rivers). Under Texas nomenclature, the Red River begins to be called that only after the confluence of its Prairie Dog Town Fork and its North Fork. Thus, a case could be made for either fork as the "main channel" of the Red River, and Texas naturally chose the northernmost fork, claiming the land between the forks as Greer County, Texas—

a claim that would be resolved only in 1896.[36] One has to wonder, though: if the shape of Texas had "always" included Greer County, would we ever have spoken of a Texas Panhandle?

With the western lands of Texas now part of the Public Domain, the United States had a huge southwestern region without effective governance. On September 9, 1850, Congress passed legislation that divided the entire area into two new territories and one new state.

The new state, California, would skip both territorial status and an enabling act, coming into official existence on that very day. (All of America had heard about the chaos of the gold rush: the region needed a government—and fast.) Its border on the north would run along the 42nd parallel as far as the 120th meridian. The eastern border would follow that meridian south, to its intersection with the 39th parallel. From that point a straight diagonal line would be drawn to the place where the 35th parallel crossed the Colorado River. The border would then run down the Colorado to the Guadalupe Hidalgo Treaty line that had given San Diego to the United States.[37]

Along the eastern portion of that treaty border, the new Territory of New Mexico would look nothing like the old Mexican state called New Mexico. It would be bounded on the south by the treaty line—up the Gila River from the Colorado, then along the L-shaped boundary still in dispute to the Rio Grande. From there the territory would follow the new Texas border, east along the 32nd parallel and then north on the 103rd meridian. The New Mexico border, though, would follow the 103rd meridian all the way to the 38th parallel, and then follow that parallel west "to the summit of the Sierra Madre" (presumably, the

CHAPTER FIVE

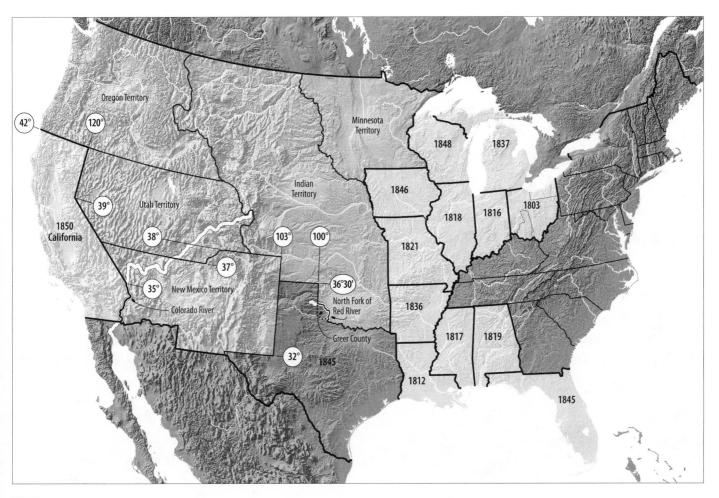

FIG 5.22

San Juan Mountains in present-day southern Colorado). Then the border would follow the crest of the mountains south to the 37th parallel, before heading west on that parallel to the border of California.[38]

The remainder of the southwest would become a new Utah Territory. (Bear in mind that the Mormons had arrived at the Great Salt Lake in 1847.) Its borders were described simply as California on the west, New Mexico on the south, the Continental Divide on the east, and the 42nd parallel on the north.[39]

(Note that the longitudes of these borders are all given with reference to Greenwich. Recall that Congress set these borders on September 9, 1850, but didn't pass the act mandating the Washington Meridian until September 28.)

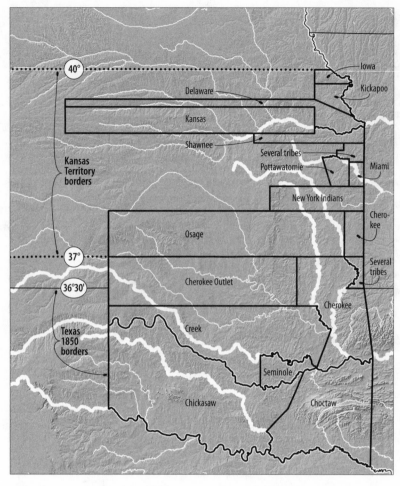

FIG 5.23

On March 2, 1853, Congress organized a new Washington Territory, with a southern border at the Columbia River west to the place where it crossed the 46th parallel. The boundary would then follow the 46th parallel to the Continental Divide.[40]

On May 30, 1854, Congress took the momentous step of organizing the bulk of what had been Indian Territory into the two new territories of Nebraska and Kansas. The Nebraska Territory would extend east and west from the Continental Divide to the Missouri River, and north and south from the 40th parallel to the U.S. border at the 49th.[41]

The border of Kansas Territory was more complex, and would be far more consequential. As a part of the Indian removal policy started in earnest under President Andrew Jackson, the land west of Arkansas and Missouri had been gradually divided into reservations for the various tribes—a conception quite alien to peoples who conceived of land as something sacred and shared. (The borders of the reservations changed constantly: figure 5.23 shows one configuration from the 1850s.) Two of the bigger reservations were those for the Osage and the Cherokee peoples. The Cherokees had been allotted a tract straddling the Neosho River, with an adjacent "outlet" to allow passage to the west. The Osage reservation was a long rectangle just north of the Cherokee lands. Their common boundary was the 37th parallel, and it was this line that Congress chose as a border for the new Kansas Territory.

But having chosen that line, Congress went one step farther. The Osage and Cherokee lands extended only as far west as the 100th meridian, and so their 37th-parallel border stopped there as

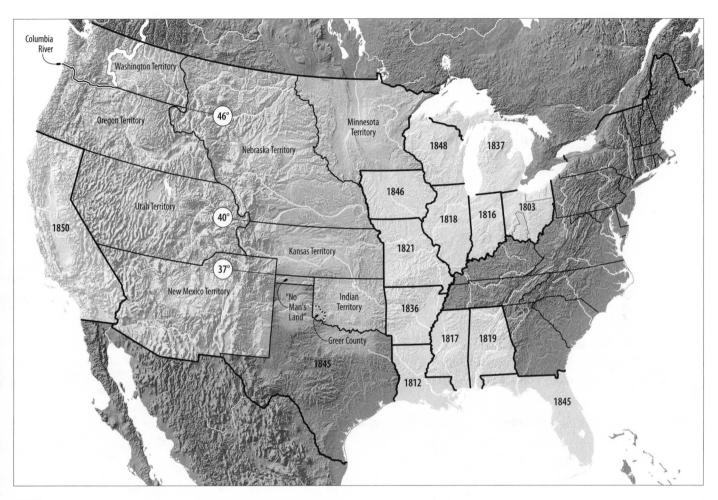

Columbia
River

Washington Territory

Oregon Territory

46°

Minnesota
Territory

Nebraska Territory

1848   1837

1846

Utah Territory

40°

1818   1816   1803

Kansas Territory

1821

1850

37°

New Mexico Territory

"No
Man's
Land"

Indian
Territory

Greer County

1836

1817   1819

1845

1812

1845

FIG 5.24

well. But Congress extended the 37th-parallel line all the way to New Mexico Territory, making that extended line the southern border of Kansas Territory. The result was that the strip of land north of the Texas Panhandle, between 37° and 36° 30', was left in no jurisdiction at all. The federal government would come to call this anomalous tract the Public Land Strip, but it was more commonly known as No Man's Land, and would remain ungoverned until incorporated into a new Oklahoma Territory in 1890.[42] And that, dear reader, is how Oklahoma got its Panhandle.

On May 11, 1858, the new state of Minnesota was admitted to the Union. Minnesota's border was to be partly a watercourse, partly a meridian of longitude; but to understand that border you need to know a little geography.

Besides the Mississippi, the two great rivers in that region are the Minnesota and the Red River of the North. The Minnesota flows roughly east to the Mississippi, but the Red River flows north, to empty into Lake Winnipeg and eventually, through the Nelson River, into Hudson Bay. At the head of the Red River's main fork (called the Bois des Sioux) is Lake Traverse; at the head of the Minnesota River is Big Stone Lake—and these two lakes, which empty respectively into Hudson Bay and the Gulf of Mexico, almost touch!

So Congress mandated that the western border of Minnesota would follow the Red River to the Bois des Sioux; then jump from Lake Traverse to Big Stone Lake; and then follow that lake to where it narrows into the Minnesota River. From that point of narrowing, a meridian line would be drawn due south to the northern border of Iowa. (Minnesota's statehood left the region west of the Red River temporarily unorganized, but Congress would rectify this condition three years later.)[43]

On February 14, 1859, the new state of Oregon was admitted. The eastern part of the Oregon Territory was not yet settled, so the state's borders were pulled west from the Continental Divide, and everything outside the new state added to the Washington Territory.

Oregon's northern border would be the Columbia River from the Pacific eastward to the Columbia's confluence with the Snake River. The border would then follow the Snake southward to its confluence with the Owyhee River. From that point a meridian would be drawn due south to the original territorial boundary at the 42nd parallel.[44]

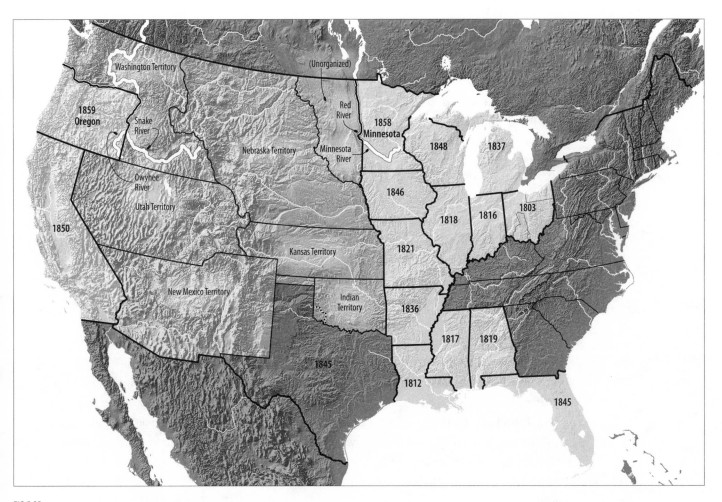

FIG 5.25

To tell you about the borders of Kansas in the space of one page, I drastically truncated the story. Let me pick up the thread here. It was the Kansas-Nebraska Act that created those two territories when it passed Congress on May 30, 1854. The Compromise of 1850 had said that the Missouri Compromise would not be extended from the area of the Louisiana Purchase into the new Southwest; the Kansas-Nebraska Act said that the Missouri Compromise would no longer apply in the Louisiana Purchase itself. The people of the new Kansas and Nebraska territories—not Congress—would decide whether to allow slavery within their borders.[45] The result was "Bleeding Kansas," seven years of open conflict between slavery and Free-Soil forces. Congress could admit Kansas as a free state, on January 29, 1861, only because Southern states, as they began to secede from the Union, recalled their representatives. The new state would keep its territorial boundaries on the north and south, but would be truncated on the west—at "the twenty-fifth meridian of longitude west from Washington" (approximately the 102nd Greenwich meridian).[46]

A month later, on February 28, Congress used that western line as a border for a new Colorado Territory. It would be a clean rectangle, stretching east to west between the 25th and 32nd Washington meridians (102° and 109° Greenwich), and north to south between the 41st and 37th parallels.[47] (Here, for you aficionados of state shapes, is the origin of Nebraska's panhandle.)

Just days later, on March 2, Congress created the Dakota and Nevada Territories. The northern border of Nebraska Territory would now follow the Missouri River up to its confluence with the Niobrara River (that's pronounced "nye-BRAIR-uh," by the way); follow the Niobrara to its confluence with the Keya Paha River ("KEY-uh PAH-uh"); follow the Keya Paha northwest to the 43rd parallel; and then run along that parallel to "the present boundary of the Territory of Washington."[48]

In 1859 a second gold rush began, this time to the Comstock Lode, northeast of Lake Tahoe; and again Congress stepped in to set up a government. The new Nevada Territory was carved out of Utah Territory (the second of many "bites" Congress would take out of the Mormon lands). Its eastern boundary was to be the 39th Washington meridian (116° Greenwich) and its southern border, the 37th-parallel boundary of New Mexico Territory, westward to

the dividing ridge separating the waters of Carson Valley from those that flow into the Pacific; thence on said dividing line northward to the forty-first degree of north latitude; thence due north to the southern boundary of the State of Oregon.[49]

And hereby hangs a tale. The borders Congress declared for California were laid out in a constitutional convention held in Monterey in September and October of 1849. The Californians knew about Lake Tahoe and the Carson Valley east of it, and they hoped that their eastern boundary lines, when surveyed and marked, might be found to fall east of the Carson Valley.[50]

In 1852 the federal official in charge of surveying in California trekked overland, set up his instruments, and "was reluctantly forced to the conclusion that the valley was from twelve to fifteen miles out of the State." In 1855 the successor

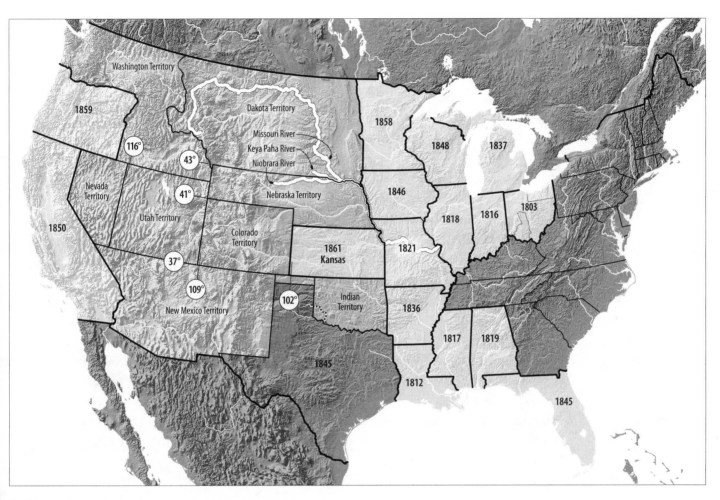

Washington Territory

1859

116°

43°

Dakota Territory

Missouri River
Keya Paha River
Niobrara River

1858

1848

1837

Nevada
Territory

41°

Nebraska Territory

1846

1818

1816

1803

1850

Utah Territory

Colorado
Territory

37°

1861
Kansas

1821

109°

102°

Indian
Territory

1836

1817

1819

New Mexico Territory

1845

1812

1845

FIG 5.26

federal surveyor had a telegraph line strung over the mountains to get an accurate time signal for determining longitude. He found, much to his surprise, that the angle between the two boundary lines fell right in the middle of Lake Tahoe!

The intent of the language about "waters . . . that flow into the Pacific" was that this disputed region would belong to the new territory—and future state—of Nevada. Congress, though, cautioned that no land within the present boundaries of California could be included within Nevada

Territory "until the State of California shall assent to the same." California would, of course, "assent" to no such thing,[51] and so the western borders of Nevada don't wiggle along mountain ridges but follow two utterly straight lines.

On February 24, 1863, Congress created a new Arizona Territory from the western half of New Mexico Territory. The border was uniquely specified—not by a declared line of longitude but by a line run due south from "the southwest corner of the Territory of Colorado." By tying the Arizona–New Mexico border directly to the southwestern corner of Colorado, Congress ensured our right as Americans to travel to the Four Corners National Monument and put one foot or hand into each of four different states.

With any westward expansion of Nevada blocked by California's boundary, Congress pushed the border of the territory eastward (and took another bite out of Utah) on March 3. Nevada Territory would now extend to the 38th meridian west of Washington (115° Greenwich).[52]

By the same legislation, Congress diminished Washington Territory to protostate dimensions and created a new Idaho Territory. Its western border would abut Oregon and follow the Snake River past Oregon's northern border to its confluence with the Clearwater River. From that intersection, the border would run on a meridian line to the 49th parallel, then east on that border to the 27th Washington meridian (104° Greenwich). The border would follow that meridian south to the border of Colorado Territory at the 41st parallel, and then follow that parallel past Colorado's northwest corner to the 33rd Washington meridian (110° Greenwich, and yet another bite out of Utah). That meridian would be followed north to the 42nd parallel, and from there back west to the border of Oregon.[53]

(You'll note that it was the creation of Idaho Territory that would give Nebraska and both Dakotas their western boundaries.)

On the last day of 1862, Congress approved the separation of West Virginia from rebellious Virginia, and its full statehood became official on June 19, 1863. The people of Virginia living west of the Allegheny and Appalachian mountains had long felt estranged from their government in Richmond. West of the Alleghenies, all rivers flow toward Ohio; and culturally and economically, the inhabitants looked in that direction. They would not follow Virginia into secession, and so a convention was held in August of 1861; the delegates resolved to form a new state and name it for the biggest of those west-flowing rivers, Kanawha. In November they convened to write a constitution (adopting, this time, the name West Virginia), which was approved in a referendum the following April. The boundary of their proposed state would be a line drawn around all the counties that had sent delegates and approved the constitution.[54] (Note that these are not the borders of West Virginia today: the section covering 1866 to 1868 tells the rest of the story.)

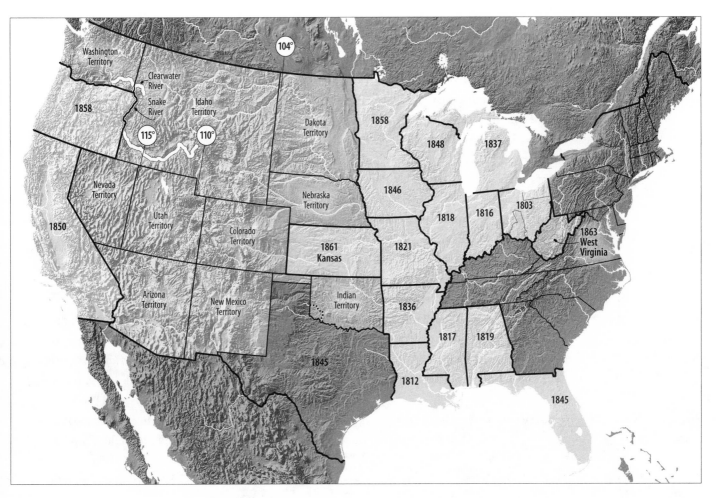

FIG 5.27

Congress passed an enabling act for Nevada's statehood on March 21, 1864. Quickly thereafter, on October 31, its statehood was proclaimed officially.[55] Note that Nevada's 1864 borders don't correspond to those we're familiar with—and indeed don't include Las Vegas.

When the Idaho Territory was created, in 1863, its capital was placed in what was then the center of mining activity, the town of Lewiston, at the far western edge of the territory, where the Snake and Clearwater rivers come together. But even before 1863, miners had begun heading to the area around Butte and Bannack and Virginia City, east of the rugged Bitterroot Mountain range and all but inaccessible from Lewiston. Almost as soon as Idaho Territory was created, the eastern miners began to ask that the territory be divided, to give them a government nearer to their digging sites.

Onto the scene from the East, in September of 1863, came Sidney Edgerton, the newly appointed Chief Justice of Idaho Territory. He got as far as Bannack, but snows blocked his route to Lewiston, so he wintered in the eastern mining region. While there he got to know the miners' interests, and so when the snows relented, he headed not to Lewiston but to Washington, to carry their message (and $2,000 of their gold) to Congress.

Also heading to Congress was a proposal from the Idaho territorial legislature to encompass both mining districts in a new state of Jefferson, whose eastern boundary would run south from Canada on the 113th meridian, then follow the Continental Divide to the 42nd parallel. In the debate, however, not only did the dividing line get pushed westward, to the 116th meridian, but the mountain border was changed as well. Idaho's southeastern border would be the 33rd Washington meridian (110° Greenwich) north from Utah up to the Continental Divide; it would follow the Continental Divide, but then split off to the west, to follow the crest of the Bitterroot range up to the 39th Washington meridian (116° Greenwich). The changed border split two important assets away from Lewiston—the boomtown of Butte and its outlying mines, and the broad, fertile Bitterroot Valley.[56]

FIG 5.28

Even though diminished by economics, Butte, Montana, is still a rough, proud mining town. For a taste of the town in its glory days, drop into one of its downtown bars.

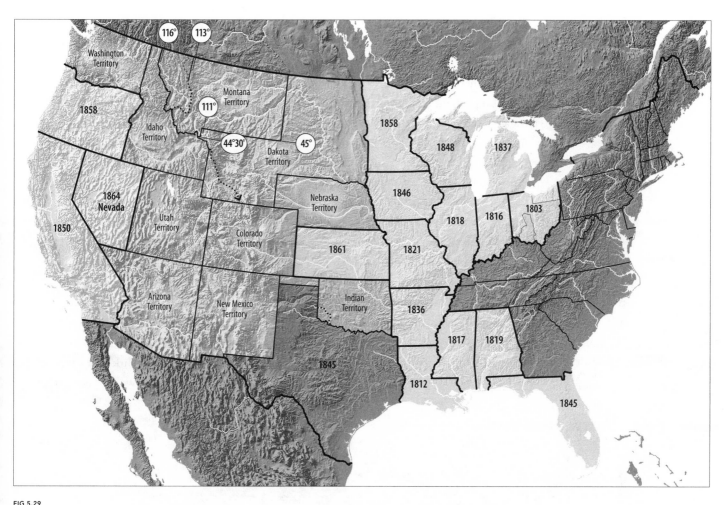

FIG 5.29

The final outcome of the debate, on May 26, 1864, gave the new Montana Territory its borders. The eastern boundary would be the one set in 1863 for Idaho Territory, the 27th Washington meridian (104° Greenwich). Its southern boundary would be the 45th parallel as far west as the 34th Washington meridian (111° Greenwich). From there the border would turn south, stopping at latitude 44° 30', then run west on that latitude to the Continental Divide, the border with Idaho.[57]

Everything south of that border, down to Utah and Colorado, would become part of the Dakota Territory. Note very carefully the area at the intersection of 111° longitude and 44° 30' latitude. You'll soon see Congress trip over the precision of the borders it declares.

In 1866 West Virginia was enlarged to its present borders by the addition of Berkeley and Johnson counties from (federally occupied) Virginia.[58] This addition gave that distinctive "hook" to West Virginia's eastern panhandle. More important, it gave the state the important trading center of Martinsburg. And at the very eastern extremity of the panhandle, West Virginia was now in possession of Harpers Ferry, the site in 1859 of John Brown's antislavery raid and his final capture.

Also in 1866, on May 5, Congress enlarged Nevada to its present borders (and took another bite out of Utah). The new eastern boundary would run southward down the 37th Washington meridian (114° Greenwich) to the Colorado River, then follow the Colorado to the California border.[59]

In 1864 Congress had passed an enabling act for the Nebraska Territory to pass into statehood, within its borders as of 1863. War matters intervened; the new president, Andrew Johnson, raised objections; and Congress finally admitted Nebraska to the Union (over Johnson's veto) on February 9, 1867. Johnson grudgingly issued the statehood proclamation on March 1.[60]

On July 25, 1868, Congress created the new Territory of Wyoming. Like Colorado, it would be a perfect rectangle, 7 degrees east to west, and 4 degrees north to south. It would have the 27th Washington meridian (104° Greenwich) as its eastern border, and the 45th and 41st parallels north and south; but most important, its western border would be the 34th Washington meridian (111° Greenwich).[61] This new boundary not only shrank Idaho to its present borders, but represents the final bite Congress would take out of Utah.

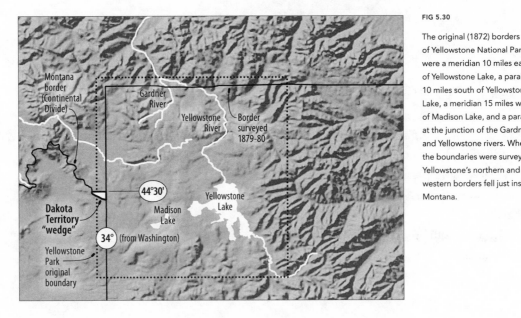

FIG 5.30

The original (1872) borders of Yellowstone National Park were a meridian 10 miles east of Yellowstone Lake, a parallel 10 miles south of Yellowstone Lake, a meridian 15 miles west of Madison Lake, and a parallel at the junction of the Gardner and Yellowstone rivers. When the boundaries were surveyed, Yellowstone's northern and western borders fell just inside Montana.

CHAPTER FIVE

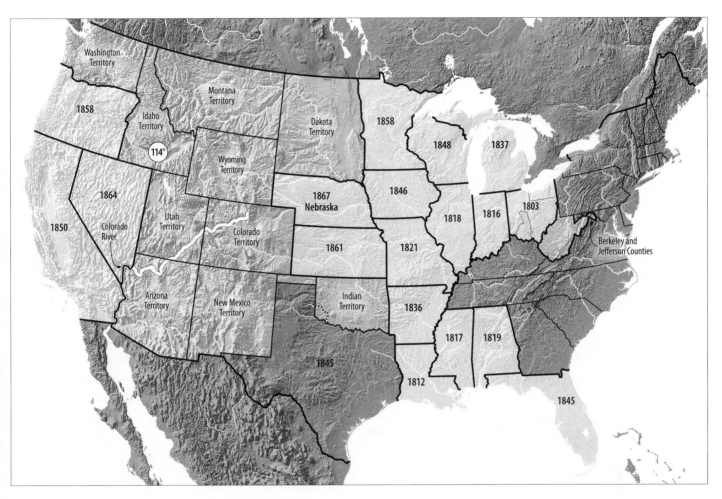

FIG 5.31

But this new border had the inadvertent effect of leaving a tiny piece of Dakota Territory wedged in between Montana and Idaho. At Wyoming's western boundary, Montana extended south only as far as latitude 44° 30', while Idaho extended north only to the Continental Divide. As you can see in figure 5.30, the two lines don't converge until just west of Wyoming.

Congress fixed its oversight in 1873, giving the tiny wedge to Montana, with the following resolve:

That all that portion of Dakota Territory lying west of the one hundred and eleventh meridian of longitude which, by an erroneous definition of the boundary of said Territory by a former act of Congress, remains detached and distant from Dakota proper some two hundred miles, be, and the same is hereby, attached to the adjoining territory of Montana.[62]

(Careful readers will note that Congress says here that Montana will get all Dakota land west of the 111th meridian, not the 34th Washington meridian as specified in the Wyoming Territory act. But since the 111th meridian [Greenwich] is east of the 34th meridian [Washington], Congress has its bases covered.)

Congress had granted statehood to gold-rush Nevada in 1864 despite a tiny population. By the same reasoning, it admitted silver-rush Colorado to the Union in 1865. The following May President Johnson vetoed the act, citing the territory's scant population and the slim majority that had voted for statehood (3,030 to 2,875 votes). Johnson vetoed a second act in 1867, but in 1875, with Ulysses S. Grant now president, Congress passed another enabling act, and Colorado was proclaimed a state on August 1, 1876.[63]

If the shape of Nebraska has looked a little odd to you in the preceding maps, it is because the state achieved its present border only in 1882. You may recall that the original northeastern border had run up the Missouri, then up the Niobrara, then up the Keya Paha to the 43rd parallel. Nebraska argued, and Congress agreed, that a better situation would be for the border simply to run up the Missouri to the 43rd parallel.[64]

CHAPTER FIVE

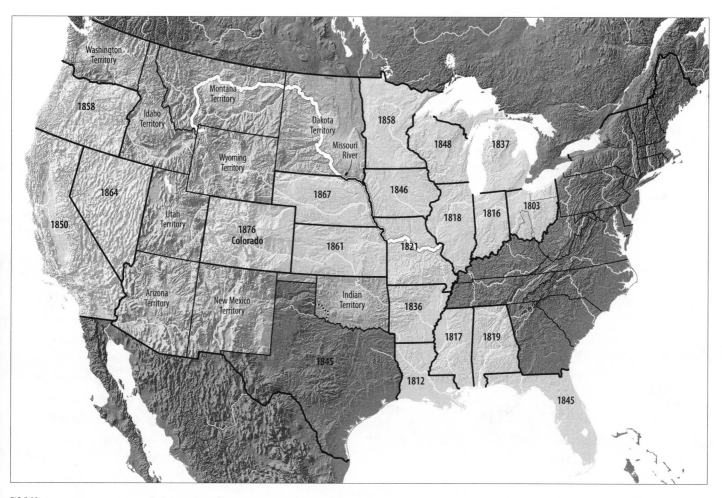

FIG 5.32

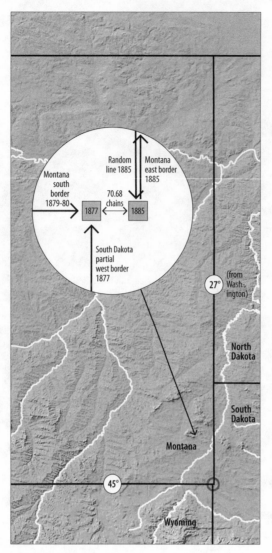

FIG 5.33

If you look very carefully at a map of South Dakota, you'll notice that its western border has a tiny jog in it. Here's why: Dakota Territory's western border was surveyed up from the south in 1877, but only as far as the southeastern corner of Montana Territory, where a monument was erected to mark the spot. Montana Territory surveyed its southern border in 1879 and 1880, tying it to the 1877 marker. Only in 1885 did Montana Territory survey its eastern border—by running a test line down from the Canadian border (what surveyors call a random line), and then checking it with a line run back up to the 49th parallel. When those surveyors compiled their calculations, they found their line ending a little more than 4600 feet east of the 1877 marker, and so they set a second monument. Eventually all parties accepted these lines, and that tiny jog resulted.

The greatest state-making date in American history is February 22, 1889. On that day (Washington's Birthday, no less) Congress passed enabling legislation for four new states: Washington, Montana, North Dakota, and South Dakota.[65] In the cases of Washington and Montana, this was merely the usual matter of advancing a territory to statehood. The case of the Dakotas is a little more complex. There was some feeling in Congress that the Dakota Territory, left undivided, would be "too big." There was also some thought that by dividing the territory, one state might go Democratic, the other Republican. Whatever the reason, the border chosen to divide the Dakotas is unique. It is not a line of latitude but "the seventh standard parallel" projected west across the width of the Dakota Territory.[66] The meaning of that opaque phrase will become clear in part 3, when we explore the Rectangular Survey; but suffice it to say for now that a "standard parallel" is the term of art for an east-west line whose function in the Survey is to assure the alignment of the rectangular parcels north and south of it—an important line in the landscape, in other words, and one that government surveyors had already begun to mark on the ground.

Though the four states were "enabled" on the same day, the effective dates of their statehood occurred over three days in November 1889. North and South Dakota were first, on November 2 (by convention, North Dakota is the thirty-ninth state, South Dakota the fortieth). Montana came next, on November 8, with Washington last, on November 11. The whirlwind of state making continued into 1890, with Idaho admitted to the Union on July 3, and Wyoming on July 10.[67]

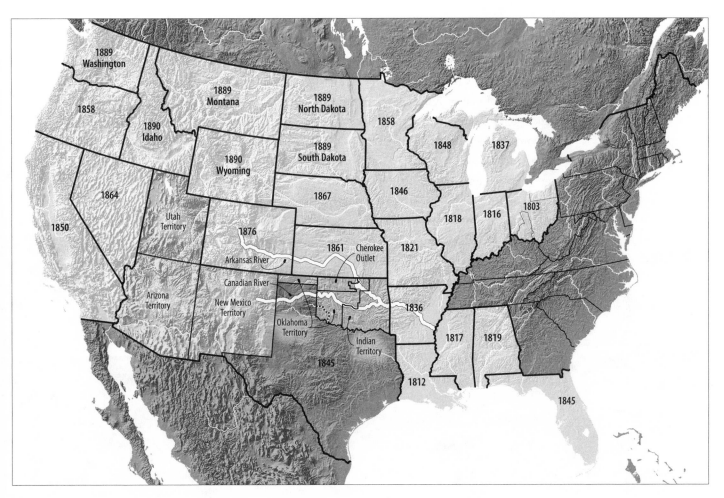

FIG 5.34

On May 2 of 1890, though, Congress took a momentous step in creating the Territory of Oklahoma.[68] Since 1825 the area north of the Red River had been reserved exclusively for Indians, off-limits to white settlement. Now Congress would reverse that policy. Its first moves were tentative: a new Oklahoma Territory was cobbled together out of lands not then assigned to particular Indian tribes, with the addition of the anomalous Public Land Strip.

On December 19, 1891, the Cherokee Nation agreed to a treaty extinguishing its rights to the Cherokee Outlet. Congress ratified the treaty on March 3, 1893, and when it became effective on September 16 of that year, the Cherokee Outlet was added to Oklahoma Territory.[69] The destiny of "Indian country" was becoming clear.

Almost fifty years after its original settlement, Congress passed enabling legislation for Utah's statehood on July 16, 1894. The reasons for the long delay are many, but the final logjam was broken when Utah agreed to renounce polygamy in its constitution. Statehood was finally proclaimed on January 4, 1896.[70]

In another dispute of almost fifty years' duration, the Supreme Court took up Texas's claim to the North Fork of the Red River as its boundary. On March 16, 1896, the court ruled that the Prairie Dog Town Fork of the Red River was the true border of Texas, and the disputed tract officially became Greer County, Oklahoma Territory.[71]

CHAPTER FIVE

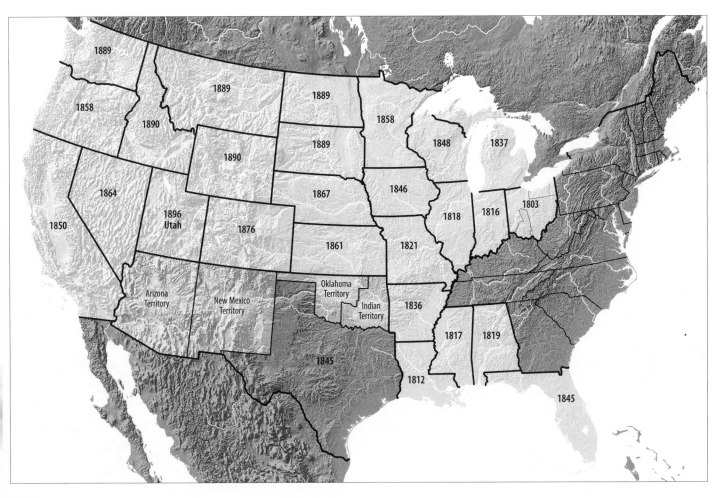

FIG 5.35

On June 16, 1906, Congress passed enabling legislation for Oklahoma, New Mexico, and Arizona
to become states.[72] Oklahoma achieved statehood
first, on November 16, 1907, with the remnants
of Indian Territory folded into the new state.[73]

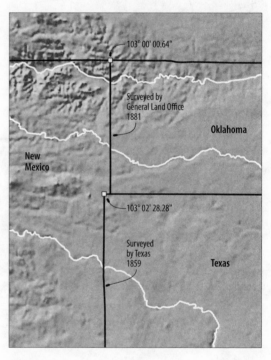

FIG 5.36

In 1859, when New Mexico was still a territory, the state of Texas decided to survey the western borders it had agreed to in 1850. New Mexico later objected, but Texas eventually prevailed in the courts. What had not been surveyed in 1859 was the border between New Mexico and the Public Land Strip that became Oklahoma's Panhandle. In 1881 federal surveyors came to the area and located the 103rd meridian with a time signal transmitted by telegraph. New Mexico accepted this as its border— and gained that tiny stub at its northeastern corner.

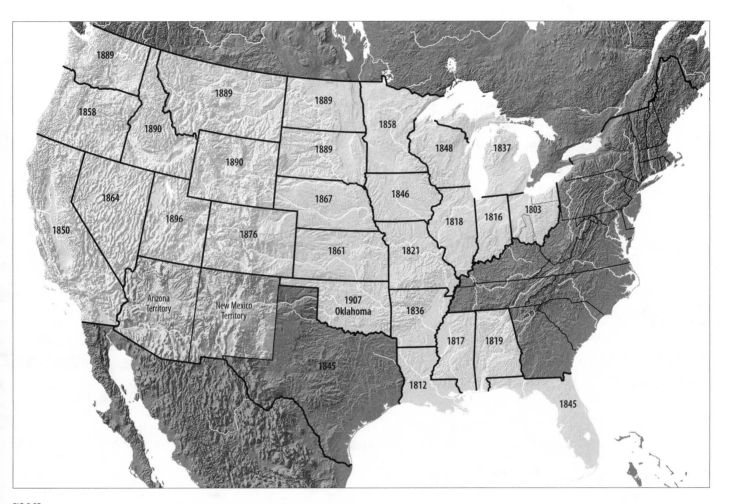

FIG 5.37

New Mexico and Arizona had to wait a little longer, but in 1912 statehood was finally granted, to New Mexico on January 6 and to Arizona on February 14.[74]

The final two states (so far) are, of course, Alaska and Hawaii, both admitted in 1959. Alaska's statehood became official on January 3, and Hawaii followed on August 21.

One Star for Each State

Being twelve at the time and history-mad in the manner of young boys, I can remember those few months in 1959 when the flag of the United States had 49 stars. They were arranged in 7 staggered rows of 7 stars each, so as to fill out the flag's blue rectangle. It was perhaps the last American flag whose stars everyone could imagine and draw correctly. The 48-state flag arranged its stars in 6 rows of 8 each—a pattern you could visualize in an instant; and the staggered 7-by-7 pattern of the "Alaska flag" was about as easy.

How, though, can you recall, to your mind's eye, the 50 stars in the current flag? When you look, do you see the stars as arranged in vertical columns and say to yourself "5–4, 5–4, 5–4, 5–4, 5–4, 5"? Or do your eyes see the stars as being in horizontal rows, which you call out as "6–5, 6–5, 6–5, 6–5, 6"? It's an undeniably handsome pattern, but try as I might, all I can truly say to myself when I watch our flag proudly wave is, "That's a whole bunch of stars up there . . ."

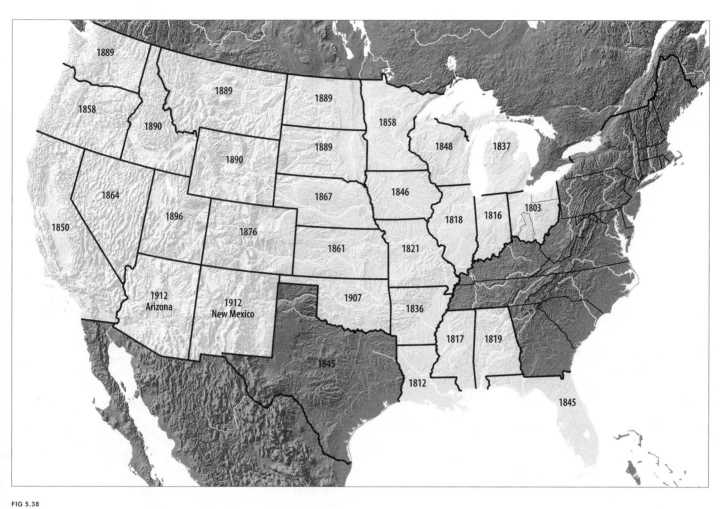

FIG 5.38

PART 3

Apportioning the States into Rectangular Parcels

OVERVIEW AND RECAP

Our story of the apportioning of the Public Domain has three main threads: the growth of the Public Domain itself, its division into states, and the apportioning of those states into parcels of land for settlers. Parts 1 and 2 traced the first two of those threads; let's now pick up the beginning of the third one.

In the spring of 1784, the Continental Congress was in Annapolis, and on its agenda was Virginia's resubmitted proposal for the cession of its western lands. Anticipating that cession, during the previous fall Congress had appointed a committee to come up with a plan for governing the lands Virginia would cede. By February of 1784, the committee had reconstituted itself with Thomas Jefferson as its chairman. They submitted their report on governance of the new lands to Congress on March 1.[1]

That same day, Congress accepted Virginia's amended offer, and consequently had jurisdiction over all the land north of the Ohio River up to Connecticut's 41st-parallel claim. The very next day, March 2, it appointed a committee, with Jefferson as chairman, "to devise and report the most intelligent means of disposing of such part of the Western lands as may be obtained of the Indians."[2]

Congress passed Jefferson's plan of governance, the Ordinance of 1784, on April 23. Two days later, Jefferson laid his plan for land disposition before the legislators. It was the practice in Congress, after the first reading of a new bill, to defer debate so the delegates could digest and discuss its provisions. During that postponement, it took up the appointment of a new minister to France, to help Benjamin Franklin negotiate commercial treaties. Congress's choice was Jefferson himself. *Amateur* of all things French, he relished the appointment and left Annapolis four days later to put his affairs in order. He stayed in America only long enough to celebrate July 4, 1784, in Boston, sailing from there the very next day.[3]

Jefferson was thus not present when his plan for land disposition came up for debate—which was perhaps merciful for him, since on May 28 Congress soundly rejected it. There matters lay, as Congress moved first to Trenton, New Jersey, for the fall, and then, at the end of December, to New York City.

On March 4, 1785, Jefferson's plan was again read and again rejected. Finally, on March 16, after the plan was rejected a third time, a new committee was formed to reconsider the question of land disposition. This committee tabled its report on April 12, and after a great deal of debate, Congress passed the Land Ordinance of 1785 on May 20, setting in motion the Rectangular Survey of the American West.[4]

SIX

Inventing a Rectangular Survey in the Ordinance of 1785

**JEFFERSON PROPOSES A
REFORMED LAND SYSTEM**

As we saw in chapter 4, Jefferson's plan for apportioning the land of the Northwest was an integral part of his plan for forming states there. As passed by the Continental Congress, the Ordinance of 1784 called for the new states to be largely rectangular, with most of them spanning 2 degrees of latitude, north to south. Almost all of the new states would thus have a north-south dimension of 120 nautical—or, in Jefferson's terms, "geographical"—miles.

In the draft of his ordinance on land disposition, Jefferson made the geographical mile the basis of his plan. Each of the new states would be apportioned into squares measuring 10 geographical miles on a side, a new unit of land measurement that Jefferson called the Hundred, since each unit would have comprised 100 square miles geographical. Since a typical state under his Ordinance of 1784 would have been 120 miles "tall," a state could be cleanly divided into Hundreds: twelve Hundreds north-to-south and as many Hundreds east-to-west as would be required to fill out the state's extent. The advantage would be that each of the new state's main divisions of private property could be made to align precisely with the state's borders on at least two and perhaps three of its four sides.

But a piece of ground 10 miles square (whether the miles are statute or geographical) was far more land than a single farmer could manage, so Jefferson's plan called for each Hundred to be subdivided into a checkerboard of one hundred Lots, each Lot a square of land measuring 1 geographical mile on a side. Each Lot would

in turn comprise 1,000 new acres (the subdivision process is shown in figures 6.01 through 6.03 on the following pages). To save you from getting out your calculator: the area of a customary acre is 43,560 square feet—or to help you visualize, a square of land measuring about 209 feet on a side. The area of the geographical acre would have been, by Jefferson's calculation, just over 37,044 square feet, a plot about 192 feet on a side.[1]

Jefferson's draft ordinance went on to mandate how this gridding of the land would be accomplished. For each new state, surveyors appointed by Congress would begin at one of the state's right-angle corners, measure 10 geographical miles along the two perpendicular borderlines, and then project the first set of dividing lines inward, north-south and east-west, continuing the process until the whole state (or least that part of it thus far ceded by the Indian tribes) had been divided into 10-mile-square Hundreds. The surveyor would mark the courses of these lines by chopping three "blazes" into the bark of every tree on, or adjacent to, the line.

Once that had been done, the Hundreds would be grouped into Districts of nine each; presumably an ideal District would be three-by-three Hundreds, so that four such Districts would stack neatly into the state's "height" of twelve Hundreds. Each District would be overseen by an appointed Surveyor, and it would be he who would accomplish the actual scribing of the lines apportioning the Hundreds into Lots. The process would work as follows.

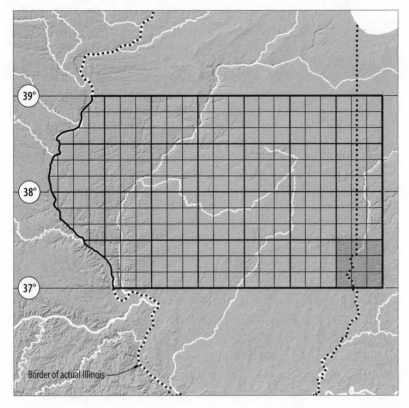

FIG 6.01

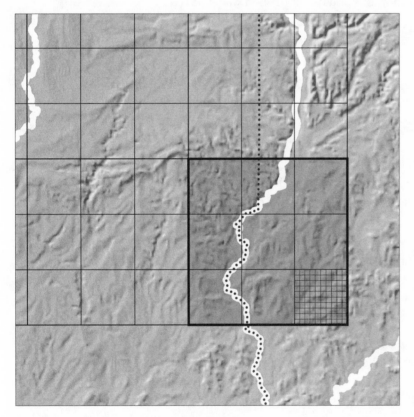

FIG 6.02

A person desiring land would have two options: he could purchase a 1-mile-square Lot as a farm for himself and his family to settle on; or he could buy a whole Hundred for resale to others. To begin the process, the buyer would go either to the U.S. treasurer or to the loan officer appointed for that particular state or state-to-be (Jefferson's plan required the loan officer to reside in that state and set up a land office there). There the buyer would hand over the per-Lot or per-Hundred price set by Congress. The amount of the payment could be offset by acreage awarded Revolutionary War veterans for their service.[2] (These bounties were graded by rank, ranging from a major general's grant of 1,100 acres, through a captain's of 300 acres, to the common soldier's of 100 acres[3]—those acreages converted, easily enough, to the new geographical acres.)

In return for his payment, the buyer would get a warrant for a Lot or a Hundred of land in that state. It would then be his job to go out into the wilderness and choose which Lot or Hundred he wanted to claim with that warrant. To register his choice and, in effect, cash in his warrant, the buyer would go next to the Surveyor in whose District his chosen land fell. Here, though, the process would get a little messy.

From the initial survey of the boundaries of the Hundreds, the Surveyor would have a map of the state gridded into 10-mile squares. If the buyer carried a warrant for an entire Hundred, he

FIG 6.03

Far left, at top is how Jefferson's state of "Illinoia" would have been apportioned into Hundreds, and then into surveying Districts. The darkened District (straddling the Wabash River, the border between present-day Indiana and Illinois) would contain nine of these Hundreds, each divided into one hundred mile-square Lots, numbered as in the diagram adjacent.

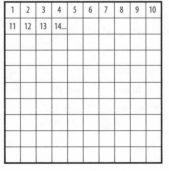

could simply point to that tract on the map, the Surveyor would inscribe the buyer's name in the square, and (if no one had previously claimed it) that would be the end of the Surveyor's responsibility for that Hundred: it would be up to the buyer to have the Hundred apportioned into smaller tracts for resale.

If, however, the buyer came with a warrant for a single Lot, the Surveyor would have a real job to do. If a Hundred was to be sold by Lots, Jefferson's plan called for the boundary lines for those one hundred tracts to be run by the Surveyor (and the lines marked by a single tree blaze, to distinguish them from the "three-chopt" exterior lines of the Hundred). On the map he would compile, the Surveyor would number the Lots consecutively from left to right, beginning in the top left—northwestern—corner, and progressing in rows eastward and then southward, in the manner of lines of type on a page. But the Surveyor was not to run these lines until he could see a pattern emerging in the desire for land (no sense surveying the interior lines of a Hundred that would be sold in its entirety; no sense in surveying, now, a Hundred that has as yet attracted no buyers). Only when the District Surveyor could see which of his nine Hundreds were drawing settlers would he set out with his instruments to grid those popular Hundreds into mile-square Lots.

As he ran the lines, the Surveyor was required to note the courses of any streams that crossed a line, as well as the positions of features like springs or salt licks that fell near it, and then record those features on his maps.[4] In this way his maps would be not just a grid for recording land ownership, but the outline for a complete geographic picture of the region.

But what about the intrepid Lot buyer who ventured into the wilds in advance of the Lots being surveyed? He would have no surveying equipment and little surveying knowledge, and thus no way even to hazard a guess as to which of the one hundred Lots in the Hundred might contain his farmstead. But there would have been some feature that attracted him to that particular plot of land—a spring, or the ford of a stream, or an opening in the trees ready to farm. Under Jefferson's plan the settler would go to the District Surveyor and identify his claim by its most distinguishing features, which the Surveyor would record so as to hold the claim until a survey might be run. In the interim the settler would continue with his farmwork, clearing the usual acre of ground each year, and hoping that when the Surveyor finally established the boundaries of his mile-square Lot, all of his improvements would be found to fall inside those borders.

This was a concern that would persist throughout the settlement period. The various land ordinances, as they evolved, all perpetuated the policy that the lines surveyed by government officials would take precedence over any others, and that farmers who settled in advance of the survey took their chances. But as long as farmsteads were small relative to the size of the tracts apportioned by the survey—which was the case well into the nineteenth century—the odds were pretty good that the lines would fall well outside the fields and woodlots the farmer had come to depend on.

As we saw earlier, in the space of the single month of May in 1784, Congress took Jefferson's plan for land disposition under consideration, sent him to Paris, and then soundly defeated it. But while the legislators were busy with their own affairs (including having to move congressional operations—twice), events were afoot in the "Old Northwest" beyond the Ohio River that would influence deliberations on a new land-disposition ordinance the following spring.

With the Americans' victory in the Revolution, the Indian tribes knew that they would soon lose the support of their allies the British. Putting aside their ancient differences, tribal delegates met during September of 1783 in a grand council on the shores of Lake Erie where Sandusky, Ohio,

now stands—shown in figure 6.04. (They could feel secure at this location, since Britain still held the forts at the lake's two ends, Detroit and Niagara.) There the Indians declared the boundary between their lands and those of the whites to be the Ohio River, now and forever.[6]

But by October of 1784, American negotiators, backed by soldiers, had split the coalition. At Fort Stanwix (near present-day Rome, New York) they induced the Six Nations—the tribes allied with the Iroquois in western New York—to renounce their claims to all lands west of Pennsylvania. (Remember that Pennsylvania and Virginia had settled a question of Pennsylvania's western border in 1779, and in 1784 that north-south line was being surveyed.)

Then in January of 1785, Americans split the coalition yet again. At Fort McIintosh, on the northern bank of the Ohio River just inside the Pennsylvania border, U.S. negotiators met with representatives of the Wyandot, Delaware, Chippewa, and Ottawa tribes from the area northwest of the Ohio River. There they got the Wyandots and the Delawares to cede all their claims to the Ohio Country and withdraw to a zone along Lake Erie between the Cuyahoga and Maumee rivers (from present-day Cleveland to Toledo).[7]

This was how matters stood when, in the spring of 1785, Congress (in New York for good, until superseded by the new federal Congress) took up once again the matter of land disposition in the Northwest. In December of 1786 the Wyandot and Delaware Indians would renounce their cessions and begin an attempt to keep whites back across the Ohio. But all that was in the future; the situations Congress had before it at that moment in 1785 were these:

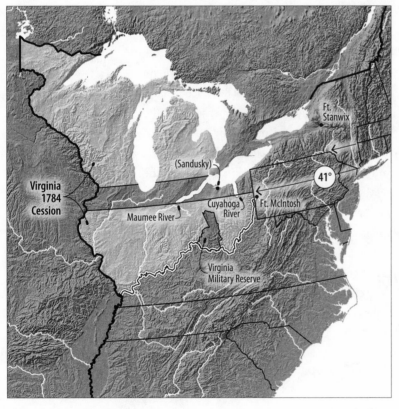

FIG 6.04

- *Settlement was pushing across the Ohio from Kentucky, Virginia, and Pennsylvania.*
- *Congress didn't yet have jurisdiction over the land along Lake Erie. On April 19, it accepted Massachusetts's cession of its western lands north of about the 42nd parallel, but Connecticut, for the time being, held fast to its claim of the strip between the 41st parallel and the lake. And overlapping Connecticut's claim were the lands allotted to the Wyandots and the Delawares in the Treaty of Fort McIntosh.*
- *Treaties notwithstanding, any area would be threatened by Indian raids until it was thickly settled. But just east of the recently surveyed Pennsylvania border was Fort McIntosh—whence defense could come, overland or down the Ohio, making that region the most secure for settlement.*

After giving Jefferson's plan the courtesy of the customary three readings, on March 16 Congress appointed a new committee to come up with a course of action the whole assembly could endorse. To help secure that approval, the committee would consist of thirteen men, one from each state, with William Grayson (like Jefferson, a Virginian) as chairman. Less than a month later, on April 12, Grayson's committee submitted its plan for land disposition.[8]

We have already seen how, in the Northwest Ordinance of 1787, Congress rejected Jefferson's grand plan for state formation and crafted a more modest law, based on conditions existing on the ground and directed at only the area over which it had jurisdiction. Grayson's committee likewise abandoned Jefferson's visionary measurement reforms, developed a system based on familiar measurements, and applied the plan to only the

region then potentially safe for settlement. But as was the of case with the Northwest Ordinance, certain features of Jefferson's grand, abstract vision managed to make it into the more modest, tough-minded replacement. Let's look at the plan Grayson's committee submitted to Congress, and see how much of Jefferson's vision survived.

True to its purpose as a consensus-building mechanism, the committee's plan opened not with a statement of what would be done but with a plan for who would do it. A team of thirteen surveyors, one from each state, would set out for the Ohio Country under the direction of a newly created officer, the Geographer of the United States. The team's task would be to apportion into Townships (Congress's first use of the term) all the territory ceded thus far by the Indians. Each Township would be 7 statute miles square.

The surveying team's work methods were precisely described (and illustrated on the following pages in figures 6.05 through 6.07). They would proceed to the point on the southern bank of the Ohio where the survey of Pennsylvania's western border had been terminated. They would then project that line due north to the opposite bank and from that point begin projecting a line due westward "throughout the whole territory."[9] Next, individual Townships would be marked out by projecting lines north and south off that Baseline at 7-mile intervals.

As the north-south lines were projected off the Baseline, each vertical row of Townships thus marked would be called a Range. Because the Ranges were numbered progressively from east to west, Range no. 1 thus became that stack of Townships directly against Pennsylvania's western border.

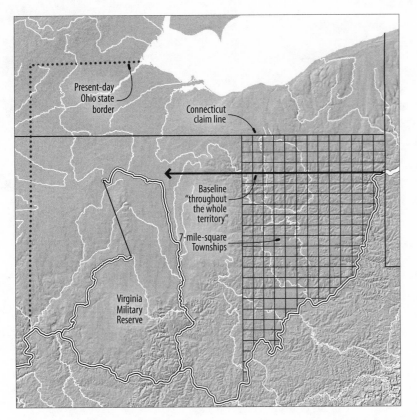

Present-day
Ohio state
border

Connecticut
claim line

Baseline
"throughout
the whole
territory"

7-mile-square
Townships

Virginia
Military
Reserve

FIG 6.05

Each Township within a Range would get a unique number. Township no. 1 in each Range would be the Township whose borders touched the Ohio River, and the numbers would progress northward, "from the Ohio to the Lake Erie." Consequently, a Township could be designated uniquely as "Township no. 7 in Range no. 3," and one would know that this was the 7th Township north from the Ohio in the 3rd Range of Townships west of the Pennsylvania line.

Jefferson's scheme had no method for differentiating Hundreds from one another. He did, however, propose a method for numbering the Lots within a Hundred, and the Grayson committee adopted it—with one telling variation.

Measuring 7 statute miles on a side, the checkerboard of a Township had forty-nine squares (not Jefferson's one hundred), and to those squares the committee gave a wholly new name: Section (the term which has endured to this day). Jefferson had numbered his Lots in the manner

of lines on a page, but the Grayson committee thought it best that the numbering system within each Township echo the pattern by which the Townships were numbered—that is, from south to north and then east to west. So Section no. 1 would be the one at the southeastern corner of the Township, with Sections 2 through 7 then progressing northward in a stack against the Township's eastern boundary.

Here, though, the committee made a departure (another that has endured). Instead of beginning the second row of Sections back at the bottom of the square, the committee mandated "beginning the succeeding Range of Sections with the number next to that with which the proceeding one concluded."[10] Not an easy phrase to fathom, that one. But imagine that you, as a mapmaker, have counted up Sections from the southeastern corner, numbering them upward. When you reach the northeastern corner, you are at Section no. 7. The only Section "next to" that one is the square immediately to the west, so only that can be Section no. 8; thus, Section no. 9 must be the one immediately south, and so on southward to Section no. 14, back at the southern edge of the Township map.

The numbering pattern thus proceeds across the Township—up, down, up—landing on Section no. 49 in the northwestern corner. There is a mellifluous word to describe such a numbering process: *boustrophedonic* (boo-STROF-e-donic). It's from the Greek and means, in effect, "as the ox plows." And thanks in part to Grayson's committee, Americans have been numbering their Townships like this for the past two centuries.

Not content with merely numbering the Sections, the committee directed that certain of them be put to specific uses. Its plan called for the Sections at the four corners of each Township to "be reserved for the United States." "The central Section" (that is, Section 25) was to be reserved "for the maintenance of public schools," and the Section immediately to the north (Section 24) was to be reserved "for the support of religion."[11]

These reserved Sections might not occur in every Township. As was the case in Jefferson's numbering scheme, if a Township was "partial"—if a portion of it was cut off by a river or a lake, or the border of an Indian territory—for whatever Sections there were in the Township, each would receive the number it would have been assigned if the Township had been "entire." By putting the school and church Sections dead center, Grayson's plan assured that if a truncated Township attained even half of the anticipated size, that Township would likely have support for its schools and churches. This is yet another concept that endured through all the modifications of federal land apportionment laws during the settlement period.

Bear in mind that this provision did not intend for the Township's school to be located in Section 25, with its church in Section 24. Rather, ownership of those two Sections would pass not into private hands but into ownership by the new state or county government, which would lease it out. The profits accrued from the arrangement would then be "applied for ever according to the will of the majority of male residents of full age" within the district.

Finally, the plan reserved to the national government a 100-acre square around any salt licks or salt springs that might be found, and allocated to the U.S. government a fraction of the proceeds of any metal or coal mines.[12]

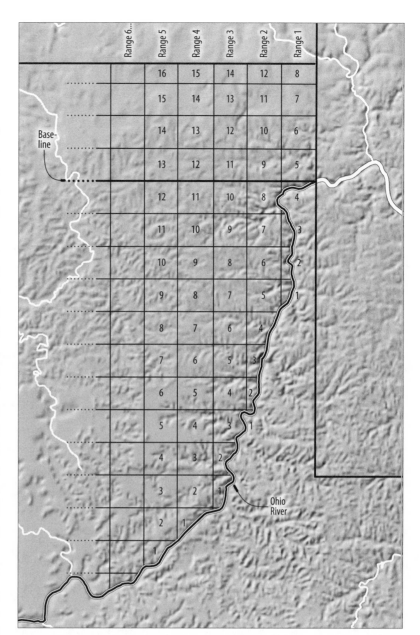

FIG 6.06

FIG 6.07

So if the Ohio Country was to be apportioned into Townships of 49 mile-square Sections, how was that apportioned land to be sold? Remember that one of the prime reasons for putting a national domain into the hands of Congress was so that it could sell the land to pay off the massive debts piled up during the Revolution. Congress had incurred the debt, but under the Articles of Confederation it could not levy taxes directly upon citizens holding wealth: the best it could do was apportion the debt among the states and then dun them for the cash. Congress, though, owed not just money but land—those bounties of graduated acreages promised to officers and soldiers who had fought in the Revolution. The Grayson committee tailored its sales scheme to these two national debts.

The plan provided that as soon as a substantial number of the north-south Ranges of Townships had been surveyed, and the maps returned to the commissioners of the Treasury, the secretary of war would draw lots to determine which Townships he would use to satisfy the veterans' claims. The remaining Townships would then be parceled out to individual states in a ratio proportionate to the most recent debt assessments. As for the question of exactly which Townships a state would get—that would also be determined by the drawing of lots. The states could then sell their allotments in any manner they chose (remitting the proceeds to Congress, of course) so long as they sold no land for less than $1.00 per acre, payable in specie. This offer to the states was not open-ended, however: if after a certain period a state had not been able to sell all the lands allotted to it, the right to sell any unsold land would revert to the Congress.

SIDEBAR

The "Mystical" Forty-Nine-Square Township

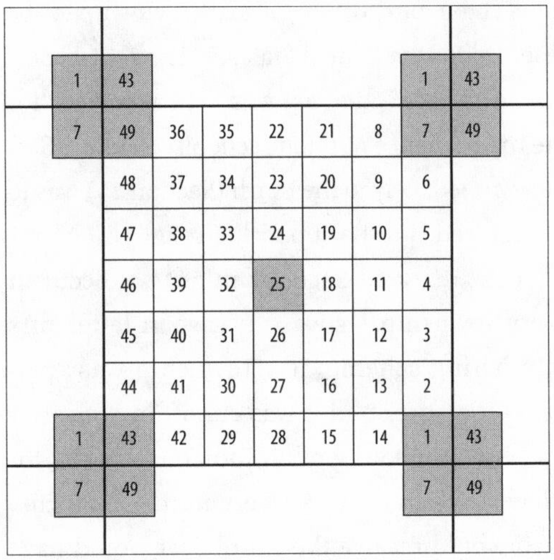

FIG 6.08

The 7-mile-square Township was never adopted by Congress, but it has a number of characteristics that lend it an almost cabalistic aura. There is first of all that occurrence of the school-supporting Section 25 occurring at dead center of the grid, 25 being half the sum of the grid's two terminal numbers: 1 + 49 = 50. Plus, if diagonal lines are drawn through that center square, the two numbers at the ends of each line also add up to 50. So will any number pairs within the grid that fall symmetrically on those diagonals (9 and 41, 19 and 31, 37 and 13, 33 and 17). These conditions are true of any numbered checkerboard—whether numbered conventionally or boustrophedonically. But only the 7-by-7 grid yields the magic number 50.

But wait: there's more. The numbers of the corner Sections reserved for the U.S. government are 1, 7, 43, and 49. Because they are corner Sections, whenever four Townships come together at a common corner, those four corner Sections will also come together—giving the United States an unbroken holding of four contiguous Sections. But note the numbers of the Sections: the positions will vary, but there will always be Sections 1, 7, 43, and 49—which will always add up to 100.

Cue the *Twilight Zone* theme . . .

CHAPTER SIX

To recount chronologically the debate on the Grayson committee's land-disposition ordinance would tax anyone's powers of concentration. So I'll organize my account around certain issues, even though the debates on those issues overlapped in time. The salient fact that pervades the debates is that in the spring of 1785, Congress was having a hard time maintaining a quorum. At almost all times, each of the states had a delegate on hand, but under the Articles of Confederation, a state's vote would count only if at least two delegates were present, and a state's vote would not count if its two delegates voted on opposite sides of the issue. Consequently, in the debate on the 1785 land-disposition ordinance— as in the 1784 debate over slavery in the West—a proposal might garner a majority of delegates present but fail to achieve legitimate "ayes" from seven states, and thus be defeated.

Some of the issues in the Grayson proposal were decided quite easily. For example, what fraction of the national domain would be set aside for military bounties? One-seventh.[13] Would a Section in each Township be set aside for the support of religion? No; not even if the language was changed to read "for charitable uses." (Significantly, only the delegates from Rhode Island and Maryland—the two states most identified with religious tolerance—voted to retain the "Church Section.")[14] And there were at least two attempts to lower the minimum per-acre price from one dollar to fifty cents. All failed.[15]

One issue, though, pervaded the debate. In the original committee proposal, it had been left up to the states whether they would sell their allotted Townships "entire" or by Sections. New Englanders at this time were accustomed to the pattern of proprietors setting up whole townships, surveying them into farmsteads, and then selling those surveyed farms to individual settlers. Southerners, by contrast, employed the practice of individual settlers obtaining warrants for a set acreage directly from the state government; the settler then was the one to stake out the boundaries of that acreage. Neither side wanted to see the settlement pattern of the other dominate in the Northwest, so each side wanted to restrain the other by writing into the ordinance a specific ratio between Townships sold whole and Townships sold by Sections. But what ratio?

The committee's second draft of the ordinance contained a provision that very much favored sale by whole Townships. This provision allowed a state treasurer to sell a Township by Sections, but he could do so only by selling them in strict numerical order. And he couldn't proceed to sell a second Township by Sections until he had sold off every single Section in the first Township—a practical impossibility.

So in a later session Grayson and James Monroe of Virginia proposed that a third of the Townships be sold by Sections. Before a vote could be taken, however, James Wilson of Pennsylvania and John Lawrence of New York upped the ante, asking that a full 50 percent of the Townships be sold by Sections.

As an amendment to an amendment, the proposal to sell half the Townships by Sections came up for a vote first. The voting can stand as an exemplar of why the nation came to feel it needed more decisive government than the Articles of Confederation could provide.

That third day in May, one northern state— Connecticut—and three Southern states—Delaware, South Carolina, and Georgia—were represented by only one delegate each, so none of their votes would count. New Jersey was absent, so in order to pass, any proposal would have to garner seven of the eight "qualifying" delegations' votes.

On the "half sold by Sections" proposal, all the delegates from New York and all the states to the south—a total of sixteen delegates—voted yes. New England, with seven delegates, voted solidly no. The vote by state, however, was five yes (New York, Pennsylvania, Maryland, Virginia, and North Carolina) and three no (New Hampshire, Massachusetts, and Rhode Island). So, in a phrase often encountered in the *Continental Congress Journals*, "the question was lost."

On the "one-third by Sections" proposal, William Ellery of Rhode Island (thinking this a fair compromise?) voted yes, splitting his state's vote and thus voiding it. Wilson (author of the "half by Townships" proposal) voted against this watered-down alternative. And Monroe and Grayson of Virginia, sensing that they might be able, later on, to get more than they had hoped, voted against their own proposal![16]

The following day James McHenry of Maryland and James Monroe of Virginia proposed a scheme calling, in effect, for the Ohio Country to be a checkerboard of the two settlement patterns: the "black" Townships would be sold whole; the "red," sold by Sections. Such a scheme would, of course, result in half the Townships being sold piecemeal, but what the plan prevented was the specter of the Townships of either pattern collecting in any one region. Because the presumably "southern" Townships would always be surrounded by "New England" Townships

(and vice versa), neither pattern could dominate the other.[17]

Due to "amendment-to-amendment" parliamentary courtesy, Congress had to spend the rest of the day debating two more proposals, neither of which explicitly promised the checkerboard of the Monroe-McHenry plan, and neither of which could muster the needed seven "qualified" states. We don't know what happened overnight, but on the following day, May 5, the checkerboard of New England and southern settlement patterns passed with only a single dissenting vote.[18]

31	30	19	18	7	6
32	29	20	17	8	5
33	28	21	16	9	4
34	27	22	15	10	3
35	26	23	14	11	2
36	25	24	13	12	1

FIG 6.09

THE LAND ORDINANCE OF 1785 IS ADOPTED

With all the questions settled, all that remained for Grayson's committee was to incorporate the language of the debates into the third and final draft of its proposal. Instead, the committee exercised its prerogative to edit the proposal, and on the very next day—Friday, May 6—it submitted an ordinance containing a change that had been neither proposed nor debated officially. And curiously, it was this change, not the revisions so laboriously debated, that "stuck," and endured to become an eternal part of the national land system.

The committee now called for Townships to be 6 miles square, not 7.[19] The *Continental Congress Journals* record no explanation for this literally overnight change, and historians ever since have been combing the delegates' private correspondence for the "smoking gun" that would definitively explain what had transpired between fifth and sixth of May. You'll certainly not catch an amateur historian like myself venturing such an explanation, other than to remark that a 6-mile-square Township would have made good sense to many people of the time. Just why it made good sense, we'll explore shortly.

The delegates discussed this and other matters for two weeks, rejecting all proposed changes to the Grayson committee plan. And so on May 20 Congress adopted the Land Ordinance of 1785—presumably by acclamation since no delegate is recorded as asking for "the yeas and nays."[20] The following is an outline of the bill they passed.

In addition to changing the size of the Township, the committee (equally inexplicably) abandoned "Section" as the name for mile-square tracts and reverted to Jefferson's term, the "Lot." (This is a change that did not stick, as the federal

Congress would later readopt *Section* as its term of choice—the term used to this day.) The boustrophedonic system of numbering the Lots was retained (as shown in figure 6.09), but since a 36-square Township has no center Lot, the committee seized upon Lot no. 16, just off center, as the one to be devoted to education—another change that endured throughout the settlement period.

The Grayson committee retained the concept of reserving four Lots for the U.S. government, but in the final ordinance those Lots don't collect at the Township corners. Instead the committee moved those reserved Lots inward from the corners to the positions of nos. 8, 11, 26, and 29 (thus depriving the central government of four contiguous mile-square tracts at the Township corners). The idea of reserving Lots for the federal government would not endure, but as the nineteenth century wore on, the number of Sections reserved for education would increase from one to two and then to four.

The committee restated its directives to the Geographer of the United States on how to conduct the survey, but added the requirement that he survey the east-west baseline personally, get a secure fix on the line's precise latitude, and, as the Townships were laid out southward from the baseline, find the latitudes of major rivers' confluences with the Ohio.[21]

Moreover, the committee inserted a clause into its instructions that set a precedent for all future state borders: "Nothing herein shall be construed, as fixing of the western boundary of the state of Pennsylvania."[22] To explain: in order to complete the survey of Pennsylvania's western border, a line would have to be run north from the Ohio River, where the Virginia-Pennsylvania survey team had stopped. As part of their mandate to lay out Townships, the confederation's surveyors would be marking a north-south line from this same position on the Ohio River, but the line they marked would in no sense be Pennsylvania's border: the only line that would "count" in this regard would be the one Pennsylvania's authorized surveyors would run. As it turned out, the Pennsylvania surveyors ran their line before the congressional surveyors. Nonetheless, an important principle was established in the Ordinance of 1785, and it held throughout the settlement period: the national government would declare the map location of state borders, but it would be the states themselves that would actually mark the courses of their boundaries on the ground.

Later we'll follow the Geographer as he marks his assigned lines on the ground, but first we must deal with two questions left hanging. Why did Congress reject Jefferson's eminently logical plan for reforming land measurement on a decimal basis and instead revert to the customary measurements? And why did Congress reject the 10- and 7-mile-square Townships and settle instead on 6 miles square as the proper size?

In 1786 Congress readily adopted Jefferson's plan for "decimalizing" the dollar. Thanks to that decision, Americans were spared two centuries of adding up 20 pence to the shilling and 12 shillings to the pound. Why, then, was there such a resistance to conducting land measurement similarly, by tens and hundreds?

One reason might be that Americans had been dealing with multiple monetary systems virtually from the first moments of settlement. In circulation all through Atlantic America were coins and certificates from all the seafaring nations of Europe. In his *Notes on the State of Virginia*, Jefferson mentions, in addition to the British pound and guinea (13 shillings), crusados from Portugal, ecus from France, ducatoons from Flanders, and pieces of eight from both Mexico and Spain.[23] American merchants and even backwoodsmen constantly had to convert these currencies one into the other, as well as haggle over what a fair conversion rate might be. When the Continental dollar was introduced during the Revolution, it was merely one more currency option to keep track of. There was, in short, no longstanding conception of "a penny" or "a dollar," much less what either might be worth.

In contrast to currency systems, land-measurement systems other than the British customary units had little penetration into Atlantic America. People there had a very firm sense of how long a statute mile was and how much ground was encompassed in an acre. To have one kind of mile and acre in the West and a different mile and acre in the established states—such a prospect must have raised, to the delegates, the specter that land measurement would become as complicated as currency exchanges had been. Why saddle the new nation with such a bother merely for conformance to a decimal idea, no matter how elegant? And if Jefferson's scheme implied the remeasurement, into the new units, of all land and distances thus far established—well, even to contemplate the task was sheer folly.

Besides, as all the delegates would have known intimately, the customary measures of the acre and statute mile, far from being "irrational," were locked together in an organic relationship that had not only familiarity to recommend it, but featured easy, "rational" calculation and conversion. So many of us live far from farms today (in mental if not actual distance) that we have little conception of what an acre is. And if in idle curiosity we pull out a conversion table, we encounter units like "furlongs" and "rods" and "chains," which have to us the same medieval air of the alchemist measuring out his potions in "drams" and "grains" and "gills." And when we look deeper into the charts and see that a rod is 16½ feet, and a chain is 66 feet, our reaction is something like, "What could they have been thinking?"

Our reaction to these measurements comes partly because decimal-based measurements are so natural to us in the twenty-first century that we can hardly imagine calculating in any other way. We Americans put up with our 12 inches and 16 ounces, and even have a defensive affection for them; but we secretly feel that the metric system is "better" because it's easier and more "rational"—which is another way of saying that its counting-by-tens basis more closely accords with how we are accustomed to doing calculations.

Decimal thinking, however, is relatively new to the Euro-American mind. What we call decimals were brand-new to Isaac Newton, and by the time of the American Revolution the concept had not penetrated much past the intelligentsia. In account books in seventeenth- and eighteenth-century America and England, you'll find numbers utterly strange to us, amalgams of the Roman numeral system and Arabic numbers, like 4^{xx}—meaning "four twenties" (which in France was spoken literally as *quatre-vingt*).[24] People in those days were accustomed to counting in multiples of a set quantity (four of the set quantity 20). To obtain small amounts, it seemed natural to think in terms of fractions of a set quantity—but not just any fractions.

Take measurements of weight, for example. If you've got a balance scale and one "set-quantity" stone (weighing 1 pound, say), you can place the stone on the left side and pour grain (or gold dust) into the right side until the two sides balance. You can then remove the stone and transfer portions of the grain to the empty left side until the sides balance again, leaving half a pound on each side. Repeat to get one-fourth of a pound, and one eighth and one sixteenth. See the logic?

Or take length. Suppose you have a "set-quantity" piece of twine—say, 1 yard. With length, you can obtain fractions not by balancing but by folding. One-half of a yard is easy, but so is one-third of a yard: fold and adjust until the three lengths are equal. (You do this every time you put a business letter into a standard envelope.) Call each of these three folds a "foot," and you have another set quantity. Continue folding in thirds and halves, and you can get any number of fractions—½ of ⅓, or ⅙; ½ of ⅙, or ¹⁄₁₂. Again: see the logic?

Simple and natural. But it's very difficult to get one-tenth of any set quantity. With the simple tools of a scales, a length of twine, and (say) a cylindrical vessel, you could find one-tenth of a set quantity, but it would take a great deal of trial and error, adjusting back and forth. And unless you recorded that one-tenth part of the standard (marking the cylindrical vessel or the twine, finding or making a new stone weight), you'd have to go through the whole laborious process the next time you wanted to find one tenth. It's so much easier to rely on (and thus think in) the fractions you can obtain effortlessly anytime you want.

That kind of "multiples-of" and "fractions-of" thinking was at work in the evolution of the customary English measurements of distance and land area, both of them growing out of medieval farming methods. As we saw in chapter 1, before the advent of the movement to enclose fields into either sheep-grazing or grain-raising plots, much of English agriculture was conducted under the open-field system. Within the extended borders of a manor, all the families attached to the lord would raise animals and crops in common.

Certain areas of the manor would be devoted exclusively to the grazing of animals: sheep and cattle would browse in the grassy meadows next to streams, which were too wet for crops; pigs would be turned out into wooded areas (which, if sustained, could also be a continual source of firewood) to root around for that mix of nuts and berries the English called mast. (A typical set of English manors is shown in figure 6.10.) The rest of the manor's lands could then be devoted to crops—but on a specific rotational system. Cows

and sheep would be kept out of a field until the crop was harvested and carried away, but then the animals would be turned into the field to graze on the stubble that was left. All during the next year's growing season, the cows and sheep would again be let into that field, to graze on the wild grasses that would sprout from windborne seeds and, significantly, to drop their fertilizing manure.

The spring after that "grazing" year, the fallow field would be turned under by teams of oxen pulling a wooden plow. (Don't think of the steel plows with V-shaped plowshares, so emblematic of midwestern and Plains agriculture: English plows were wooden, and the plowshare threw earth in only one direction.) As the plow was pulled forward by the oxen, its blade dug into the earth, wedged a continuous strip of earth upward, and rotated it over to the right, in the manner shown in figure 6.11. What this accomplished would be obvious to any farmer, but needs to be explained to most urbanites.

First, plowing killed the wild grasses. Grass grows when its blades can reach sunlight and its roots can draw water from damp soil. Turned over by the passage of the plow, the strip of earth buried the grasses in darkness and exposed their roots to drying by the sun. The plow also spread out the previous season's deposits of manure and mixed them into the soil. And for the soils of England, which are often heavy and prone to waterlogging, plowing loosened the soil and aerated it.

Upon this prepared field the manor's farmers then spread their grain seeds broadcast. That is, they planted not in the perfectly parallel lines of corn or alfalfa that is our American image of a

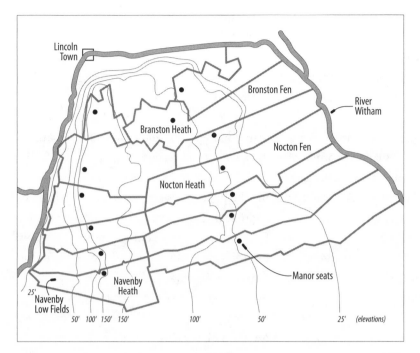

FIG 6.10

Here are the manors south of Lincoln, England, enclosed by the River Witham. The central farmsteads (and the track connecting the manor seats) sit on the cusp between the fens (lowlands) and the higher heath lands. For a sense of scale, the Nocton manor is about 6 miles long.

Redrawn from C. Stewart Orwin and C. Susan Orwin, *The Open Fields* (Oxford: Oxford University Press, 1967).

FIG 6.11

Courtesy of the Museum of English Rural Life, University of Reading.

farm crop. They threw their seed grains so as to distribute them, as best they could, evenly across the whole plowed field, grabbing handfuls of grain from a shoulder sack and strewing it onto the field in carefully placed, nonoverlapping arcs.

In this open-field system, each family nonetheless owed the lord of the manor a portion of its animal and vegetable produce. We can easily imagine how a family kept track of its animal increase, with ear nips or branding marks, combined with the obvious teat attachment of newly born young to the mother. But how to differentiate the vegetable produce among families if all crops were grown on common fields?

The answer is that the "common" fields were actually divided into individual plots, with each family assigned a proportional number of plots by rotation. A family would cultivate the plot assigned it for that year, and it would be that family which owned that plot's production (subject to the lord's share, of course). The problem was that though a single family could handle the task of cultivating their plots during the long growing season, the tasks of plowing and harvesting required more resources than any single family could muster.

As any nonindustrialized farmer knows, crops don't become ready for harvest on a predictable schedule. On any typical manor, the lay of the ground would determine how much sun and water a field would get; and soils would vary from field to field, so that even for one crop, the best time to harvest family plots might come all at once, or be spread out over weeks. The perpetual dilemma: harvest too early, and the crop will be so damp it will rot when put into storage; harvest too late, and the crop will be so dry it will fall from the stems to the ground and be unrecoverable. Harvest time on a commonly held farm was an activity in which all might be called to lend a hand.

The amount of effort needed to plant the common fields was more predictable, but also demanded more resources than a single family could command. Oxen were indispensable during the plowing season, but a family could not afford to maintain a pair for a whole year merely for the return they would give pulling the plow for a few days in the spring. Moreover, in the heavy soils in some parts of England, it could take two or more pairs of oxen to pull a plow through the mat of even one year's growth of grass. So oxen were invariably owned in common. How, then, to divvy up the work of a brace of oxen held in common by all the families of the manor?

The solution gradually arrived at was that the commonly owned oxen would be allocated by days of plowing. That is: however much land the manor's ox team could plow in a day, that much land would be one of the rotating shares of cropland allotted to each family. How, though, to mark out the allotments year to year?

Here English farmers turned a whole series of conditions to their advantage. First, they knew that a team of oxen—especially a team of two or more pairs—is tremendously difficult to turn. Best to let the oxen pull the plow in a long, unbroken line through that heavy, root-encrusted soil for as long as they could pull before becoming winded and needing a rest. While they caught their breath, the farmer would wheel the team around until it was facing in the opposite direction. He would then guide them down the line previously cut, aligning his plowshare next to the slice of earth just thrown over so as to plow up only new earth. When he got to the end of the first line he plowed, the oxen would again be winded, so he'd maneuver them again to plow a line outside the first line he had plowed—a maneuver depicted in figure 6.12. He would then continue, plowing around and around his original line, until the sun began to set and the oxen would have to be put out to pasture to be ready for the next day's work.

The result of plowing in this way was, as might be imagined, a long, capsule-shaped plot. But because each pass of the plow threw a strip of earth inward of the furrow it made, over a span of years those strips—thrown farther inward with each year's plowing—would begin to pile up, and the center of the capsule-shaped tract would begin to rise. After only a few years of plowing, a shape like a very low but very long circus tent would rise from the earth, its ridge being the line of the initial plow cut, its sloping flanks recording the successive passes of the plow throwing turf strips uphill. Figure 6.13 shows the landscape this kind of plowing produced.

Farmers of course could have prevented this, merely by reversing the direction of their plowing each year, but the crowning of the fields had two salutary effects. First, in a land where the water table was high and groundwater often threatened to waterlog the roots of seed crops, a crowned field would help lift those roots above the level of the water, and drain off rainfall as well. To cover the manor's croplands efficiently, these crowned plots were laid out side by side, giving the common field a ridge-and-trough, corrugated aspect—which led to the second salutary effect.

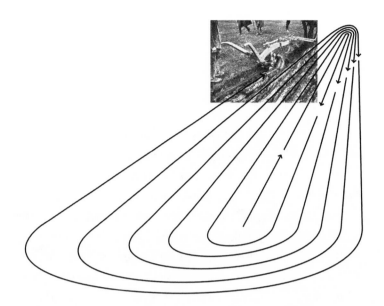

FIG 6.12

If you trace the route of the plow in this diagram, you'll see that the plow always throws earth inward of the furrow it makes in the soil.

Courtesy of the Museum of English Rural Life, University of Reading.

FIG 6.13

After centuries of plowing, throwing the earth inward results in landscapes like these.

Courtesy of Helen French.

At the end of every fallow year, the boundaries of every long plot would be readily apparent: even under a cover of grass, the valleys would show where one plot ended and another began.

English farmers had a word for the plots: they called each of them a *furrow-long*, a long furrow. And here at last we get into land measurement. There was, of course, no consistent length or width to these long furrows. Where soils were light, the oxen might plow quite some distance before becoming winded; and in truly heavy soils they might be able to cut through only a few dozen feet before demanding a rest. But in the manner in which customary measurements evolve, a consensus gradually arose concerning the typical English furrow-long; and by about the year 1400, the standard was considered 4 perches wide by 40 perches long, those 40 perches totaling one-eighth of a mile in length. What, though, is a "perch"? You are entitled to ask: it is what we today call a rod, 16½ feet. So a standard furrow-long was 66 by 660 feet, and an English mile was eight of those furrows-long laid end to end, equaling 5,280 feet. (From this length we get a term heard today only in horse racing—the furlong, ⅛ of a mile.) But to think of furrows-long or furlongs or miles in terms of *feet* is to throw together two systems of measurement meant for different purposes.[25]

To the English of that time, feet (and its adjuncts, inches and yards) were the units used to measure *length*: that system would have been applied to portable objects and fixed objects (like houses) of a middling size. In contrast, English people would have seen the mile (and its adjuncts, furlongs and rods or perches) as the proper units to use when measuring *distance over*

the earth. They would have conceived of the mile as comprising 8 furlongs, with each furlong containing 40 perches. They would not have thought of the mile as "containing 5,280 feet." That weird number comes about only because the perch and furlong have so fallen out of our consciousness. For an English farmer to imagine long distances in terms of feet would be like us imagining speed in terms of, say, yards per minute (60 mph is, for example, 1,760 ypm: sound weird?). If you can keep "feet thinking" out of your conception of distance and think instead in terms of furlongs and perches, you'll be thinking like a Briton or American of the eighteenth century, and you'll be able to appreciate the elegance of the customary system of land measurement.

The centerpiece of that system was the *acre*. What tied the acre to the system of long-distance measurements was the fact that it was defined as the area of land encompassed in a furrow-long. Now the standard furrow-long may have been 4 perches wide and 40 perches long, but it was never intended that "an acre" be a plot of ground with those precise dimensions. An acre is an area of ground, in any shape, that contains 160 square perches. Perches being so difficult for us to imagine, try thinking in terms of feet: a shape 1 foot by 10 feet has an area of 10 square feet, but so does a shape 2 feet by 5, or a shape 3.16 feet on a side. If you give that 10-square-foot area a unique "set-quantity" name, you have the concept of the 160-square-perch acre.

Remember too that the 1:10 ratio of the furrow-long's sides was not an outcome of decimal thinking: in the fifteenth century the base-ten counting system (invented in India and passed westward by the dynamic Muslims) had not yet penetrated to the isles off Europe's northwestern coast. One-by-ten was merely the nearest whole-number approximation that captured the evolved dimensions the oxen of England produced in a day of plowing.

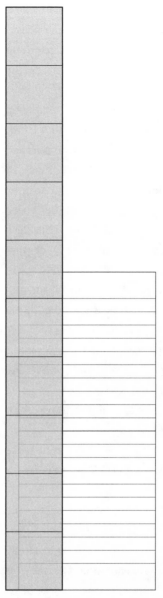

FIG 6.14

If you have trouble visualizing a furrow-long acre, here is the shape—1 chain wide by 10 chains long—superimposed on a football field. The *chain*, as a unit of measurement, is explained in the next section.

MEASURING IN CHAINS,
AND IN METES AND BOUNDS

The decimal idea eventually penetrated England, and in the early 1600s Edmund Gunter seized on the fortuitous 1:10 ratio of the furrow-long as a means to make surveying more precise by making it decimal. Through a conceptual leap, Gunter looked at the 4-perch narrow side of a canonical furrow-long and reimagined that dimension as new a unit of measure to be called a *chain*. No mere theoretician, Gunter made an actual chain of equal links. His original links were thin metal rods with their ends bent back into circles. It must have taken him countless trials, given the instruments of the day; but eventually Gunter fabricated one hundred of these links, each the exact equal of the others, and precise enough that when linked end to end, the whole assemblage would stretch out to precisely 4 perches in length[26] (see figure 6.15).

How does Gunter's chain make surveying more precise? If an acre is 4 by 40 perches, it is, in Gunter's terms, 1 by 10 chains. The area of an acre is thus, under Gunter's new unit, 10 square chains. Out of that fact comes the true elegance of the customary system of land measurement—in the following terms.

To find the area of a rectangle, we naturally multiply the lengths of two adjacent sides. When we measure the standard furrow-long rectangle in chains, we find one side to be 10 chains in length, the other 1 chain, giving an area of 10 square chains. Since that area equals 1 acre, we can convert 10 square chains to 1 acre by simply dropping a digit.

To apply that principle, suppose you have used Gunter's chain to measure two sides of a rectangular plot of land and have found them to be, respectively, 20 and 40 chains in length. Multi-

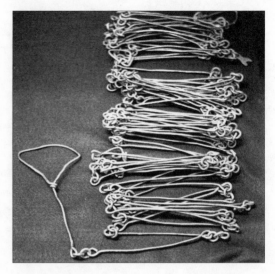

FIG 6.15

This is a Gunter's chain used in the Rectangular Survey of Michigan in the mid-nineteenth century. By that time surveyors had added circular loops between the links, allowing the chain to fold (and unfold) more easily.

plying the lengths of these two sides, the area of the plot is 800 square chains or—dropping a digit—80 acres in area.

"Dropping a digit" is, you will recall from grade-school math, equivalent to moving a number's decimal point one place to the left. So if your plot had measured 21 by 39 chains, it would have had an area of 819 square chains or—moving the decimal—81.9 acres. But because Gunter's chain has 100 links in it, you can measure whole chains and then hundredths of a chain. Our imagined plot, measured more precisely, might be 21 chains plus 10 links on one side and 38 chains plus 85 links on the other: 21.10 chains times 38.85 chains equals 819.735 square chains, thus 81.9735 acres.

A simple, natural, rational system. Of course, all this good sense evaporates if you insist on thinking of the acre as 43,650 square feet and the chain as 66 feet and the link as 7.92 inches. But chains and acres and miles, and their intimate interrelationships, would have been quite present in the minds of the delegates to Congress, even those from "urbanized" New England. The delegates were all men of some means, and there were very few fortunes in colonial America that lacked some connection to the land. Today only farmers and surveyors have a clear conception of what an acre is, but a man of consequence at the end of the eighteenth century would likely have considered the ability to "think in acres" as one of his most indispensable skills. Little wonder that the delegates to Congress were loath to abandon that system of land measurement and adopt another system under which, for a time at least, they would be reduced to the level of novice.

What is curious is that while the delegates held tenaciously to the customary system of measuring the area of a parcel of land, they abandoned the customary system of marking out the boundaries of such a parcel. That system of boundary marking had just as much common sense to recommend it as did the measuring system, but it also had some serious drawbacks. To understand how these drawbacks impelled Congress to seek a new system for marking boundaries, you need to have one more lesson in the system of land allocation worked out first in England, and then in America, in the centuries before the Revolution.

From time immemorial in England, the boundary of a piece of land had been described by a method that came to be known as *metes and bounds*. A landowner would declare that the boundary of his land ran from, say, "the large oak tree east of my home to the red-colored stone in the midst of the meadow, to the ancient grave marker on the edge of the copse of maples, to . . ."—that is, from a landmark (a mete) down a line (a bound) to another landmark.

The system worked in a rough-and-ready fashion, but could break down if any of the landmarks were to disappear or be moved. Most English landholdings were small, the patterns of cultivation were ingrained on the land, and the patterns of agriculture and husbandry resulted in most of the land being revisited frequently enough that any lost landmark would be quickly noted and its position readily reestablished from remembered cues in the surrounding landscape.

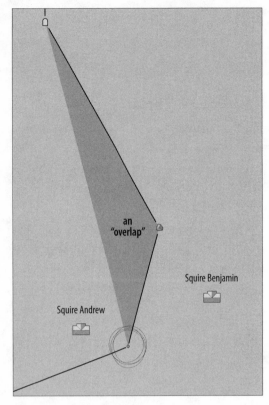

FIG 6.16

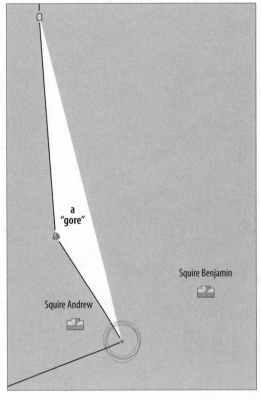

FIG 6.17

But even if a landowner knew his land's boundaries with precision, the system worked only if adjacent landowners "knew" the same boundaries. Look at figures 6.16 and 6.17, for example. Squire Andrew may assume *his* land is bounded on one side by that "oak tree . . . to stone . . . to grave marker" boundary just mentioned. But Squire Benjamin, on the plot immediately adjacent, may believe that the border of his land runs directly from the oak tree to the grave marker. If the position of the central red stone landmark is such that the tree-stone-marker line forms an angle pointing toward Squire Benjamin's holdings, there will be a triangular overlap in their claims. Conversely, if the middle stone is located such that the tree-stone-marker angle points toward Squire Andrew's holdings, there will be a triangle of land that neither claims as his own—a *gore* between claims, in surveying parlance.

Either condition, once discovered, could lead to litigation. An overlap would make both squires feel they had been denied their ancient rights, but the gore would likewise tempt them to grab at an ambiguous possibility. But overlapping and gored land claims are endemic whenever individual landowners describe their own boundaries. The ideal situation would be if every landowner described his boundaries by means of the same metes and bounds that every abutting landowner used to describe his boundaries.

If such a condition could be brought into being, there would exist a true *cadastre*—a net of straight lines, running between recognized landmarks, covering the entire region, with every such polygon ascribed to an undisputed owner. But how would that cadastre be recorded on a piece of paper that all could see and endorse?

The system that evolved (and is still in use today) was to describe the periphery of a tract by citing a sequence of straight lines between landmarks, the sequence beginning at one landmark and then circling around to end at the same landmark. That circling-around between landmarks had been done traditionally: what was new was to note the length and the direction of each of the lines. This surveying terms of art are the *distance* and *bearing* of a line between *monuments*.

The boundary description of Squire Andrew's property under this new system might be: "Begin at an oak tree marked by a triple blaze; on a bearing 15° east of north proceed a distance of 10 chains, 20 links to a red stone; from there proceed on a bearing of 30° west of north a distance of 20 chains, 50 links to a grave marker; from there proceed . . ." in like fashion to, say, a tree marked by a brass nail, then to a wooden stake, and finally from that stake "proceed on a bearing 20° north of east a distance of 15 chains 25 links to the point of beginning." Figures 6.18 and 6.19 illustrate these terms.

Such a sequence of words may be cumbersome, but it describes Squire Andrew's property with geometric precision. There is the extra benefit that should any one of the monuments be destroyed (that red stone, say), its location can be found by "retracing the lines"—measuring out the set distance on the prescribed bearing to (as surveyors say) *reestablish the monument* with some new marker.

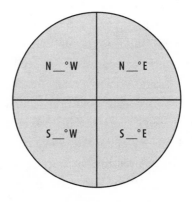

FIG 6.18
The four quadrants at left convey how surveyors have standardized the notation of bearings and distances. Any line can be considered as directed from the circle's center into one of the quadrants: the notation of the quadrant determines how that line's bearing will be expressed.

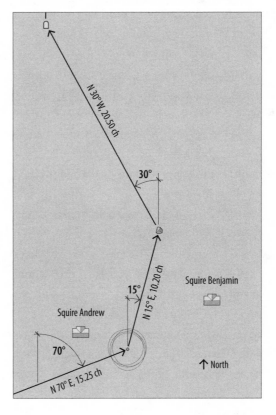

FIG 6.19
Here is the notation system in use. If it's easier to imagine, you could read "N 30° W" as "Face north, then rotate 30° to the west"; or more simply, "30° west of north."

It's difficult to place a beginning date on many of these developments in land description, much less a date of general acceptance; but the technique of metes and bounds recorded on maps by bearing and distance was in place in Atlantic America by the period of the Revolution, ready to be employed in the apportionment of the West. Also available was a large cadre of surveyors trained in at least the rudiments of these techniques. (We saw how both Washington and Jefferson had spent time looking through the sights of surveyors' instruments.) For men of lesser station, if they had a head for numbers and enjoyed the outdoor life, surveying could be an almost ideal career, rewarding as well as remunerative. Many such men took on the minimal training to qualify for the work.

So it would have been possible, in theory, to apportion the West by allowing farmers to stake out their own claims to land in the wilderness. A farmer could find a favored stream, a clearing in the woods already made (probably by Indians conducting their own agriculture), perhaps even a spring. He would then chop blazes on tree trunks to mark the boundaries of his claim, trying to enclose approximately the acreage stated on either his warrant or his bill of sale. Farmers around him would of course do the same. Virtually all of the back-country South was settled by such a process, and so might the West have been. Eventually the gores and overlaps could have been negotiated away, mutually agreed-upon descriptions of each property arrived at, and a cadastral survey map of each region prepared. That was, after all, what was being undertaken, with some success, in England following the enclosure movement. But the American situation was decidedly different.

In most cases in Britain, surveyors were essentially taking note of evolved patterns of ownership and then recording them through the new medium of geometric description. Where disputes arose, there would often be evidence of unchallenged use or occupancy by one of the contending parties, precedents which, in a last resort, the courts could use in rendering their decisions.

By contrast, in America the claims would have not evolved over centuries of negotiated use, but come about all at once. Worse, the only "evidence of occupancy" on a claim would likely be a small farmstead near the center of a claimed tract: the tract's periphery—where overlaps between claims would most naturally occur—might hardly have been visited by the contending parties since their initial reconnoiterings of the area. This meant that boundary disputes might not occur until decades after the initial claims had been filed, when the operations of the claimant farmers (or their sons) finally expanded enough to come within shouting distance.

Compound those problems with the fact that as a thrown-together society, America was, even then, a litigious place. You may think there are too many lawyers in the United States now, but in the eighteenth century there were, per capita, an equal proportion of people who called themselves lawyers; and a great chunk of their business came from boundary disputes between abutting landowners.

There was hardly an American landowner alive in the 1780s who had not been involved in some dispute, litigated or not, over landownership. The disputes occurred all across the nation, but more frequently in the South and the middle states, less so in New England. The reason why New Englanders went to law less frequently was because the northern states had gradually adopted a principle that has come to be called *survey prior to settlement*.

We saw that in medieval England the ideal agricultural manor was one that contained within its boundaries three distinct landscapes:

- *a wooded portion where pigs could forage, where small game might be hunted, and from which timber for fuel and construction could be cut;*
- *open fields (most likely forest land cleared and enlarged through previous woodcutting) where grain crops could be grown in rotation with manure-producing cattle grazing the fields in fallow years; and*
- *grassy meadowlands on the verge of streams, where cattle could be grazed, with the further benefit of huntable waterfowl, catchable fish, and the possibility of boat transport to carry any produce in excess of the manor's needs downstream to market.*

Farmers who emigrated from England to America naturally looked for tracts that could provide these three landscapes, and because the East Coast is not so very different from England (in its topography and hydrography at least), English farmers were able to find tracts in this new land that would fructify in response to their ancient patterns of agriculture and husbandry.

In the southern and middle colonies, with a tradition of noninterventionist governance that continues to this day, farmers were largely left free to carve out wilderness tracts that seemed, to their own personal judgment, to encompass (or promise) those three essential landscapes, with the resulting confusion and litigation of tract boundaries. New England, by contrast, was founded on a strong belief in governmental control over settlement. When settlement was new and still concentrated along the coast, control could be exercised by the authorities themselves; but when the sheer force of population growth mandated westward expansion, the most expedient method for perpetuating the tradition of centralized control was to place that control in the hands of men who had proved themselves to be supporters of the status quo.

For favors enacted or money rendered to the colonial governments, these proprietors would be granted all the land within a newly created township whose borders would be surveyed by the colonial government. The proprietor's task would then be to survey the township into tracts, which he would sell or grant to people who would actually settle and farm there—with a sizable tract often reserved for the proprietor and his heirs. The ideal was that the proprietor would be not a "lord of the township" but eventually, as the plots were gradually alienated from his ownership, just one more landowner (even if the leading landowner) as the township evolved from a land-selling corporation to a governmental entity.

The colonial governments imposed few rules about how the proprietor would apportion his township. It was assumed that he would follow existing conceptions about what constituted a three-landscapes ideal farm, and divide his holdings into tracts that would meet that criterion—how else could he hope to attract settlers who held the same conception of a tenable farmstead? The proprietor looking at his newly granted landholdings usually saw something like the following.

The landscape of his township would likely be characterized by ridges running very roughly parallel across his holdings. Between these ridges

would flow streams that gradually converged into rivers which flowed to the sea. The ridges were largely wooded by a continuous hardwood forest that thinned as the land fell toward the streams, the tree cover gradually giving way to open meadows in the floodplains of the streams, with wetlands at the verge of the stream banks.

The proprietor would know from his own experience that roads which run on ridges will drain themselves, and that farmers would want their houses near the roads. So he would instruct his surveyors to mark onto the land a series of long rectangles, each roughly equivalent in area, each extending from a projected "Ridge Road" down to the stream that roughly paralleled the ridge. In this way, as the purchaser of the plot cleared the trees upward from the floodplain meadow, uphill toward the ridge, he would achieve, through his own efforts, the ideal three-landscapes farm he had been taught to expect.

Surveyors would blaze trees in the forested portions of the ridge slope, and drive wooden stakes in the marshy stream verges. Thus, they would mark out—prior to settlement—a rack of roughly parallel, roughly equivalent farmsteads from which an emigrant farmer could make his pick. And because the surveyors would map each of the tracts by that distance-and-bearing system, there would be no gores, no overlaps, no ambiguity about who owned a piece of land.[27]

The shape of the townships granted to proprietors went through a definite evolution between the mid-1600s and the late 1700s. From the earliest times of settlement, it had been the practice to set new townships directly adjacent to previously established ones. Given the randomly indented shape of the New England shoreline, the initial townships, directly along the coast, took no typical shape. But as settlement pushed inland, into territory more or less unbroken by insuperable geographic features, it became easier to form more nearly "ideal" townships, shaped not by geography but by what the earliest colonists had learned about governance and settlement in this new land.

Remember that it was presumed that these proprietary townships would gradually evolve from the status of one man's property into an entity governed by its landowners. Governance in seventeenth-century New England was conducted not by elected representatives (as it was in the South) but by the people themselves, acting through two bodies. There was the township government itself, whose major decisions were made in periodic assemblies of all the township's men in a town meeting. But there was also the church, which all in the township were expected to attend on holy days, to receive instruction not just on the salvation of their souls but on the conduct of their lives, in both the private and civic spheres.

The requirement for an ingathering of the townspeople was enshrined in law, but those laws were not merely dictates imposed by stern Puritan divines. They were more nearly the reflection of a consensus among settlers that, unless people in the wilderness got a bit of "society" from time to time, they would go a little mad. (Later frontier experience—from the hollows of Appalachia to the Great Plains—must engender some respect for that view.)

Travel being what it was then, a township could not be so extensive as to prevent its residents from making the trip from their farmsteads to the meeting and back in the course of a single, dawn-to-twilight day. That was one factor in determining the ideal township's size, but there were others. How small could a township be before the number of its parishioners would be insufficient to support a minister in an appropriate style? How small could it be before its able-bodied men, formed into a militia, would be insufficient to defend the town against Indian attacks? How big could it get before it had more fences than the town inspector could check in the traveling season between "last mud" in the late spring and "first snow" in the early autumn? At what point indeed would a township's extent be too great for the residents to be able to imagine it as an organic entity?

All these factors and others went into the determination of a township's "ideal" size. Some factors pulled in the direction of tiny townships; others favored vastness. But after some trial and error, by the late 1600s New England had settled upon a township approximately 6 miles square as just about right.

So when, for example, officials in Massachusetts set up the new town of Concord in 1635, they mandated that it be a square, not oriented to the cardinal directions but tilted to follow features in the local landscape,[28] which you can see in figure 6.20. Concord was being carved out of then-huge Newtowne (which, after John Har-

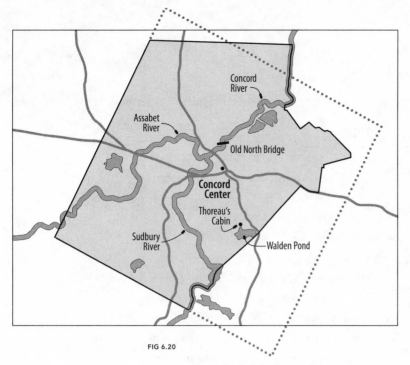

FIG 6.20

vard's gift, was renamed Cambridge in recognition of its new status as a seat of learning), and the colonial government thought it best that the township be centered on the Concord River and its tributaries. The rivers meander in that area, but their general flow is toward the northeast, so two of the square's sides run in that direction. The square was further positioned so that the nascent village of Concord Center would indeed be near the township's center. Concord's borders have changed since then (only two of the square sides remain), but it is one of those serendipities of arbitrarily declared borders that the boundaries of the town of Emerson and Alcott should also contain the shores of Walden Pond.

Don't think, however, that the 6-mile-square township was seen as the definitive model; and "squareness" itself was only one of the strategies deemed proper for shaping townships. If other factors (prior borders, coastlines, or river courses) seemed relevant, townships would be given wholly other shapes with no thought of having fallen short of an ideal. In 1635 squares were an expedient alternative, but not yet a model.

All through the mid-1600s, New England governments continued to plant new townships, reaching a total of seventy-two by 1692, with square ones gradually coming to dominate. By the late 1600s, though, town establishing took a leap: farmers wanted to skip over the hilly land immediately inland from the early towns and settle in the fertile valley of the Connecticut River. The governments of New Hampshire, Massachusetts, and Connecticut complied with these wishes (in New Hampshire's case actively encouraging settlement, partly to thwart New York's claims to the area), setting up new proprietary townships. The pattern quite naturally was to use the river as a township border, and draw township boundaries perpendicular to the river's course. The Connecticut River, like its twin the Hudson, has a remarkably straight north-south course in that part of New England, with only a few slight swings east to west. So as settlement progressed, the valley gradually acquired a ladder of townships aligned to the gentle undulations of the Connecticut River. But once established, the idea of a ladder of townships took on a life of its own.

The most striking instance of such a ladder pattern is in what is now southern and central Vermont and New Hampshire, as shown in figure 6.21. There in 1761, Governor Wentworth of New Hampshire ordered an unprecedented sixty-eight new townships to be laid out, all of them 6 miles square.[29] (The proffered reason was to provide bounty lands for soldiers from the French and Indian War, just about to end; but we can be certain that thwarting New York was again a motivation for so bold a move.) Because there were to be so many townships, there would need to be rank on rank of them back from the river. The most reasonable, expedient way to lay them out would therefore be as a grid. In the area centered around Hanover, New Hampshire (home of Dartmouth College), the grid rotates to the east to use the river as one of the grid lines. But in the area of Vermont toward Bennington, the river—and thus the grid—runs north-south, producing 6-mile townships almost aligned to the compass points.

What makes this area of New England so fascinating to map lovers is that the resulting township grid is actually visible on conventional maps of governmental borders. When the whole of the West was apportioned into 6-mile-square "Congressional" Townships (as they were called in parts of the West), seldom did the Township become a unit of government; and so even though counties in the West are often agglomerations of such Townships, those county boundaries only hint at the 6-mile grid underneath. Governor Wentworth's grants to proprietors, though, were explicitly meant to become real towns, real units of government. Almost all of them did, and even though some of the borders have changed since, the original 6-mile grid still can be seen on maps—one of the few places in the United States where this occurs.

But even if you can see the original grid of townships on political maps of New England, what you will not see is the feature that virtually defines the built landscape of much of the Midwest and West: those roads that run only north-south and east-west. We know that roads run that way because they follow the edges of the squares into which the invisible Congressional Townships were divided. But fly over the gridded townships of New England, and you'll see roads that angle and curve without reference to the rectangular township boundaries you see on maps of the region. The reason, of course, is

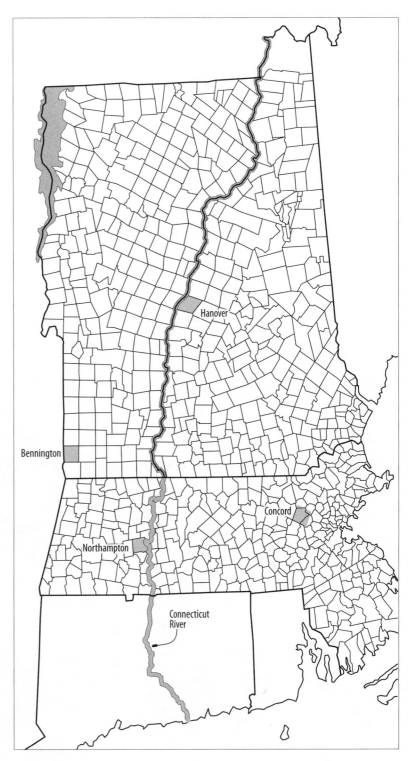

FIG 6.21

that even in the compass-aligned township grid of southern Vermont, the apportionment of the township interiors was left to their proprietors and settlers, and invariably they followed the same ridge-to-vale, "ideal-farm" principles that had proved themselves in that region over the course of 150 years.

Land Apportionment in a Square Township

My colleague Scott Tulay has constructed "transects" of the landholdings in his hometown of Northampton, Massachusetts, at fifty-year intervals. The map in figure 6.22 (redrawn from his) shows the farmsteads as they existed around 1700. You can see from figure 6.21 that the much larger Township of Northampton was about 6 miles square, aligned with the general run of the Connecticut River.

Within those geometric boundaries, though, the land was apportioned to produce long, "three-landscapes" farms on the local topography. The township authorities chose not to lay out farms from the wooded heights down to the Connecticut River (such large farms would become practicable only with the mechanization of the nineteenth century). Rather, they surveyed tracts up from the banks of the Connecticut's gentler tributary, the Mill River, flowing into the oxbow loop abandoned by the bigger river in one of its periodic floods.

There are great variations in the farm plots, but a precept of farm size was that a single man cold qualify for 4 acres, whereas a married man was due 6 acres. You can see this general trend in the smaller farms: they are approximately 2 furrow-longs deep by 2 furrow-longs wide—say, 1,300 by 130 feet. The larger "family" farms are, by contrast, 3 furrow-longs deep by 2 furrow-longs wide—about 2,000 by 130 feet.

This is nowhere near the size of the manors south of Lincoln in England, but both the family farmsteads and the baronial estates follow the same "three-landscapes" model of an ideal farm.

Connecticut River

Town center

Connecticut River floodplain: held in common

4-acre farms

6-acre farms

Mill River

Connecticut River oxbow

FIG 6.22

From this story of English land measurements and New England land settlement, we can see why Congress might have rejected Jefferson's scheme for "decimalizing" the measurement of land. For four centuries Englishmen and then English Americans had conceived their land—and calculated its value—in terms of the customary acre, a unit held so habitually as to be an almost unconscious mode of thought. To cast out that system and adopt another would not just overturn comfortable habits, but render hard-won skills of judgment obsolete, and for what? "Rationality"?

We can see too why, when casting about for a basic unit with which to apportion the Northwest, Congress might have rejected the "rational" 10-nautical-mile square and its alternative the 7-statute-mile square in favor of the 6-mile-square township that had shown its usefulness over the course of a century.

So both the acre (and its integral system of measurement by chains) and the 6-mile township had precedent to recommend them. What was wholly without precedent was the idea of dividing the Township into 1-mile squares of 640 acres (80 chains square, thus 6,400 square chains, thus 640 acres). Nowhere on the continent had there been a tradition of 640-acre, square-shaped farmsteads. And nowhere on the planet had there been a tradition of mile-square farmsteads aligned to the points of the compass without any adjustment to local geography. What could Congress have been thinking?

We must here make some allowance for ignorance. Not that the delegates were ignorant of farming conditions: most of the southern and middle-state delegates were, like Jefferson, land-

owners; and even the "urbanized" delegates from New England had intimate contact with agricultural conditions. Their ignorance was about the vast Northwest, which almost none of them had seen—and by implication, the even more unknown Southwest that would fall their way once Georgia and the Carolinas surrendered their claims. Who could say what might be a proper farmstead for those strange territories? (As it turned out, the mile-square tract would be found to be too big for the corn-and-pig farms of Ohio, Indiana, and Illinois; too small for the slave-cultivated cotton plantations of Mississippi and Alabama; and somewhere near correct for the mechanized wheat fields of the eastern Plains. Not too terribly bad for a shot in the dark made in 1785.)

To ignorance as a reason for the mile-square farmstead, we can probably add a measure of desperation. Deep in debt from the Revolution, Congress needed money, and quickly. It was determined to have its parcels surveyed prior to settlement, but it couldn't afford the precious time to send scouts into the Ohio Country to reconnoiter the lay of the land; follow those scouts with surveyors to mark out the ridge roads and ridge-to-stream ideal farmsteads, and record those surveyed lines on plats; allow potential farmers into the wilderness to judge the farmsteads on offer; and then conduct auctions of those plots.

Such a scheme might delay land sales for a decade or more. As for the alternative of selling large tracts to proprietors and having them survey the farmsteads, there were two objections. First, headstrong southerners would never accept the judgment of any person—government surveyor or proprietor's surveyor—about what constituted a proper farmstead. Second, the delegates to Congress harbored the hope (not entirely unreasonable) that the bids on many small plots might be a greater sum than the bids they could get on those plots offered as "entire" agglomerations. Offering mile-square plots, marked out not by the biased judgment of a hired agent but by the arbitrary (and therefore unbiased) application of a blind, disinterested system, would placate the southerners' hope for freedom from authority (and perhaps tempt their fondness for a crapshoot) while still putting desirable properties onto the market.

Thus a third factor in Congress's deciding as it did: regional and factional distrust. Already in the air was the peculiarly American answer to mistrust of faction: a trust in *system*—the belief that if reasonable people can fabricate a system that all can agree seems logical, then all those who sign on to that system will feel that they had been treated fairly, whatever outcome that system might produce.

This is the boiled-down essence of the U.S. Constitution, and the Federalist Papers' arguments for it. That the papers convinced the electorate, and that the Constitution was adopted, is testimony to the extent to which this idea was already current in the American mind. That John Rawls could make much the same argument in his *A Theory of Justice*, and have it so widely respected, is testimony to this idea's enduring hold on the American psyche. Whatever passions Jefferson the man might have stirred among members of Congress, the delegates' final acceptance of his mile-square farmstead must be evidence of their shared belief that this system—native to no region or faction—might,

when applied, produce a result that all could accept as a fair.

We'll never know whether the force of Jefferson's accomplishments "sold" Congress on the idea of a mile-square farmstead, or whether his formulation of the idea merely struck them as "right." Nonetheless, of all the land-allocation ideas Jefferson offered, the only one that garnered assent—in the Continental Congress and throughout the nineteenth century in the federal Congress—was apportioning the western lands not by the judgment of persons or groups acquainted with the terrain, but by the disembodied operations of division and redivision along lines of latitude and longitude, those lines unknown by anybody before being found by surveyors, and thus unmanipulable to the interest of anyone.

Better to trust a blind "system" than the machinations of bureaucrats or corporations. From the beginnings of the Republic, some Americans have seen government as the best vehicle for determining "how things should be." Also from the beginning, other Americans have seen the marketplace as the best testing ground for deciding what to do. The idea of surveying the wilderness into rectangles offered a conception favoring neither marketplace determinism nor bureaucratic fiat. Both would have to give way to rectangles. And that solution has, for two centuries, seemed right and proper—again, manifest—to Americans.

SEVEN

Putting a Rectangular Survey on the Ground in Ohio

1785: CONGRESS PUTS THE SURVEY IN MOTION

On May 27, seven days after approving the Land Ordinance of 1785, Congress appointed Thomas Hutchins overseer of the new survey called for in the ordinance. Both Congress and Hutchins anticipated that the Virginia-Pennsylvania boundary surveyors would extend their line north to the Ohio—the designated beginning point for the survey—in time for Hutchins to use that starting point to complete some of his surveying work before the snow began to fall.

Thomas Hutchins, born in 1730 in New Jersey, had been a loyal officer in the British colonial service from 1756, serving the Crown in the French and Indian War and in the early stages of the Revolution. In 1778, though, he found himself stationed in London, and from there he deserted, escaping to France, where he made his way to Benjamin Franklin. You will recall that Congress had sent Franklin to Paris to persuade the French to aid the American rebels. Franklin was impressed enough by Hutchins to send him back to America with a letter of introduction to Congress. In 1781 Congress concurred with Franklin's assessment and appointed the new patriot as one of the first two Geographers of the newly United States. On April 1, 1785, as debate began on the Land Ordinance, Congress called Hutchins to its seat in New York to advise it on the finer points of the act, and to stand ready to begin to carry out the provisions. The call to duty came when he was given a new three-year term as, this time, the sole Geographer of the United States.[1]

While Hutchins was beginning his preparations, the reconstituted Virginia-Pennsylvania surveying team had taken to the field. On June 6, 1785, the axmen of the party began creating a vista northward by clearing away brush and trees from the surveyor's line of sight, with Rittenhouse's Transit Instrument not far behind. Two months later they reached the Ohio River, and on August 20, one of the crew ferried across the Ohio and drove a stake into the flats on the river's northern bank. From this stake Hutchins would begin his survey.[2]

THE TASK AT HAND
The Surveyor's Compass

FIG 7.01

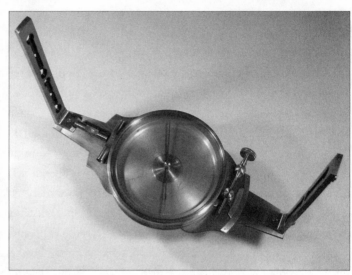

FIG 7.02

Thomas Hutchins would carry a variety of instruments with him, but prime among them would have been the surveyor's compass.[3] Less accurate than Rittenhouse's Transit Instrument, the typical surveyor's compass was both more portable and more durable, and perfectly adequate (in the hands of a skilled and conscientious surveyor) to the task of dividing land into rectangles. Virtually all the land surveying done between the Revolution and the Civil War was accomplished with this versatile instrument. Sometime around 1860 a studio photographer posed two nattily attired gentlemen with their compass. Figure 7.01 shows that then, as now, surveyors "clean up nice."

Figure 7.02 shows the author's own surveyor's compass, a reproduction from the good folks at the Ames Instrument Company in Ames, New York. Its parts were cast from original molds, then milled with modern tools. It's a hefty piece of work: you can sense its size from the first illustration, but be aware as well that its weight is around 15 pounds.

The instrument was constructed of brass so as to have no magnetic effect on the compass needle, and all its calibrations were etched onto its surfaces (no painted markings that might chip off with use). Its main component was, of course, a glass-faced compass, which sat on a flat plate; from this plate two sighting vanes projected upward. It was these vanes that turned a mere compass into a precision surveying instrument.

CHAPTER SEVEN

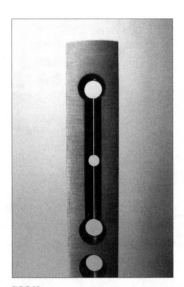

FIG 7.03

In instruments like the astronomical transit we saw earlier, you looked through the whole aperture of a telescope to locate what you wanted to sight, then used the crosshairs in the aperture to align the scope with precision. With the surveyor's compass, the pair of sighting vanes does the work of the telescope: figure 7.03 shows a close-up of one of the vanes. You would do "wide-angle" sighting by positioning your eye to look through a pair of these holes, then rotating the compass around until the object desired for sighting was roughly centered in one of the holes. You'd next lower your head so that your eye looked through a pair of slits, then gently nudge the compass around until the object fell in the center of that much narrower field of view. Figure 7.04 diagrams that view through two slots.

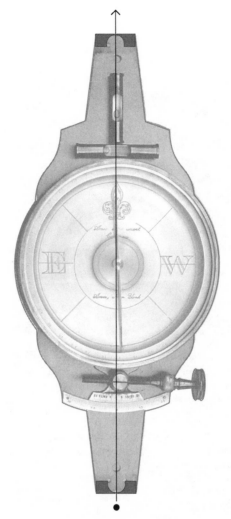

FIG 7.04

FIG 7.05

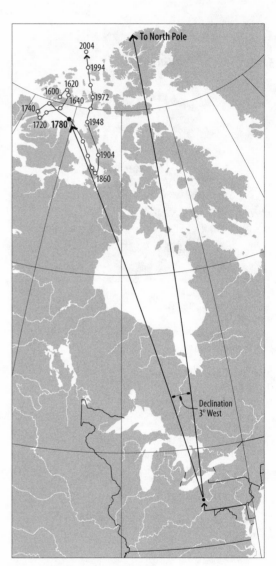

FIG 7.06

You doubtless know that a compass needle aligns itself not to true north but to the lines of force of earth's magnetic field. Those lines surround the planet like the sections of a tangerine, and for as long as humans have used compasses, those lines have dived into the earth at places *other than* the two ends of the earth's axis of rotation. Figure 7.05 shows a "compass needle" aligned with one of those imagined lines of force.

What is more, the magnetic field wanders slowly over earth's surface. Figure 7.06 shows how the north magnetic pole has migrated during the past four centuries. You can see that in Ohio, in the years around 1780, the north magnetic pole would have made a compass needle point about 3° west of true north. Surveyors would express this condition by saying that the needle has a "*declination* of 3° west."

It was possible to use the surveyor's compass to find north by sighting on Polaris, but the techniques were complicated and prone to error. In long-settled areas, surveyors would instead rely on published tables of declinations. In frontier places like Ohio, the official in charge of the survey might find the declination with precision instruments and then mandate that all the crews in his charge use that figure in their work. Many surveyor's compasses had a feature that could compensate automatically for magnetic declination; the images in the next five figures show how that feature worked.

Figure 7.07 should make you aware, first of all, that we are talking here about a really *big* compass, almost 7 inches across.

The compass had to be big, since only a big circumference would spread the necessary half-degree marks far enough apart to be useful. Figure 7.08 shows both the etched marks and the magnetized compass needle.

FIG 7.07

FIG 7.08

Figure 7.09 shows how the compass would look with the sights aligned to true north, and the needle pulled 3° to the west.

But on this instrument, the compass case can rotate on the base. With this thumbscrew and the finely calibrated scale next to it, the case can be rotated the exact amount of the magnetic declination for that region. (Figure 7.11 shows the author's fingers doing just that.) For Ohio in 1784, the surveyor would have rotated the case 3 degrees counterclockwise.

With the compass case rotated, as in Figure 7.10, when the needle points to the magnetic pole, it matches the north point on the compass face, and the sights are aligned true north–south. Clever. But why are the E and W on the compass face reversed?

Not all of a surveyor's work involved sighting to the four compass points. Suppose you, as surveyor, have to make a sighting toward N45°W—that is, 45° west of north. You are standing behind the "bottom" end of the compass in Figure 7.10. You rotate the whole compass counterclockwise, and as you do, the needle starts to point "west" of north. When the needle aligns with the 45-degree mark between the N and the W—N45°W—your sights are pointing in the required direction.

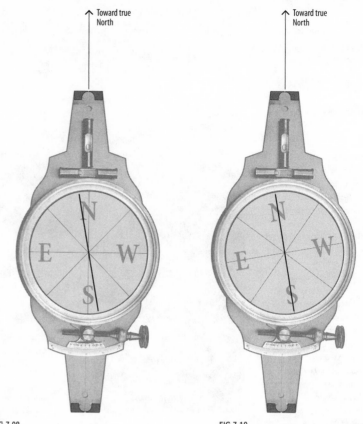

FIG 7.09

FIG 7.10

FIG 7.11

○ Range pole in distance

THE TASK AT HAND

Projecting a Line

Stake beneath
plumb bob

FIG 7.12

FIG 7.13

●

FIG 7.14

↑ Foresight

Second
stake

↓ Backsight
○
First stake

FIG 7.15

In almost all cases, the work of a land surveyor involves projecting a line out from some already determined point on the earth—from a marker set in some earlier survey, or (as is the case here) from a stake set by the surveyor himself. In figure 7.12 the "surveyor's assistant" (my colleague Scott Tulay) spreads and closes the legs of the tripod that will hold the compass so that a plumb bob will fall directly over the stake.

Once the compass is mounted on the tripod, the stake will be directly beneath its center, as indicated in figure 7.13. The job in this case is to project a line from the stake directly northward, so you swing the declination-corrected compass around on its mount until the sights are aligned north–south. You then have your assistant carry forward a straight wooden rod called a picket or range pole.

As in figure 7.14, your pole man holds the rod off the ground so that it hangs plumb, and you look through the slits of your sighting vanes. You signal for your assistant to move the pole from side to side, keeping it hanging plumb. When the pole comes into the center of your narrow field of view, you have him carefully lower it directly into the ground. You now have two markers in the earth, aligned directly north–south.

To project your line further, you go to the range pole just driven and substitute a stake for the pole. As before, you set up your compass directly above the stake, as indicated in figure 7.15. But this time you walk around to the other side of your compass and look back at the first stake. You have your pole man hold his rod above that stake, and this time you rotate the compass until you get the pole in your sights. You are now looking directly down the line you just projected, and if you walk back around the compass and sight in the other direction, you can project that line forward to yet another range pole. Surveyors call this technique "backsighting and foresighting": this is the task the surveyor's compass was designed for.

FIG 7.16

FIG 7.17   Courtesy Nebraska State Historical Society.

As you backsight and foresight your line northward, right behind you and your pole man comes a second crew, the chainmen who will do the job of measuring distances on your line. Picture the field of their operations. Because of how you did your work, from every point marked by your stakes, the "next" point on the line can be seen (as well as, often, the "previous" point). The axmen on your crew will have cut away some of the underbrush, meaning that a chainman at a point on the line can see clearly to the next marker—even when squatting. We'll see in a moment why this is important.

We've talked about Gunter's surveying chain in the abstract. Figure 7.16 is a close-up of a chain used in the survey of Michigan, but the chains carried by the Ohio surveyors would have been similar. They were made entirely of thick steel wire, round in cross section, the wire too stout to be bent by bare fingers but thin enough that pliers could do the job. By the time of our Ohio survey, it had been discovered that the chain would be less prone to kinking if the links were connected to each other by a pair of circular loops made of the same wire.

In addition to the chain, the chainman also carried a set of eleven chaining pins. The fierce-looking gentleman in figure 7.17 is Robert Harvey, Surveyor General for Nebraska, photographed at the end of his career with a lifetime of survey-ing tools—including, beneath his forearm, a set of chaining pins. A chainman would usu-ally knot a short rag to the wire circle at the top of each pin, and then stick his pins into a loop attached to his belt.

The chainmen do their work in pairs, with one man desig-nated the head chainman, the other the rear chainman. The rear chainman holds the chain (like the one shown in figure 7.18), folded up in an hourglass shape, and he squats behind the first stake, facing forward. The front chainman stands in front of the stake, carrying the full set of eleven chaining pins in his belt loop. He takes out one of the pins and sticks it into the ground next to the stake. Then, with his right hand, he pulls the forward handle of the chain out of the bundle. Next he takes a second pin—with (say) a bright-red rag tied on—puts the pin into his right hand along with the chain handle, and turns his hand so that the rag is above his hand. Holding that hand away from his body, he walks toward the range pole in the distance, frequently looking back at the rear chainman.

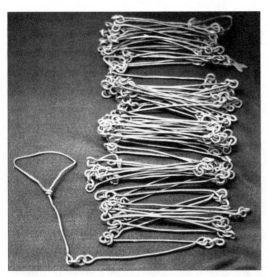

FIG 7.18

The rear chainman, squatting down behind the pin, manipulates the chain bundle so that it pays out smoothly. He sights simultaneously on the range pole in the distance and the rag on the lead man's pin, trying to keep the rag perfectly aligned with the pole. The head man inevitably veers left or right as he advances, but the rear chainman guides him back onto alignment with hand signals and shouts.

The rear man also watches to see how much of the chain is left in his hands. Just before the last link is pulled from his bundle, he shouts "*Halt!*" to the head man. The head man then stops and gives the chain one final shake to loosen any kinks, and between them the two men stretch the chain taut. With his end of the chain precisely positioned against the first chaining pin, the rear man shouts, "*Stick!*" The lead man pushes his pin into the ground at the end of the chain and shouts back to the rear man, "*Stuck!*"

The forward end point of the chain now securely marked, the rear man pulls his pin, inserts the pin into the loop in his belt, and walks toward the head man, folding the chain back into its hourglass-shaped bundle as he goes. When he gets to the head man, the two repeat the process, only now the pin that the head man inserted in the ground serves as a new starting point; and the head man takes another pin from his loop and carries it forward with the chain.

After another "*Stick!*" "*Stuck!*" exchange, the rear man pulls the second pin, inserts it into his belt loop, and again walks up to the head man. When they meet, the rear man will know—from the two pins now in his belt loop—that he has advanced 2 chains from his starting point. As the process repeats, the rear chainman, who is responsible for an accurate count, thus always carries in his belt loop a true record of the number of chains he has paid out and rebundled. So when the head man puts the eleventh and final pin in his set into the ground while the rear man pulls the chain taut, with those two pins in the ground, the rear man must have nine pins in his belt loop. The protocol of surveying called for the head man then to shout, "*All out!*" or "*Tally!*"— meaning: "*I'm out of pins!*" and "*Tally up your pins as you come to me!*"

Pulling the rear pin, the rear man advances as before, bundling up the chain, but also counting up the pins now in his possession. If he finds 10 pins, then the count of chains is confirmed and the point marked by the 11th pin is a full 10 chains from the starting point—that canonical eighth-of-a-mile derived from the ancient furrow-long. With eight of those "*Tally!*" calls, the chainmen know that they have advanced a full mile along the line.

Late in the summer of 1785, Thomas Hutchins set out from New York City to fulfill his commission as Geographer of the United States, beginning the first survey of the Ohio country. He arrived in what is now Pittsburgh on September 3, but he barely stopped there. The very next day he was at Fort McIntosh, down the Ohio almost at the Pennsylvania border, to consult with the fort's commander, Colonel Josiah Harmar. Hutchins had heard rumors of Indian unrest in the area just west of where he would be surveying, and he hoped for assurances from the colonel that those reports were unfounded—or at least exaggerated. Harmar replied that the Geographer could "very safely repair"[4] to the survey area; and so Hutchins, probably not entirely assured, went back up the river to Pittsburgh to buy supplies, hire workmen, and meet with the assistant surveyors sent by each of the states as required in the Land Ordinance.

As it turned out, only eight delegates showed up (some of whom we'll meet later). Nonetheless, on September 20 Hutchins began to move his crew down the Ohio, and on September 30 he (or an assistant) positioned a surveyor's compass over the wooden stake that had been driven into the northern bank of the Ohio on August 20 by a crewman from the Virginia-Pennsylvania boundary survey.

As he rotated his compass 90° west from his calculated north meridian, Hutchins knew that the terrain of this part of the Ohio Country would make surveying difficult. As a British officer serving at Fort Pitt, he had made treks both overland to Lake Erie and on the Ohio River down to the Mississippi; the map he compiled from those journeys was considered one of the era's best. He would have known that 10 to 20 miles back from the river was a plateau of rolling hills, but the general elevation of this plateau was hundreds of feet above the river. Consequently, the streams falling from the high plateau had cut it into deep valleys.[5] (If you have visited Pittsburgh, you have seen a terrain similar to the region of the first surveys.)

To add to his trepidation, he would be doing his work without the protection of Continental soldiers. We saw that part of Congress's reasoning in having the survey commence at that particular location on the Ohio was the nearby presence of Fort McIntosh. But forts like these were set up not to continually fight Indians but to induce them to sign treaties, and during September the Delawares and the Wyandots agreed to a treaty conference to be held on the northern bank of the Ohio at the mouth of the Great Miami River. Accordingly, Colonel Harmar mustered all the troops on active duty at Fort McIntosh, loaded them onto boats for the trip down the Ohio, and on the very evening before Hutchins began his work, floated past the surveyors' camp on the riverbank. (You can trace Colonel Harmar's long voyage in figure 7.19.)

Despite this abandonment, the Geographer and his crew set out the next morning to project their first line west. What Hutchins did not know was that despite the impending negotiations, four days earlier, on September 26, a band of Indians had attacked a trading post near the Delaware tribe's village of Tuscawaras, about 50 miles west of the surveyors. One of the two white traders at the post had been killed, the traders' merchandise carried away, and—in a sign of defiance—the door of the trading post and many

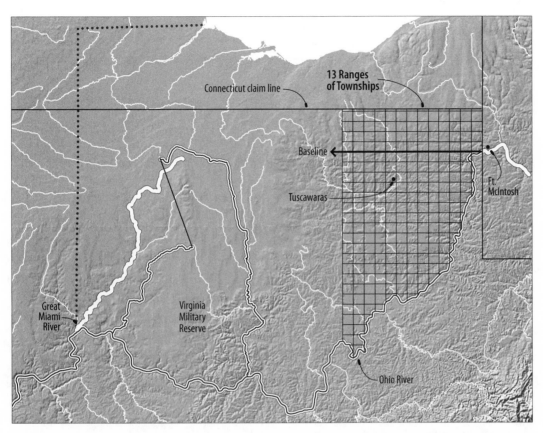

FIG 7.19

surrounding trees had been splashed with red paint. News of this raid reached Hutchins during the first week of October. Little wonder that on the 8th he broke camp and took his crew across the Ohio back into Virginia, determined not to take the field without assurances of protection. Hutchins and his men had spent little more than a week on the job and had completed only about 4 miles of the projected 13-mile Baseline.[6]

"Point of Beginning"

Historians and surveyors have given, to the location where Thomas Hutchins began his work, the name Point of Beginning of the Rectangular Survey. The name is suitably grand for the starting point of the greatest continuous surveying effort on the planet. The phrase itself, though, only contributes to the myth that the continentwide pattern of rectangles began at this single point and spread unbroken until it reached the Pacific. In fact, as we shall see, the Rectangular Survey, although pervasive, is radically discontinuous. It actually spreads out from thirty-eight official Points of Beginning and dozens of other unofficial points. It would be truer to anoint this the "Point of Beginning of the *Method* of Rectangular Surveying," but no one should be forced to get her mouth around that inelegant locution.

The true Point of Beginning is now unfortunately lost to us. Because of the New Cumberland Dam downstream, the water level on that stretch of the Ohio has risen, and the site of the wooden stake is underwater. But local historical societies and historically minded surveyors banded together in 1960 to give at least some indication of where the true point lay.[7]

In 1881 Pennsylvania and Ohio had undertaken to resurvey their common border, and the two states marked the northern and southern termini of the recalculated meridian line with identical 4½-foot-tall granite obelisks, both standing on squat bases engraved "Ohio" and "Pennsylvania" on their western and eastern sides.

Now, many rivers have a terrace, a level of land above the river at some distance back from it, from which a slope descends to the river's floodplain. The Ohio River has such a terrace at the Pennsylvania border, and the 1881 surveyors placed their monument on the northern bank just at the foot of its slope. Unfortunately, in the twentieth century the land around the monument was bought by Youngstown Steel, and the company used the site as a dump for the slag left over from its steelmaking processes. Respectfully, they refrained from dumping slag directly on the monument, but by 1960 that restraint had left the obelisk standing in a deep, U-shaped trench whose "legs" extended into the Ohio River.

Local groups got the monument moved directly north, out of the trench and up onto the terrace, into a new park on the south side of the nearest road paralleling the river. Thus the obelisk, with its east and west inscriptions, still marks the Ohio-Pennsylvania border; but since 1960 its west face bears an additional inscription that begins "1112 FEET SOUTH OF THIS SPOT WAS THE POINT OF BEGINNING FOR SURVEYING THE PUBLIC LANDS OF THE UNITED STATES."[8]

1786: SURVEYING CONTINUES

Thomas Hutchins and his crew waited in Virginia, but with no assurances of protection forthcoming, they soon returned to Pittsburgh. Eventually all the delegate-surveyors departed for their homes (these were "gentleman surveyors" who could not be expected to winter at an army camp in the wilderness), with Hutchins leaving last, en route to New York City to report to Congress. Late in December he laid before the legislators a map showing the pitiful 4 miles surveyed, but he augmented it with a glowing (and verbose) narrative of the fecundity of the region. Despite his bad treatment by the army, Hutchins asked to be authorized to continue the survey.

Before it had to decide the matter, two pieces of good news encouraged Congress to hope that completion of the survey might be possible. One was the successful outcome of the conference Colonel Harmar had taken his troops to. On January 31, 1786, Harmar and the Indians signed the Treaty of Fort Finney by which the Chippewas surrendered their claims.

The second piece of good news was Colonel Harmar's establishment of a fort (named for him) on the Ohio at the mouth of the Muskingum River (see figure 7.20). The site of this new Fort Harmar was just west of the area to be surveyed. With the area now bracketed northeast and southwest by two army posts, and with the Indian cessions seemingly secure, Congress felt enough confidence to send Hutchins out into the field again.

Congress (perhaps not entirely assured by these hopeful developments) diminished the area Hutchins would be required to survey. Originally, his mandate had been to cover all the land west of Pennsylvania from the Ohio River to Lake

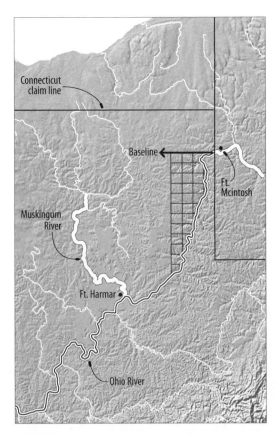

FIG 7.20

by this cavalier dismissal of ancient and proven surveying principles. Congress's instruction, though, must be considered in context. Most of the 6-mile-square New England townships had not been surveyed to the true meridian, and those "untrue" Townships had not suffered unduly from that fact. Plus, as we've seen, by a happenstance of the migration of the north magnetic pole, in the area of the first surveys the needles of the surveyors' compasses would point to true north with a deviation of only about 3 degrees— "good enough for government work," Congress might have said, and certainly good enough to serve as a basis for selling off 1-mile-square Lots and 6-mile-square Townships to pay down a crushing national debt.

So in June Hutchins was on his way to Pittsburgh, and by the latter part of July he had been joined there by delegate-surveyors from all the states but Delaware. He had every hope for success. His plan was to push the Baseline farther westward, and at each 6-mile interval, to send one of the delegate-surveyors south with a team of crewmen. Each team would project a line southward and then, at 6-mile intervals, project a line back toward the east. That eastward line would then close upon the north-south line marked by the previous team sent south, and in that way the four sides of every Township would be traced. Even though the requirements as to "north" had been eased, the surveyors were still required to blaze trees, note natural features, and monument each mile point on their lines. They would not, however, project those 1-mile lines across those Townships which were to be sold in 1-mile-square Lots. In the interest of getting the survey done, those lines would merely be *protracted*—drafted

Erie, but that assignment (made in May of 1785) had come in the wake of Congress's acceptance of Massachusetts's April 19 cession of territory. There was reason to hope, then, that Connecticut might cede its claim as Massachusetts had done—with no strings attached. By the spring of 1786, though, it was clear that Connecticut was going to insist on its Western Reserve (to which Congress would accede on May 26), meaning that Connecticut, not Congress, would survey the territory along Lake Erie above the 41st parallel. So on May 9 Congress told Hutchins to confine his work to the area south of the east-west Baseline he had begun.[9]

On May 12 Hutchins got a second boon from Congress: he was told that in the interest of time, he should run his lines not by the true meridian but by the "north" given to him by his magnetic compass.[10] Surveyor-historians are scandalized onto the plats the surveyors would draw up but not marked on the land. Nonetheless, when the purchaser of a Lot came to take possession of this land, he would have those 1-mile markers as starting points for a survey he would conduct.[11]

As Hutchins set out, everything seemed set for a productive surveying season. He had hopes of getting his full quota of thirteen Ranges of Townships surveyed before the snow flew, but Indian troubles would again intervene to dash his hopes.

First came Colonel Harmar's reluctance to supply the surveyors with a military escort (a reluctance overcome only by a direct order from the secretary of war), which delayed the start of work until early August. All seemed well as the team carried the line west and then sent delegate-surveyors south—until the morning of September 18. By that time the Geographer had been able to send seven surveyors south, and the previous night his crew had erected a range pole on the Baseline well into the 8th Range. But when the crew awoke that morning, they found their farthest pole snapped off—a clear warning to them from the Indians. That very afternoon a rider tore into camp with intelligence that Indians were planning to sweep up from the southwest, swing around behind Hutchins's crews and their escorting soldiers, and cut them all off from their base back at Fort McIntosh. To save his men from the impending attack, Hutchins now had to locate all his surveyors—scattered southward over the seven Ranges—and get them all back across the Ohio to safety.

With his crew assembled across the river, Hutchins determined that the best he could do would be to complete the survey of the four east-ernmost Ranges—the ones that intruded least into what the Indians seemed to consider their territory. With military escorts the surveyors finished marking the four Ranges by mid-November; but in view of all they had endured, they said, in effect, "That's it for the year." The first nighttime frosts had come, and the troops, "barefoot and miserably off for clothing," were suffering more than the surveyors. So the soldiers decamped to their winter quarters at Fort Harmar, Hutchins dismissed the surveying teams, and he stayed behind as a guest in a settler's home in Virginia to work on drafting the plats of what his teams had surveyed.[12]

15-inch white oak

N58°W, 28 links (about 16½ feet)

8-inch wooden post

S6°W, 27 links (about 16 feet)

10-inch white oak

FIG 7.21

Marking Boundaries for Settlers

Thomas Hutchins's crews had no precise instructions about how to mark the corners of the Townships: the surveyors were expected to exercise their best judgment and make use of whatever materials were at hand in the vicinity of each corner. In Ohio there would usually be rocks that could be piled together to form a small cairn at the corner, or a small tree that could be felled and its trunk axed into a post to be driven into the corner point. But surveyors, land settlers, and even Congress knew that merely monumenting the corners would not be enough: these markers would be a mile apart in a virgin forest. Imagine the chances of a settler finding the corners of his property merely by traversing the forest and keeping a sharp lookout for unnatural-looking piles of rocks or rough-hewn posts! No, his only real chance for finding his corners would be to walk, as best he could, in the footsteps of the surveyors. To make this retracing possible, Congress mandated that the surveyors compile **field notes** and copy them out as an adjunct to the Township plats they would draw.

Field notes "note" two kinds of things: the natural features the surveyor observes, and the marks the surveyor leaves. The format of the notes derives quite naturally from the manner in which a surveyor of long straight lines works. That is: the very purpose of the survey was to trace the outer boundaries of the Townships and mark those boundaries with a monument every mile. So as the chainmen measured the line in segments between range poles, someone on the crew would be carefully keeping an accumulating total of chain measurements since the previous mile marker. When that total accumulated to 80 chains (and thus 1 mile), the crew of axmen would monument that spot, and the chainmen would proceed with their line, beginning again at "0 chains, 0 links." Since the record keeper always knew how far, in chains, he was from the previous monument on the line, whenever he encountered a feature or left a mark, he could locate it by giving its "chain position" on the 80-chain line. As might be expected, the form of the surveyor's notes evolved, becoming codified only in the nineteenth century; but the field notes from the very first survey give a sense of how they worked. A paraphrase of the notes by two surveyors for the second mile of Hutchins's Baseline tells us,

- At 13.75 chains there is a white oak 18 inches in diameter, now carrying two notches on its east and west sides.
- At 35.82 chains is another white oak, this one 24 inches in diameter, now marked with two notches on its east and west sides.
- At 43.37 chains, we crossed the summit of a narrow ridge 90 feet high, bearing N45°E.
- At 49.97 chains, we crossed a brook running S20°W.
- At 64.37 chains, we topped another ridge about 170 feet high.
- At 77.37 chains, we regained level ground.

- At 80.00 chains, we set a wooden post in the ground. There is a 15-inch-diameter white oak 28 links from it on a bearing from the post of N58°W, and a 10-inch-diameter white oak 27 links away on a bearing of S6°W. [Figure 7.21 illustrates this oak post and the two marked trees.]

To this record of observations and actions, the surveyors then appended their opinion that the land seen along the line in the last mile seems particularly good for the cultivation of wheat.[13]

You can easily see how, by looking at the plats and reading the field notes, a prospective buyer of land in the Ohio Country might get some very rough ideas about the parcels on offer. And you can see how a buyer who actually consummated a purchase could track into the wilderness with a redrawn copy of the plats and the field notes (and a hand compass, of course), navigate from blazed tree to stream bed, and eventually come upon the wooden post marking a corner of his property.

When he arrived back in New York in late February of 1787, Thomas Hutchins found Congress disillusioned with the slow progress of his survey. They had seen the Land Ordinance as a way to survey the Ohio Country and bring it quickly to market. Now two years had passed, and they didn't even have ready for sale the minimum amount of land the ordinance had specified (remember that the law had called for a minimum of seven Ranges, so that one-seventh could be allotted as bounties for soldiers). So on April 21, a mere three days after Hutchins submitted his plat of the four Ranges, Congress decided that it would go ahead with the sale of whatever land had been surveyed, and authorized that the four Ranges Hutchins had surveyed be offered for sale at auction.[14]

Anxious to expedite the process, Congress abandoned the cumbersome process it had adopted in the more optimistic times of 1785—that the thirteen states would draw lots for lands proportionate to their tax assessments, sell what lands they could, and then return the rest to the confederation government. Under the new plan, the army would draw its one-seventh bounty allotment, and then all the remaining land would be put on sale at one time and in one place, at the seat of government in New York City.[15] The sale would be five months hence; and prior to it, advertisements would be run in at least one newspaper in every state.[16] To encourage sales, Congress backed down from the provision that a buyer must pay for his purchase in full at the time of the transaction, and instead allowed buyers to put one-third down, with the balance due within three months.

Hutchins's report had counted 675,480 acres in his four Ranges,[17] meaning that, with the army's share removed, Congress would have on offer some 500,000 acres in 27 whole or partial Townships, 15 of which were to be for sale in mile-square Lots. The auction began on September 21, but by October 9 only about 108,000 acres had found buyers, and so the auction was closed down.

It had been a disappointing affair. In the auction's nearly three-week run, only thirty separate sales agreements were concluded, with total sales of $176,090—and out of that would have to come the more than $14,000 the survey itself had cost.[18] It must have looked to the delegates in Congress as if the project of selling off the Northwest in 6-mile-square and 1-mile-square parcels would never get them the revenue they needed.

Even before the auction, though, Congress had become increasingly disenchanted with the land apportionment system mandated by the Ordinance of 1785—primarily because the survey was so slow and expensive. By the time Hutchins had laid his paltry four Ranges before them, Congress was looking at the possibility of selling off the West in huge chunks to speculative companies—thus getting a quick infusion of cash and saddling the speculators with the cost of surveying. We'll look at two of these companies in a moment. Meanwhile let's jump back to April, when Hutchins was in New York awaiting his orders from Congress for the coming 1787 surveying season.

1787: THE SEVEN RANGES ARE SURVEYED

At the conclusion of its debate, Congress directed Hutchins to confine his fieldwork to the completion of the seven Ranges he had begun surveying, the obvious hope being that a delimited survey might be platted faster and thus put to auction sooner. Sensing that he might never be granted the chance to become "The Man Who Surveyed the Ohio Country," Hutchins asked to be relieved of his field duties. (Massachusetts and New York had finally negotiated their common border; Hutchins wanted to survey that line.)[19] Congress granted his request, and so during the 1787 season, work would not be done by the majestic movement of delegate-surveyors under the direction of the Geographer of the United States. The three remaining Ranges would be surveyed by whichever of the delegates could get in the field first and stake a claim to a portion of the work (and, not incidentally, a knowledge of the terrain that could be profitably imparted to future settlers or land speculators—more about this later).

During the winter of 1786–87, two of the surveyors had stayed in lodgings on the Virginia side of the Ohio, and the home of another was just across the Alleghenies in York, Pennsylvania. All three were in the field by April, not even waiting or indeed asking for a military escort. The Indians found them, of course, and by mid-May the surveyors were asking for help from the military. Colonel Harmar once again delayed taking the field; but once he deployed his troops, the surveyors and their escorts brought the survey of the three remaining Ranges down to the Ohio River by early July. They could not know it then, but the expansion of these Townships westward would not resume until 1798,[20] and the Town-

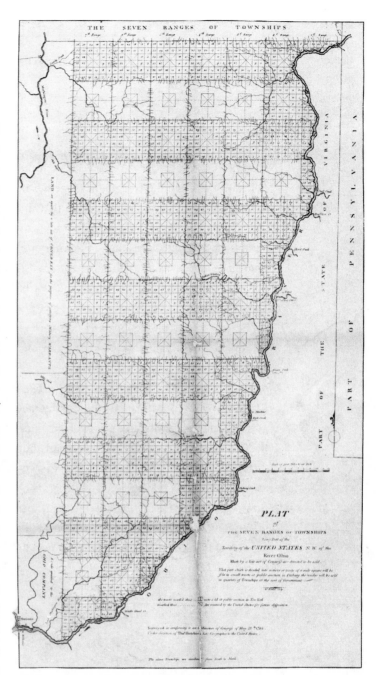

FIG 7.22

Library of Congress, Geography and Map Division.

ships they did survey would ever after be known as the Seven Ranges, with the Baseline at the top of the Ranges called *the Geographer's Line*.

The three surveyors stayed in the area until the end of August, compiling their notes; they then headed back to New York. With both the field surveyors and the Geographer together in the city, it might have been expected that the plats for the final three Ranges would be quickly completed. But squabbles among the surveyors, Hutchins's preoccupation with his private surveying, and his determination to complete a map of all seven Ranges (though not requested by Congress) delayed the completion of the report until July of 1788. (His map is shown on the previous page in figure 7.22; you might want to compare it to a map of the lines actually traced by the surveyors, in figure 7.35.) By this time the Continental government was just about to go out of business: the Continental Congress would essentially shut down in September, with the new federal government to take over on March 4, 1789, the date of the new president's inauguration). But the Board of Treasury—itself about to be superseded—took delivery of the documents, merely holding on to them until its mandate would expire.[21]

In Ohio the Continental and federal Congresses tried out a whole series of schemes for selling off the Public Domain. They experimented with both 6-mile-square and 5-mile-square Townships. They sold great swaths of territory to land companies, idealistic and speculative. They sold farmsteads to individual families, first by whole Sections and then by increasingly smaller fractions of a Section. The manner of numbering Sections evolved: at first they were numbered northward from the Ohio River, then north from a Baseline, and finally outward in all directions from a single point. And the techniques of surveying evolved as well, sometimes in the direction of more precision, sometimes in the direction of more speed.

By the 1820s, as the last of the Indian tribes surrendered land in Ohio and the survey extended across the entire state, most of the practices that today constitute the Public Land Survey System were in use. Figure 7.23 is a map of boundaries resulting from all this trial and error. As a systematic survey, it's far from elegant, but it serves as an indelible record-on-the-land of the ideas of officials in Washington and surveyors on the ground. For the rest of this chapter, when I describe some campaign of Ohio surveying, the map of that project will be in the same position on the page as this map so they can be "flip-compared" like the earlier national maps.

Now back to our chronology.

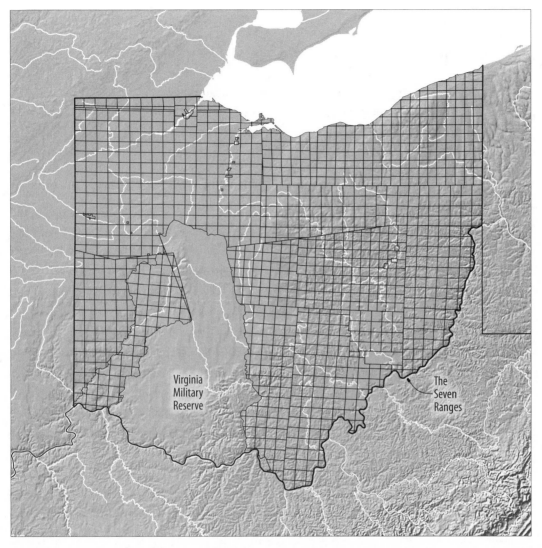

FIG 7.23

Even before the first land sales in 1787, Congress had already been approached by lobbyists from joint-stock companies with an alternative idea about how to sell off the Northwest to pay down the debt. "Instead of auctioning off the land in mile-square or 6-mile-square parcels" (they hinted to the delegates), "let us buy contiguous parcels of a million acres or more. We will pay the minimum price you set for Ohio land, $1.00 per acre. But because we will be buying so much land, and because our contiguous tracts will necessarily encompass both good land and bad, we ask for a discount: let us pay you in Continental dollars and bonds."

This bargain tempted Congress. During the Revolution and in the years since, it had been issuing both currency and bonds backed not by specie but only by a promise to redeem the notes upon demand at some future date. The convertibility of these notes (their "exchange rate," in today's terms) had fluctuated all during this period, mostly downward, to the point that all Continental notes were worth considerably less than their face value. So if Congress took up the companies' offers, it would be getting only a fraction of what it might get if it sold the same lands piecemeal for hard currency. But various issues of bonds were coming due, and a quick infusion of cash, even devalued cash, presented a solution to its dilemma, at least for the short term.

On October 22 (eleven days after the disappointing sale of the first four Ranges) Congress passed resolutions that would very roughly set the terms of its negotiations with land companies.[22] The key to that plan was the Geographer's Line, specifically its assumed extension west through the Ohio Country. Congress hoped that the Indian tribes might be induced to cede all the land south of the line and thus open that land to sale and settlement. A broad north-to-south swath of that land was not Congress's to sell—the Virginia Military Reserve, which had been withheld from cession—but that exclusion still left huge tracts east and west of the reserve.

The eastern boundary of the Virginia Reserve, the Scioto River, had not yet been mapped, so Congress had no way of knowing how many acres the United States possessed between the reserve and the Seven Ranges. Its plan was to use the lands along the northern edge of the tract, directly against the Geographer's Line, to satisfy its own military bounties by drawing an east-west boundary far enough south of the line that a sufficient acreage would be encompassed. With that one exception, Congress declared in its resolutions that it was willing to entertain proposals to buy any of its remaining lands, in tracts of "not less than one Million of Acres in One body."[23] First off the mark, even before Congress had finalized its plans, was the Ohio Company of Associates.

Benjamin Tupper had been a delegate-surveyor on the Four Ranges; he was thus one of the first white men to gain acquaintance with the Ohio Country. In the winter between the first two surveying seasons, he was one of the prime participants at a meeting on March 1, 1786—at a Boston tavern called The Bunch of Grapes—of people interested in investing in Ohio land. Two days later the terms of investment were settled, and the Ohio Company of Associates opened its books for subscriptions in the venture.[24]

The charter envisioned the sale of eight thousand shares, of which each individual investor could buy a maximum of five. The incorporators envisioned settlement in something like the traditional New England pattern: each share entitled the investor to a town lot and nearby farm acreage. Each share carried the option of paying either $125 in gold or $1,000 in Continental currency[25]—an indication that at least in Boston the exchange rate of "Continentals" was something like eight to the gold dollar. Sales began, and soon after that, so did the 1786 surveying season—which saw Winthrop Sargent, delegate-surveyor for New Hampshire, joining Tupper as an unofficial scout for the Ohio Company. You'll remember that the start of surveying was delayed that year, and Sargent seized the opportunity to venture down the Ohio to the Muskingum River to reconnoiter the Muskingum's lower valley, an area that would form the heart of the Ohio Company's lands.

After a year on the market, about a quarter of the company's shares had been sold, so the directors of the syndicate felt prepared to make a petition to Congress. Their request was for the land scouted by Sargent, a tract north of the Ohio and east of the Virginia Military Reserve (see figure 7.24 on page 237), whose northern boundary would be an east-west line positioned to enclose 1.5 million acres. Their request was put before a congressional committee, which issued a favorable report on July 14, 1787, exactly one day after the Northwest Ordinance was adopted.[26]

Congress debated the sale sporadically through the summer and into the fall, long enough for the auction of the Four Ranges (September 21 to October 9) and its own land-for-sale resolution (October 22) to occur. Eventually, a sales agreement was hammered out, and representatives of the Ohio Company signed it on October 27. The agreement showed that Congress really wanted to make the sale but also wanted to extract some concessions out of the bargain.

The company had hoped to pay only 50¢ an acre for its land. Congress held out for its Four Ranges minimum of $1.00 an acre, but made an allowance on the grounds that a third of the company's land might be unsaleable—effectively reducing the per-acre price to 67¢. Congress further sweetened the deal by allowing the company the option of making its payments in either specie or Continental certificates. To seal the deal the company would have to pay $500,000 immediately; that much again when the boundary of its tract was surveyed (and its total acreage thus ascertained); and the balance afterward in six equal payments.

Despite the problems encountered in the Four Ranges, Congress was not yet willing to give up the principle of survey prior to settlement, so it mandated that the system of 6-mile-square Townships and 1-mile-square Lots be extended into the Ohio Company's lands—and that those lands were to be surveyed by the company at its own expense. The idea of reserving five Lots in every Township was also continued: Lot 16 was again to be set aside for the support of education, and the four Lots just inward from the Township corners—nos. 8, 11, 26, and 29—were to be reserved also, as before. Lots 8, 11, and 26 were to be held, as in the Seven Ranges, by Congress; but Lot 29 would be held by the Ohio Company for support of religion. (You will remember that in the debate over the 1785 Land Ordinance, it had been New Englanders like the investors in the Ohio Company who favored preserving a Lot for religion.) And in an entirely new provision, Congress required that two of the company's Townships be devoted to the support of a university or seminary.[27]

Other land speculators were importuning Congress that summer of 1787, one of whom was John Cleve Symmes of New Jersey. Symmes hoped for 1 million acres just west of the Virginia Military Reserve, between the Great and Little Miami rivers. On October 3 Congress approved his request, offering him essentially the same terms the Ohio Company would accept later that month.[28]

Symmes's scheme differed from that of the Ohio Company, both in how he raised money and how he planned to parcel out his land. His was a classic speculative scam. Before taking title to his land, Symmes invited both settlers and speculators to make claims on the land he hoped to buy. To secure those claims, he would naturally demand a down payment of "earnest money." Then with those funds he would satisfy the conditions of his purchase contract with Congress. Like many such schemes, Symmes's speculative pyramid eventually collapsed under its own weight, but it lasted long enough for the land to be surveyed into Townships and Lots whose boundaries endured on the ground even after his syndicate folded.

If the speculative nature of the Symmes Purchase was what caused it to fail, the undoing of the Ohio Company was its hopeful idealism. Even though the company had been born in a tavern, the incorporators hoped to continue in Ohio the venerable New England tradition of people centering their lives in close-knit towns: every share in the company entitled its holder to a town lot where the settler would build a home, and acreage outside of town where he would do his farming. But as early as 1787, a newer agricultural pattern was emerging in which settlers had no house lot in town, only an isolated farmstead out in the country. The Ohio Company found that a sufficient number of people couldn't be persuaded to buy town-and-farm shares. But because of the congressional mandate that parcels be surveyed in advance of sale, the company's surveyors monumented those "out-of-town" rectangular parcels, and the resulting boundaries endured on the land as a framework for later settlement— even as the company that had marked them fell out of existence.

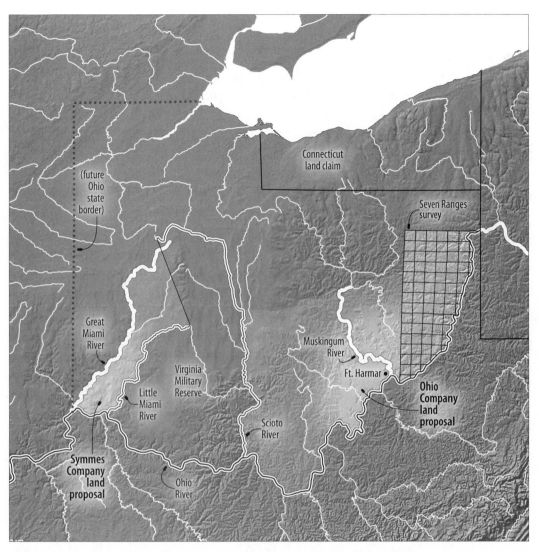

FIG 7.24

Though the sales agreements with the Ohio and Symmes companies specified that they would survey the Townships within their purchases, Congress reserved the right to mark the outer boundaries of the companies' tracts—a task which fell naturally to the chief surveyor of the nation, Geographer Thomas Hutchins. By late September of 1788, he was at the mouth of the Scioto River (see figure 7.25), ready to survey the potential border of the Ohio Company's lands.[29] You will recall that Congress had reserved to itself an east-to-west band of territory south of an extended Geographer's Line to satisfy Continental military bounties. Hutchins's task was to survey the southern boundary of that tract, which would later be known as the U.S. Military Reserve. His plan was to map the course of the Scioto to a point about 80 miles north of its mouth at the Ohio, and from there to drive a line due east to the western border of the Seven Ranges. Hutchins was free to conduct the Scioto survey, because in July he had finally turned in his detailed Seven Ranges plat to the Board of Treasury in New York.

Only 12 miles up the river, Hutchins gave the work to his two deputies, Israel Ludlow and Absalom Martin (both veterans of the Seven Ranges survey), and returned to Pittsburgh. There he fell ill, and in the following spring of 1789, he died.[30] By that time any power of appointment had passed to the new federal government, and no successor to Hutchins was chosen as Geographer of the United States.

Once Hutchins placed the survey in the hands of his assistants, they decided to split the work between themselves. Martin would do the meticulous work of mapping the Scioto, while Ludlow would go back up the Ohio to the Seven Ranges, proceed north up its western border to the latitude specified by Hutchins, and then survey that line westward back to the Scioto. When Martin and Ludlow compiled their survey notes in June of 1789, they were able to calculate the area of the Ohio Company tract as containing just under 5 million acres.[31]

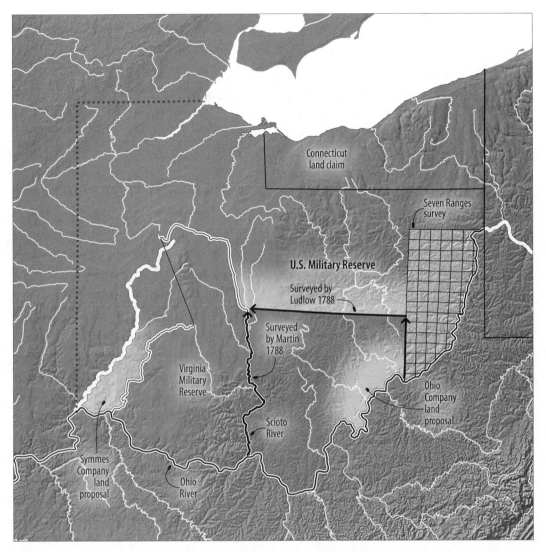

Connecticut
land claim

Seven Ranges
survey

U.S. Military Reserve

Surveyed by
Ludlow 1788

Surveyed
by Martin
1788

Virginia
Military
Reserve

Scioto
River

Ohio
Company
land
proposal

Symmes
Company
land
proposal

Ohio
River

FIG 7.25

With Congress's military reserve now bounded, the Ohio Company could ascertain the northern boundary of its own million-and-a-half acres. It decided to survey ten new Ranges of Townships west of the original Seven, these likewise running north from the Ohio.[32] The northern border of the company's tract would be an east-west line two Townships north of Marietta, the settlement it had founded in March of 1788. Its lands thus formed a rough right triangle (shown in figure 7.26), with the Ohio as hypotenuse; when finally mapped, the triangle was found to contain just under 1.8 million acres.[33] But because the company could not sell enough shares to make the payments its contract required, it had to surrender the northern fringes of its tract—which accounts for the jagged borders of what is now known as the Ohio Company Purchase. The boundary around the purchase was never surveyed as a boundary: the jagged border is, rather, a line drawn later by mapmakers to encompass the lands eventually controlled by the Ohio Company.

Similarly, the border of John Cleve Symmes's purchase was never surveyed specifically. After his work on the Ohio Company's boundaries, Israel Ludlow definitively joined his interests with Symmes's, becoming chief surveyor for the syndicate. By 1794 Ludlow had mapped the Little Miami (the eastern border of Symmes's tract) up to its source, surveyed a parallel of latitude from there to the Great Miami, and mapped that river down to its confluence with the Ohio. Because the source of the Little Miami was much farther south than had been assumed, this "Between the Miamis" tract contained not the anticipated 1 million acres but (by Ludlow's calculation) scarcely half that.[34]

In any case Symmes, strapped for cash, was in no position to pay for even a half-million acres, and so the U.S. Treasury simply declared the border of his purchase to be a parallel of latitude run between the rivers that would enclose the roughly 300,000 acres he could pay for. The line was not run, but by convention the border of Symmes's "Miami Purchase" is designated on maps by an east-west Township line near that mandated parallel—the one shown in Figure 7.26.

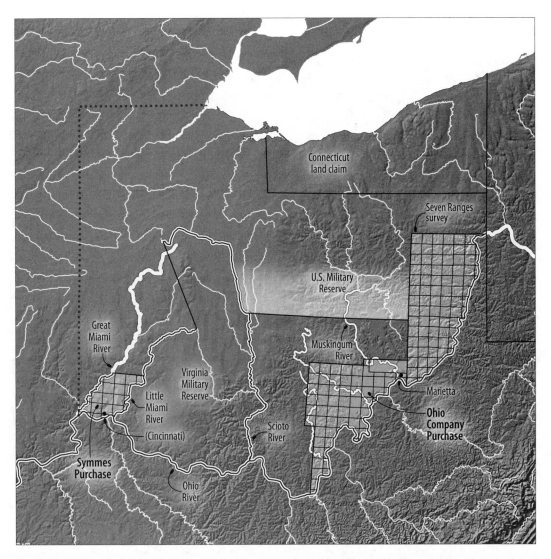

FIG 7.26

In its role as successor government to the Continental Congress, one of the first tasks of the new federal Congress was to decide which of its predecessor's acts it would carry forward. In its judgment, the Northwest Ordinance of 1787 was a wise statement of public policy, and it readopted the ordinance with only those changes necessary for conformance to the new Constitution. The Land Ordinance of 1785, however, presented problems: its method of selling off the Public Domain in presurveyed parcels clearly had not worked. The 1787 auction at New York had shown that. But the alternative strategy of selling off million-acre tracts to private syndicates had likewise not yet shown itself to be a preferable alternative. Congress thus opted to do what prudent Congresses, faced with conflicting indicators, have often done—that is, nothing. It allowed the 1785 Land Ordinance to lapse.[35] From 1789 onward, the United States had no policy about how frontier lands would be apportioned to settlers.

But in the absence of a grand policy decision, contractual obligations still had legal force, and the agreements with Symmes and the Ohio Company mandated that their lands be surveyed according to the dictates of the 1785 Land Ordinance. Contracts, however, can be renegotiated, and we can imagine that a federal Congress so indifferent to the idea of prior rectangular survey might have entertained a request from the proprietors that they be allowed to revert to the pattern being followed right next door in Virginia's Military Reserve—claims made and described by the settlers themselves. Certainly such a procedure would have been cheaper for the proprietors: they would be spared the costs of surveying the tract

prior to sale. Nonetheless, both the Symmes and Ohio companies went forward with surveying their lands into the Townships and Lots of the 1785 Ordinance. Why?

Remember that the chief surveyor for the Symmes syndicate was Israel Ludlow, a veteran of the Seven Ranges survey. He knew not only how to enact the surveying provisions of the 1785 Ordinance but how to exploit them, helping families find the best Lots available (for a fee, of course). The Ohio Company had even closer ties to the Seven Ranges survey. The initial surveys of the its lands were conducted by three of the original delegate-surveyors, and the military and governmental leader of the settlers was General Rufus Putnam, who had been Massachusetts's original delegate to the surveys. (He had a prior commitment as Surveyor General for Massachusetts's land in Maine and so sent a substitute to Ohio.)[36]

Administrators as well as surveyors, Ludlow and Putnam didn't just repeat the procedures used in Seven Ranges but made two important innovations in how surveyors would be hired and in how a survey in virgin territory would be commenced.

In surveying the Ohio Company's lands, Putnam used the numbering system begun in the Seven Ranges in that the Townships in each successive Range were numbered upward from the Ohio River. Because of how the company was selling land, his surveyors didn't just monument the 1-mile intervals on the exterior borders of the Townships but drove those lines into the Townships to mark off the individual Lots and in some cases to subdivide the Lots, a practice that would be followed later in the federal surveys.

Putnam's real innovation, though, came in how he contracted with his surveyors for their services. In the survey of the Seven Ranges, the implicit assumption had been that the delegate-surveyors would be men of some means enacting a public trust, serving as a kind of "Congress in the field" to assure folks back home that the job was being done to no state's or region's advantage or disadvantage. Their pay—$2.00 per mile surveyed—was intended to cover the expenses of hiring and maintaining their crews. But the per-mile amount proved so inadequate that the delegate-surveyors had constantly to petition Congress for redress, which Congress granted, making the surveyors' pay in effect an open-ended reimbursal system.

Putnam decided in 1788 to hire entrepreneurs, not gentlemen, as his surveyors. Under his system, a precisely described surveying task would be offered to any qualified surveyor at a stated per-mile rate. The rectangular system made it easy to calculate the miles of lines to be run, and thus the total payment due to the surveyor. Out of that amount, the surveyor would pay all the expenses he incurred and keep any remainder for himself as profit.[37] In such a system the obvious temptation to the surveyor would be to run the lines in a quick, shoddy manner to decrease his expenses. As this contract system evolved, a succession of safeguards against such practices were put in place. At a minimum, payment would not occur until an acceptable plat was filed; at a maximum, no payment would occur until the lines run in the field were inspected. There was, of course, a countering temptation that the official offering the contracts would skimp on the per-mile rate, but the system itself worked against that. Offer the contracts at a rate insufficient to cover overhead and some profit, and there would be no takers.

From 1788 through 1795, surveyors worked for the Ohio Company under this contract system, gradually pushing their rectangles outward from the initial settlement at Marietta. The survey of Symmes's Miami Purchase was neither as systematic nor as well administered, but it too followed the 1785 system of numbering Townships in Ranges north from the Ohio. Israel Ludlow, though, had a problem: the boundaries of Symmes's tract were three river courses, so there were no straight lines from which to extend a survey.

Ludlow's solution (shown in figure 7.27) was to lay down two intersecting lines—one a parallel of latitude, the other a meridian of longitude—and then have his surveyors project their Township borders outward from those lines. In that region, the Ohio River swings slightly northwest before continuing its generally southwest course, so Ludlow found the bend in the river that pushed farthest north into the Symmes tract and began his two intersecting lines just north of the bend. He later extended the latitude line to the Little and Great Miami rivers, the eastern and western borders of this survey; still later he extended the longitude line north to its intersection with the Great Miami.[38] (As it turned out, Ludlow's two lines are not perfectly aligned to the cardinal points but rotated slightly clockwise.)

As the Rectangular Survey evolved, it became the pattern to originate all surveys at the intersection of a meridian of longitude and a parallel of latitude, with the longitude line gaining the grandiloquent title of **Principal Meridian**, the latitude line being called a **Baseline**, and their point of crossing named the **Initial Point** of the survey. In those later surveys not only would the laying

out of Townships begin at that intersection, but also their numbering. Ludlow, working from the mandate of the 1785 Ordinance, didn't use his two control lines that way (he numbered his Townships up from the river), but the method he used to establish his initial point would become the model followed in all surveys from the nineteenth century onward.

Because all the borders of the Symmes tract were rivers, Ludlow must have sensed that any 6-mile grid, no matter how it was positioned on the tract, would yield partial Townships, interrupted by the rivers, at its edges. Even if a perfect 6-mile grid could be drawn as an overlay on a perfect map of the tract, and the grid shifted back and forth over the map, no alignment could claim to present a better alignment with the rivers than any other. Therefore, a grid started at a certain point would be as efficacious a means of apportioning the tract as a grid started at any other point. That being the case, why not start the survey at the point most convenient for the survey effort? For Ludlow's endeavor, the most convenient beginning place was that point from which an east-west line would not cross the Ohio. For later surveyors, other factors would be decisive. In chapter 10, we'll see that the Initial Point of the Montana survey was determined by the distance through which the two lines could be projected without encountering mountain peaks.

But for all future surveyors, the fact that drove Ludlow's decision would drive theirs—namely that with the defeat of Jefferson's plan of land-measurement and state boundaries linked to longitude and latitude, the apportioning of the land had become fundamentally unhooked from the determination of state borders. No matter how

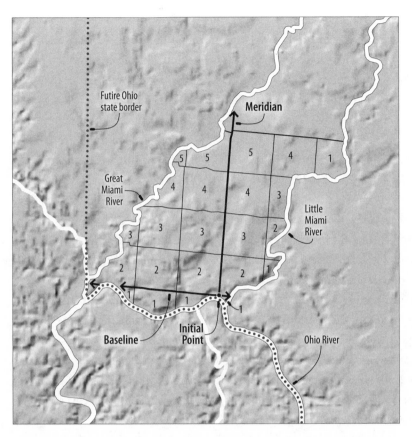

FIG 7.27

state borders were established—whether by lines projected from river mouths or by meridians of longitude or parallels of latitude—the grid of 6-mile-square Townships, when marked out on the ground, would never coincide perfectly with the borders of a state. Even if projected perfectly from one straight-line state border, by the time the survey crept across the state, its collision with the opposite borders would inevitably produce Townships just as "partial" as those resulting from the grid's collision with a river course.

The principle that Ludlow demonstrated, even if he did not fully realize it, was that for the Rectangular Survey to be an accurate apportioner of

land, it needed to be controlled at various points by Principal Meridians and Baselines; but for it to be an *efficacious* apportioner of land, the location of those controlling lines was irrelevant. Leaving their location to the surveyor's judgment would produce a result no worse than relying on any other basis. And such a strategy—if it made the surveyor's work easier, faster, and more accurate—might actually produce a better result.

Even after allowing the 1785 Land Ordinance to lapse, the new federal Congress didn't entirely ignore the question of how to apportion the West into parcels suitable for settlement. It was more a matter that the issue seemed moot, given that marauding Indian tribes were at least temporarily preventing white settlement from expanding outward from its base just north of the Ohio River. Plus, why act on such a difficult issue when the lands into which settlement might expand were not even under Congress's control? Due to the actions of the Continental Congress, most of the land close to the river had been either sold to Symmes or the Ohio Company or left in the hands of Virginia.

Congress did hold out hope for the renewal of land sales at some future date, and so it asked Treasury Secretary Alexander Hamilton to study and report on the matter. Hamilton's 1790 report was never adopted, but it's worth looking at for a moment, both because some of its ideas were adopted later, and because its tenor shows how precarious the Rectangular Survey was at this time.

The report reads as if the record of congressional land legislation were entirely blank. There would be three classes of people who want to buy land in the West, Hamilton determined: the individual farmer who wants his own plot, the small group of farmers who want to settle together on a contiguous parcel, and the speculator who wants to buy up a large tract and then subdivide it for resale. That being so, the government should apportion the West into some large number of tracts, with each tract limited to only one of the three types of land sale. In the tracts for individual farmers and groups, the settlers themselves would locate their lands—up to a maximum of 100 acres for individuals and 500 acres for groups—and choose their lands' boundaries in whatever manner in they saw fit. Speculators could choose their lands also, but would have to take them in 10-mile-square Townships (a curious echo of the plan of Hamilton's rival, Jefferson). In all cases the boundaries would be marked by government surveyors, with the costs of the surveys to be paid by the buyers of the land.

All of this would be overseen by a new General Land Office, under the direction of a Surveyor General. But because of the differing needs of the three groups, the General Land Office would maintain not just a headquarters in the nation's capital but two regional land offices: one in the Northwest, the other in the Southwest (the lands south of Tennessee, which Congress had every hope of receiving from South Carolina and Georgia). The General Land Office would deal with the speculators, the regional offices with the groups and individuals.

However bought, the lands would be sold not at auction but for a fixed price of 30¢ an acre in specie. (In 1790 a dollar of gold could be bought with something like four to seven paper dollars; an acre would thus cost, in paper money, between $1.20 and $2.10.) An individual or group would pay the entire price, no credit, to the regional land office. Speculators, on the other hand, would deal with the General Land Office, where, since they were making a big investment, they'd have to put up only one-fourth of the purchase price, with the rest due in two years.[39] (For the worker, cash on the barrelhead in a flea-bitten office; for the wealthy, easy credit in a posh suite.)

When rendered into proposed legislation, Hamilton's ideas were rejected, but the debate over them revealed a surprising shift in Congress's sympathies. In the debate over the 1785 Ordinance, it had been southern delegates who fought for their traditional practice of "indiscriminate location." By 1790 there had been nearly a decade's worth of experience with veterans of the Revolution locating their bounty warrants in the backcountry. From its inception, this system had always generated overlapping, ill-defined claims, and many lawsuits; but in the colonial period, when new settlement was tentative, the price in litigation had seemed sustainable when balanced against the ideal of the freedom to choose one's own land. But with the post-Revolution rush for land, some threshold of tolerance was crossed—and the opening of the Northwest and Southwest promised an even greater land rush in the future. As various representatives rose to speak of their states' experience, it became clear that Congress would never support a land allocation system in which individuals chose the boundaries of their own farmsteads. It would endorse only a land system based on settlers choosing from among presurveyed tracts.

But who would do the surveying? In 1790 the Symmes and Ohio companies were both running lines across the lands they had bought. Both syndicates were having problems, but Congress could still hope that with more realistic management at the Ohio Company and less corruption in the Symmes operation, a land policy of selling off large tracts might yet work. That hope would be abandoned by 1796.

1794: THE GREENVILLE TREATY OPENS OHIO FOR SETTLEMENT

By the beginning of the 1790s, white settlement was beginning to spread inland from its base along the northern shore of the Ohio River. Indian raids were incessant, and military expeditions authorized by President Washington proved disastrous. The situation was made precarious by the fact that the British still occupied Fort Detroit, which gave the tribes a secure zone from which to attack, and into which they could retreat.

General "Mad" Anthony Wayne broke this pattern of feint and retire by attacking directly into the Indians' zone of safety, with a decisive victory at the Battle of Fallen Timbers (near Toledo—shown in figure 7.28) on August 20, 1794. Wayne delayed opening negotiations, instead withdrawing southward to Fort Greenville, which he had established the previous year.

While Wayne waited, the British agreed, in October, to abandon their U.S. forts—which fact Wayne was able to impart to the tribal representatives when negotiations began at the fort in the summer of 1795.[40] Under the Treaty of Fort Greenville, the tribes agreed to surrender all claims to the lands south and east of this line:

- *starting at Lake Erie and running up the Cuyahoga River, then over the short rise of land to the Muskingum River, and then down the Muskingum to an important crossing at Fort Laurens;*
- *from Fort Laurens westward to a trading post called Loramie's Store;*
- *from there westward to Fort Recovery near the head of the Wabash River;*
- *and from Fort Recovery southward to the point where the Kentucky River flows into the Ohio.[41]*

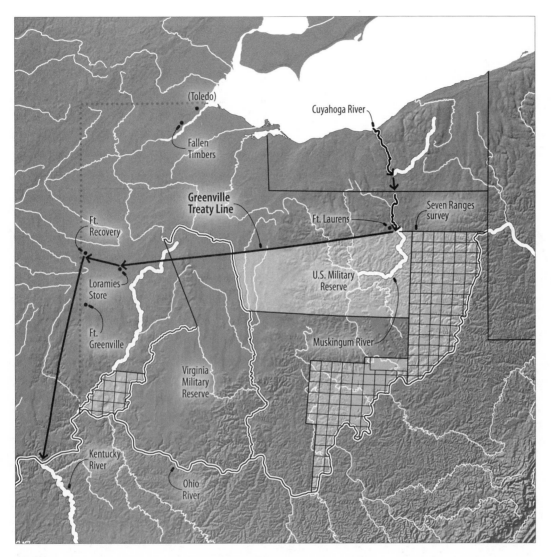

FIG 7.28

By 1796 events had shown Congress that large-tract sales were so inextricably entwined with corruption and mismanagement that they could never serve as a means for apportioning the Public Domain. The first evidence that put a blot on the large-tract idea came not in the nation's lands in the Northwest but in the vast Southwest still held by Georgia on the basis of its ancient colonial grant. Like all former colonies burdened with Revolutionary War debts, Georgia had looked to any source that offered a fast payoff; in 1789 and then again in 1795, it sold, to a whole series of companies, lands comprising almost all of present-day Mississippi and much of Alabama. In Congress's eyes, it was bad enough that these grants stood in the way of extending the Public Domain to include the Southwest. What was worse, some of the land companies had obtained "title" to land through direct negotiations with Indians—a prerogative the national government reserved exclusively to itself. Worst of all, allegations arose that company officials had offered bribes to Georgia legislators, a scandal that caused Georgia to rescind the grants in February of 1796.[42]

Just the previous month the House of Representatives itself was rocked by a land-company bribery scandal. In late 1795 Robert Randall had been lobbying Congress to favor his syndicate's plan for gaining title to what is now the Lower Peninsula of Michigan. Lobbying is now and was then a perfectly legal activity, but then Randall's business partner Charles Whitney stepped over the line by offering shares in the company to members of Congress.[43]

So land companies were in a decidedly bad odor when in early 1796 Congress took up the question of allocating land in the Ohio Country. The matter had some urgency, since by the signing of the Greenville Treaty the nation now had secure title to a huge chunk of territory. But if events had convinced members that neither grants to land companies nor indiscriminate location by settlers would be sound policy, what alternative was there?

The obvious alternative—obvious to us today—was the system embodied in the Land Ordinance of 1785. As enacted, the Land Act of 1796 mandates an almost identical Rectangular Survey, but in its debates Congress acted as if the old law, and the tracts successfully conveyed under it, had never existed. Not only was the 1785 Ordinance scarcely mentioned, but the exemplar of a Rectangular Survey most frequently cited was the system New York was using in its territory west of Syracuse.[44] It's surprising to us that the 1785 Ordinance wasn't lodged more deeply in the institutional memory of Congress, but this only goes to show the precarious existence of the Rectangular Survey in its early years. Bear in mind too that the Land Act of 1796 was not in any sense a future land policy for the National Domain. Its provisions applied only to lands south and east of the Greenville Treaty line. A Rectangular Survey would become firmly established as national policy only in the first decade of the nineteenth century.

As adopted, the 1796 Land Act again called for 6-mile-square Townships, each divided into thirty-six 1-mile-square tracts now to be called (as ever after) Sections. And as before, only every other Township was to be subdivided and sold off in Sections; the rest were to be sold "entire." This time, though, the corners of the Sections would actually be monumented—after a fashion. Fearing the cost of marking all the interior lines of a Township, Congress concocted a scheme under which surveyors would run every other interior line, north to south and east to west. With lines at 2-mile intervals, each Section would have at least three of its corners monumented, from which the fourth corner could later be projected.

This 2-mile-interval system facilitated a provision unique to the 1796 law: rather then reserving the four lots just back from the Township corners, Congress reserved for its disposition the four Sections at the Township's center, bounded conveniently on all sides by the running of those 2-miles-apart interior lines.

As the Continental Congress had done, the federal Congress mandated a numbering system for the Sections within each Township. The 1796 Act called explicitly for boustrophedonic numbering, but mandated that Section no. 1 be in the northeastern corner, with the numbers running first east to west then west to east—the system in use to this day.[45] Taking a leaf from Hamilton's report, Congress called for a Surveyor General to oversee the surveying effort (Hamilton's General Land Office would come into being in 1812[46]). It also took the Treasury secretary's advice on land sales: whole Townships would be sold at the national capital, but Sections would be sold at land offices on the edge of settlement—for the first sales in the Ohio Country, there would be one office in Pittsburgh and another in Cincinnati.

As for the mode of sales, Congress debated the idea of setting a fixed per-acre price, but eventually reverted to the practice of selling the land at auction, with only the minimum per-acre price fixed. With the signing of the Greenville Treaty, possession of the land would now be secure from Indian raids, so Congress felt justified in raising the minimum price from $1.25 to $2.00 per acre.

During the debate the new federal Congress carried forward one more idea from the old Continental Congress: a Military Reserve of land out of which Continental Army veterans could redeem their land-bounty payments for their service. Congress decided to utilize the area set apart by the Continental Congress: the area north of the line surveyed by Israel Ludlow in 1789. The northern border, however, would not be the Connecticut Western Reserve (as had been anticipated in 1789) but the Greenville Treaty line farther south.[47]

A problem facing Congress was the graduated-by-rank acreages of the bounties. As we saw earlier, the grants were in multiples of 100 acres: major generals received 1,100, captains 300, and common soldiers 100. The acreages possible under the 1-mile-square Section system, no matter how subdivided, would not yield tracts of the proper size. So Congress devised a system, unique to Ohio, that might have pleased the decimal-loving Jefferson in its numerical elegance.

The basis for surveying in the U.S. Military District would be a 5-mile-square Township,[48] a dimension wholly without precedent in American experience. But at 80 chains to the mile, 5 miles is 400 chains, and if you think in chains rather than in feet or miles, you can appreciate the numerical elegance of the system. The whole 400-by-400-chain Township comprised 16,000 acres. Divide it in half, by a north-south or east-west line, and you get a tract 400 by 200 chains, or 8,000 acres. Halve it in the other direction, and

you get a 200-by-200-chain tract—4,000 acres. Keep repeating the process, and you get acreages of 2,000, 1,000, 500, and so on. Those unique 5-mile Townships are shown in figure 7.29, as are the rectangular Townships of the lands reserved by Connecticut—likewise 5 miles square.

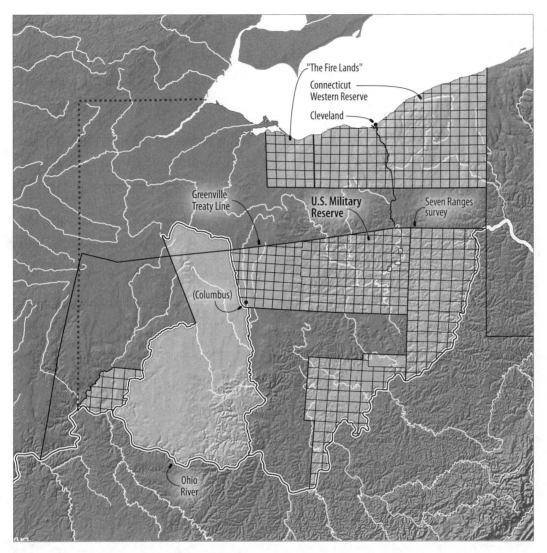

FIG 7.29

"The Fire Lands"

Connecticut
Western Reserve

Cleveland

Greenville
Treaty Line

**U.S. Military
Reserve**

Seven Ranges
survey

(Columbus)

Ohio
River

The official act that would set the survey in motion was the appointment of the first Surveyor General, and in the fall of 1796 President Washington chose Rufus Putnam, the man who had headed the survey of the Ohio Company lands. Putnam took office on November 5, and by the spring of 1797 he had set up his field office in Marietta, the town he had laid out for the company.[49]

Putnam knew that a prime task in the survey would be the marking of its outer boundary, the Greenville Treaty line. For this task he could think of no better man than Israel Ludlow, who had surveyed the southern border of the U.S. Military Reserve. Using the system that had worked well for him, Putnam drew up a contract under which Ludlow would survey the treaty line and adjacent Townships. Putnam himself journeyed north to the eastern end of the line to locate its starting point, and during the 1797 survey season Ludlow took the line as far west as Loramie's Store. Other teams worked that season between the treaty line and Ludlow's earlier Ohio Company line to mark out the 5-mile-square Townships of the U.S. Military District.[50]

During the 1798 season Ludlow worked to finish up the Military Reserve surveys, then turned to the area west of the Symmes lands. Again the question arose: where to start the grid? Just as had been the case when Ludlow surveyed the Symmes tract, no latitude or longitude lines had yet been run, so a controlling meridian line would have to be established. But rather than start the line in an arbitrary location, why not start it at a point that would become important? The 1787 Northwest Ordinance had mandated that the border between the "middle state" and

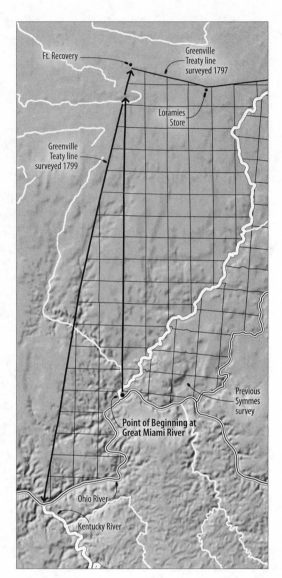

FIG 7.30

the "eastern state" be a meridian line projected north from the mouth of the Great Miami at the Ohio. So in October of 1798 that is where Ludlow began his Principal Meridian, as shown in figure 7.30. The line of course became the boundary between Indiana and Ohio, but it also became, in 1819, the First Principal Meridian.[51] (It's a confusing fact that the "first" meridian to govern the Rectangular Survey would actually be the tenth one to be given an official name—coming four years after the Fifth Meridian; but all will become clear soon.)

Ludlow carried his meridian and its dependent Townships northward through the 1798 season, to the point that in 1799 he could pick up his treaty line at Loramie's Store, extend it to Fort Recovery, and finally carry it down to the Ohio opposite the Kentucky River's mouth. By the end of the 1800 season, virtually all the Rectangular Survey below the treaty line had been completed, including the extension of Ludlow's own survey beyond the Symmes track north "between the Miamis."[52]

It was, all in all, a remarkable four years of surveying, even if more remarkable for its speed than for its accuracy. Because the Ohio surveys originated from so many disconnected points, the rectangles of each crash into others in haphazard ways. Especially disturbing is Ludlow's "between-the-Miamis" survey, which perpetuates the clockwise-rotated grid of the original Symmes tract. Besides the many starting points, inaccuracies came from Surveyor General Putnam's rather cavalier attitude toward true north. But if Putnam cannot be said to have advanced the art of Rectangular Surveying, he must be credited with setting its administration on a sound footing. From 1797 until 1910, virtually all government surveying was done by Deputy Surveyors under strict, cost-per-mile contracts.

There was one part of Ohio south of the Greenville Treaty line that was not surveyed into rectangles. You will remember that Virginia had reserved from its cession all the land between the Scioto and Little Miami rivers. By accepted practice, this border would have been surveyed by mapping each of the two rivers to its source and then marking a straight line between those two points. When Virginia staked this claim, it had been assumed that the courses of the Little Miami and the Scioto ran roughly parallel and were of roughly equal length, so that a line between their source points would tilt not too far from east to west. Ludlow's survey of the Little Miami showed, however, that its diminutive sobriquet was well deserved, whereas surveys of the Scioto revealed that its course not only wandered well north of the Little Miami's source but ended at a point some miles west of that source. The obvious solution was to start a line at the source of the Little Miami and carry it north and west to the source of the Scioto.

The man chosen for the task was William Ludlow, the son of Israel. In 1802 he set up his instruments at the head of the Little Miami and directed them toward the latitude-longitude position of the source of the Scioto, as you can see in figure 7.31. By this time, the east-west part of the Greenville Treaty line had already been surveyed by his father, in 1797. Consequently, when Ludlow's projected source-to-source line intersected the treaty line, he stopped his work: the land north of the treaty line was Indian land and thus closed to claims by Virginia war veterans.

Virginia, though, challenged Ludlow's line. In a pattern learned in border disputes among all the colonies whenever geographic realities showed chartered boundaries to be nonsensical, Virginia offered a "solution" that was clearly untenable but which had just enough legalistic sense to serve as a bargaining position. It proposed that the western boundary of its Military Reserve be a line drawn from the source of the Scioto to the mouth of the Little Miami at the Ohio River—a line whose course would fall entirely west of the course of the Little Miami and thus invalidate many of the purchases made from Symmes.

The federal government responded in 1812 by having Charles Roberts survey a true line from the source of the Scioto back to the source of the Little Miami. It turned out that this line, when surveyed, angled ever so slightly west of Ludlow's conjectural source-to-source line. The final agreement—not reached until 1818—was to accept Roberts's line north of the Greenville line and Ludlow's line to the south, the two to be connected by a short jog along the treaty line.[53]

The boundary drawn around the Virginia Reserve also marks the extent of rectangular surveying in Ohio. Within the reserve, land was apportioned according to Virginia practice: settlers armed with warrants for specific acreages staked out the boundaries of their own farmsteads. With the setting of the borders of the Virginia Reserve, the extent of the Public Domain in Ohio was also set.

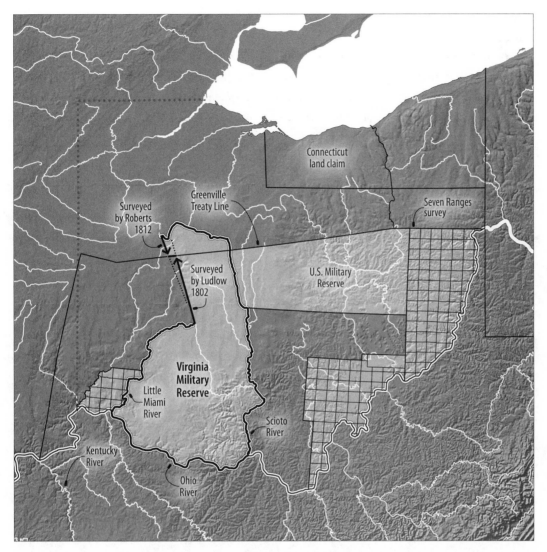

FIG 7.31

After such an inauspicious beginning, it's a wonder that the idea of a Rectangular Survey wasn't simply jettisoned. Yet the system prevailed and, after its readoption in 1796, has endured ever since. One reason might be the simple fact that in the survey of Ohio, a workable system of surveying procedures gradually emerged; and a system, regardless of its flaws or risks, will sometimes seem better than no system. This is not to say that all the procedures evolved by around 1800 worked well—clearly some did not. But in the course of pushing rectangular lines through the woods of Ohio, it became clearer, to all involved, how things should be done "next time."

One of the procedures that worked was the habit of blazing trees to mark the lines of the survey. Chopping notches into, or axing patches off of, tree bark to mark a trail or line is a practice so intuitive that it arose independently in many woodland cultures. The practice builds on the universal wayfinding principle of **intervisibility**: if I can stand at one blaze and see the following one (or see it soon enough), I will know where to go next. But the surveyors of Ohio continued the American surveying practice of systematizing the blazes and chops. They saw, by the end of their work, that a good system would consist of at least this:

- *If a tree falls to one side of a line, blaze the quadrant of the tree's trunk that faces the line.*
- *If the tree falls on a line (and the tree is too big to cut down), blaze it on the two opposite quadrants that are "looking" down the length of the line—as if the line, in passing through the tree, had itself marked its passage.*

- *Corners of parcels should be marked by a monument at the corner, and adjacent trees should be blazed as "witness trees" to the monument; the trees' bearings and distances away from the corner should be recorded in notes that would accompany the drafted plats.*[54]

As the Rectangular Survey evolved during the settlement period, the system of chops, notches, and blazes became ever more systematic, to the point that corner posts came to bear so many notches and markings as to appear almost hieroglyphic.

For as long as surveyors have run lines, they have made notes to recall to themselves what they saw and did in the field, but seldom had field notes been as systematized as those in the Ohio surveys. And almost unprecedented was the idea of compiling those notes in tabular form and making them available to members of the public for their use. This was a second feature that endured and became more methodical as the survey evolved.

A third precedent begun in Ohio was the scale to which the plats of the Townships would be drawn. At the time, "40 chains to the inch"[55] was the standard, so an 80-chain mile thus equaled 2 inches on paper. Thus, the plat of a full Township would be a square 12 inches on a side. The practice that evolved was to draw the Township at the left edge of a rectangular sheet in "landscape" orientation and fill the right half of the sheet with notes.

Of all the practices adopted in Ohio, this is the one that has endured with the least change. The content and format of the surveyors' notes evolved, and the sheet size eventually settled at the 16 by 20 inches shown in figure 7.32. If you

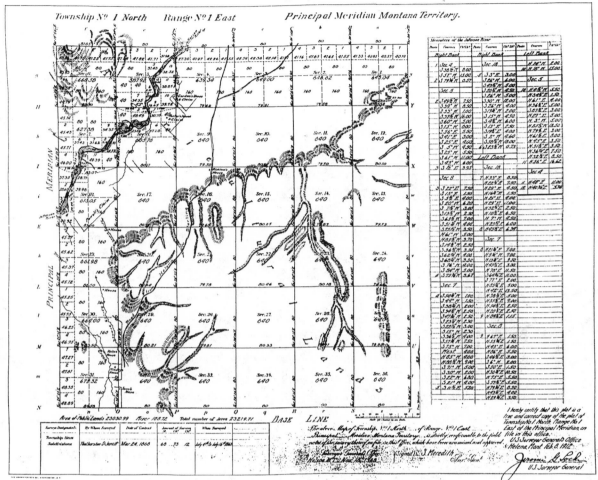

FIG 7.32

Here is a plat from the survey of Montana, drawn up in 1867. The tables at the right of the sheet tell the direction and bearing of the *meander lines* that run around lakes and streams within the Township.

Bureau of Land Management, United States Department of the Interior.

go to a public land archive and ask to pull out a Township plat (which is your perfect right, by the way), you will see the Township and its Sections recorded at the scale of 2 inches to the mile.

It was also in Ohio that the system of numbering Sections was worked out. You will recall that Congress described the enumeration of the Lots in its 1785 Land Ordinance. The description is in words, without a diagram, so there's some ambiguity about what Congress intended—if indeed it knew. It had called for the Lots to be numbered in the bottom-up, right-to-left manner in which the Townships were numbered, as shown in figure 7.33. But as we saw earlier, other language in the ordinance implied a "down-up-down" boustrophedonic pattern, as in figure 7.34. In his plats of the Seven Ranges, the Geographer of the United States chose to follow the "bottom-up" numbering system in both Townships and Lots. Only with the 1796 Land Act did Congress describe explicitly the boustrophedonic numbering system that has been used ever since.

However numbered, the data for drawing the plats for the Seven Ranges was supplied by the gentleman surveyors sent into the field by the various states. From the time of the first land sales in 1787, when the surveyors turned in their invoices, it was plain to all that the delegate system had been a mistake. It took a while, but the federal government would eventually adopt a system of hiring Deputy Surveyors—surveyors certified as competent in the skills of the profession, who would contract with the government to survey a specified area for a set fee. It would be up to each Deputy Surveyor to hire and pay his assistants, and purchase supplies, out of that fee. This contract system would continue until 1910, when Progressive Era reforms put the survey into the hands of civil service employees—the system that continues to this day.

Another negative lesson of the Seven Ranges survey was the folly of allowing surveyors to run their lines by magnetic north rather than the true meridian. The requirement was restored by the federal Congress in the 1796 Act, and subsequent laws have mandated that requirement in the centuries since. And a good thing too: as the survey moved deeper into in the Old Northwest, into what is now Michigan, underground iron deposits would render magnetic compasses almost useless.

A further defect was not visible on the carefully rectified plats Thomas Hutchins submitted, but it would become obvious when settlers took serious possession of their land and discovered the locations of the lines shown in figure 7.35. Often, where four Townships came together at a common corner, the corners did not meet at a single point on the land. Some Townships did not even close into complete rectangles.[56] This was a direct consequence of the manner in which surveyors had extended the survey south from the Baseline Hutchins ran. The instructions to the delegate-

36	30	24	18	12	6
35	29	23	17	11	5
34	28	22	16	10	4
33	27	21	15	9	3
32	26	20	14	8	2
31	25	19	13	7	1

FIG 7.33

31	30	19	18	7	6
32	29	20	17	8	5
33	28	21	16	9	4
34	27	22	15	10	3
35	26	23	14	11	2
36	25	24	13	12	1

FIG 7.34

surveyors were essentially: "Run 6 miles toward south, as indicated on your compass; monument that point and turn 90° to the east; then run 6 miles in that direction, stop there, and monument that point." There was no mandate for a surveyor to tie his corners with those established by the surveyor working immediately to the east. Such confusion about boundaries was precisely what a Rectangular Survey was supposed to avert. Hutchins's technique of sending successive teams south and then east through a Range of Townships would never be used again.

Finally, after Ohio it became an accepted principle that surveys should be projected from Principal Meridians and Baselines, and that those lines should be drawn with very special care. Though surveyors have given to Hutchins's base the honorific name of the Geographer's Line, they don't honor it for its accuracy. Pride of place for accuracy in eighteenth-century surveys goes, without question, to Pennsylvania's southern and western borders, and especially to the southern border run primarily by Mason and Dixon. Granted, the boundary surveyors carried precision instruments like Rittenhouse's Transit, while Hutchins and his crew carried only surveyor's compasses. Plus, the Geographer and his men were pushing their line through a virgin forest across rugged terrain. Nonetheless, the far end of the Geographer's Line—7 Townships, or 42 miles, west of where he began it—is about 1,500 feet south of its Point of Beginning.[57]

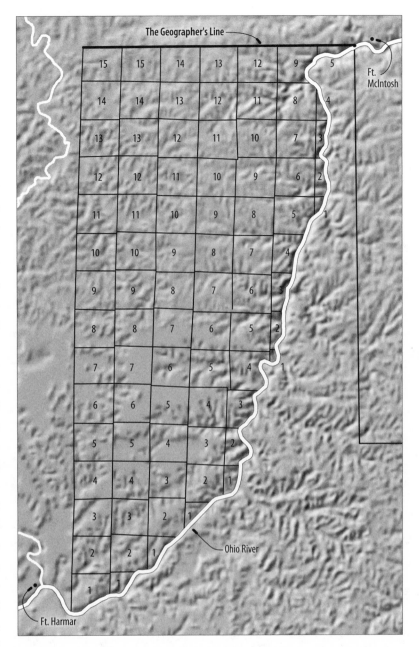

FIG 7.35

EIGHT

The Rectangular Survey Evolves into Its Final Form

1800: DIVIDING SECTIONS IN HALF

In 1798 it was determined that the Ohio Country had crossed the threshold of the five thousand free males necessary for forming a government; the next year an assembly was convened, not just to set up that government but to send a delegate to Congress. The Ohioans' choice was William Henry Harrison: at last Congress would have in its midst a true voice from the West. Though he could not vote, Harrison convinced Congress to approve a bill allowing land to be sold in Half Sections of 320 acres—still far bigger than the farmsteads settlers were used to, but all that could be obtained at the time.

Yet Congress still hesitated. As with the 1796 law, the Act of May 10, 1800, would apply only to the lands below the Greenville Treaty line. And within that territory, only in the lands west of the Muskingum River would half the tracts be sold in full Sections, half in Half Sections.[1] It was a tentative start on the road to providing settlers with farmstead-sized tracts, but once Congress began accommodating farmers' needs, it continued on that course until well into the twentieth century.

The bill also helped farmers by authorizing the four new land offices shown in figure 8.01—at Cincinnati, Marietta, Chillicothe, and Steubenville, towns just back from the edge of settlement. This was another precedent Congress would follow throughout the settlement period, one which gives us the indelible image of the just-founded western town consisting of little more than a general store, a saloon, and a land office.

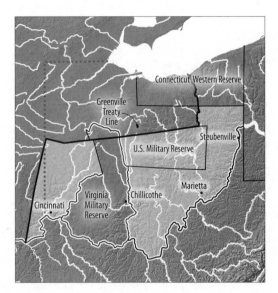

FIG 8.01

For our story of apportioning the land, the most important provision of the 1800 Act is not so much the idea of halving Sections as the precise manner in which that halving would be carried out. The act gave precise instructions about how surveyors were to mark out the Half Sections. As they ran the Section boundaries at 1-mile intervals, east to west and north to south, they were to place monuments every mile in the north-south lines but every half mile on the east-west lines. Thus, the only division allowed in a full Section would be a north-south "vertical" cut like the one shown on the next page in figure 8.02: if a farmer wanted to buy 320 acres of land from the government, he could get it only in the form of a vertical rectangle.[2]

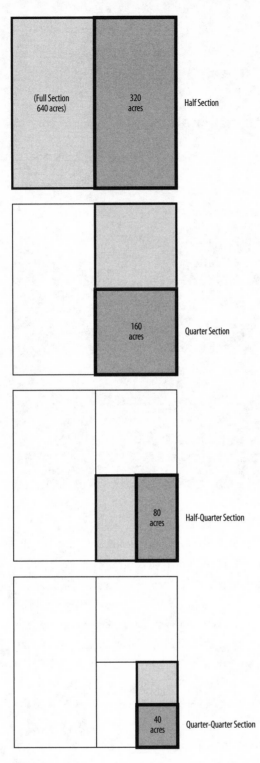

(Full Section
640 acres)

320
acres

Half Section

160
acres

Quarter Section

80
acres

Half-Quarter Section

40
acres

Quarter-Quarter Section

FIG 8.02

Congress would expand this mandate in 1805, when it allowed the sale of Quarter Sections—but only Quarter Sections formed by running an east-west line across the middle of a tall Half Section. Once set, this precedent controlled subsequent legislation. After 1805, with only a few extraordinary exceptions, the United States would sell land only in plots achieved by perpendicular halving. As the settlement period wore on, Congress eventually shrank the minimum parcel size to 40 acres—one-fourth of a Quarter Section, known as the "Quarter-Quarter," or more familiarly, the "Forty." The Quarter-Quarter completed the range of offerings the government would make to settlers:

- the Section: 640 acres;
- the Half Section: 320 acres;
- the Quarter Section: 160 acres;
- the Half-Quarter: 80 acres; and
- the Quarter-Quarter: 40 acres.

If a settler wanted to buy land, it came only in those sizes and only in those shapes. Still does, in fact. And it all began when in 1800 Congress decided that it, and not the settler, would be the one to divide the Section. One more provision of the 1800 Act set a precedent which endures to this day. It may seem obscure to easterners, but it's a principle known intimately by people who own farms or ranches on Public Domain land.

Bear in mind that what Congress wanted to offer to the public were parcels with precise acreages—640-acre Sections and 320-acre Half Sections—and it wanted to be done with each parcel as soon as it was sold. (The term of art for the sale of a tract of government-owned land is that the parcel has been "alienated from the Public Domain.") The last thing Congress wanted was for some buyer to come back years later with the complaint, "I decided to sell my Half Section farm, so I had it surveyed by an expert, and he tells me it contains only 300 acres. That's 20 acres less than I was told it contained. Land around my farm is now selling for $50 an acre, so instead of the $16,000 I expected, I can now ask for only $15,000. You, Congress, owe me a thousand bucks!"

A shaky argument, perhaps, but one not hard to imagine occurring in a sale between two individuals. To prevent this from ever happening, Congress created a legal fiction. The first part of the fiction was to declare that no matter how many actual acres a Section or Half Section might be found to contain later, the title for the parcel would read "640 acres" or "320 acres," and the buyer could forever after sell it as containing that full acreage. Every Section would have a face value of 640 acres, and the owner could take that to the bank—and still can.

This legal fiction presented a problem, though. If the measurement of the outer boundary of a Township was not precisely accurate, yet Congress declared it to contain exactly 23,040 acres (36 Sections of 640 acres each), it would not have an accurate measure of the Township's area—or, eventually, of the states and the Public Domain itself. So Congress said that if a Township were found to have sides that exceeded or fell short of exactly 6 miles,

the excess or deficiency shall be specially noted, and added to or deducted from the western and northern ranges of sections or half-sections in such townships

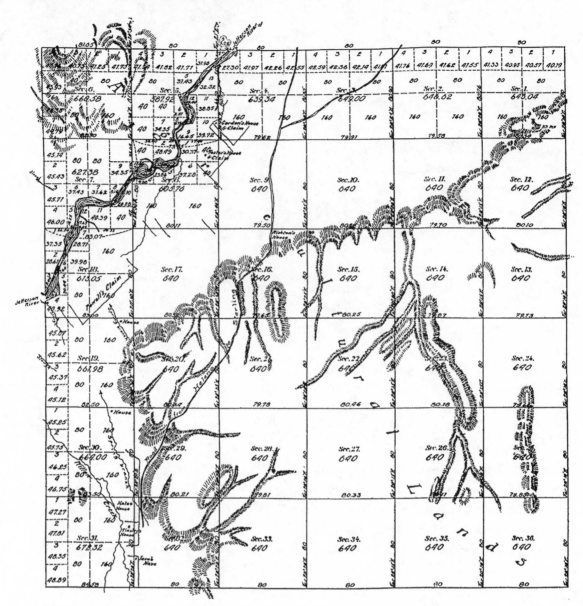

FIG 8.03   Bureau of Land Management, United States Department of the Interior.

. . . the sections and half-sections bounded on the northern and western lines of such townships shall be sold as containing only the quantity expressed in the returns and plats, respectively, and all others as containing the complete legal quantity.[3] [Figure 8.03, the Township map from figure 7.32, shows this idea in operation.]

Later in this chapter I'll show how the surveyors accomplished this addition and deduction, and assigned nonstandard acreages to the Sections on the northern and western borders of Town-ships. Let's return, though, to our farmer who wants to sell his land. We now know that even though it "actually" contains only 300 acres, he can sell it as containing 320 acres; but what if he wants to sell off, say, half of his farm? Does each half contain 160 acres? Yes, it does—or rather, it will, under the **aliquot principle** Congress was to adopt in 1805. But to continue the story of how the Rectangular Survey developed, we first have to go back to Congress and then into the swamps of Mississippi.

You'll remember that in 1802, Georgia at last ceded its western lands to the United States. During the previous year the Choctaw tribe had been induced to surrender a tract of land bordering the Mississippi whose apex was Vicksburg and whose base was the 31st-parallel border with West Florida (a shape shown in figure 8.04). There was a great desire for land in this region so close to New Orleans, and so in the Act of March 3, 1803, Congress authorized an extension of the Rectangular Survey to the new Mississippi Territory. This time not just half but all the Townships were to be surveyed into Half Sections. To oversee the survey, President Thomas Jefferson appointed Isaac Briggs as Surveyor of the Lands of the United States South of the State of Tennessee.[4]

Briggs had specific instructions from Treasury Secretary Albert Gallatin. Since the 31st-parallel boundary line had been surveyed by Andrew Ellicott in 1798 and 1799, Briggs was to use that as a Baseline from which to project his survey. To control that projection, he was to choose a point on Ellicott's line and from there project a Meridian northward.[5]

All this was as might be expected, but then Gallatin gave a novel instruction about numbering Townships. He followed the old pattern of calling a vertical stack of Townships a Range but rationalized the system of numbering the Ranges and the Townships within them. You will remember that in the Seven Ranges, each Range had been numbered in sequence from east to west; and, even though the Townships within those Ranges were projected southward from the Geographer's Line, the Townships had been numbered upward from the Ohio. In the Symmes tract Israel Lud-

FIG 8.04

low had marked a Baseline and Meridian from which to project his Townships, yet he began the numbering of his Ranges at the western border of the tract, with the Range numbers increasing west to east. Within those Ranges he continued the practice of numbering the Townships upward from the Ohio.[6]

In the surveys west of the Symmes tract, Ludlow projected his Townships off a Meridian at the mouth of the Kentucky River and numbered his Ranges away from the Meridian: "1st Range east of . . . ," 3rd Range west of . . ." But within each of those Ranges he again followed the practice of numbering the Townships upward from the Ohio.

The problem with numbering Townships upward from a wiggling border like a river was that the Township numbers in any two adjacent Ranges would have no discernible relationship to each other. We saw this in the Seven Ranges where, for example, immediately west of Township 9 in the 2nd Range is Township 12 in the 3rd Range (see figure 7.35: both Townships are just below the Geographer's Line).

Briggs's instructions at last provided a comprehensible, universally applicable numbering system. The Ranges would be numbered outward from the Meridian—"east of" and "west of," as Ludlow had done—but the Townships within each Range would be numbered upward from the Baseline,[7] in the manner shown in figure 8.05. In such a scheme, immediately east of "Township 5 in Range 3 East" would be "Township 5 in Range 4 East." It would thus become possible to speak of Townships as arranged in east-west *Tiers* as well as in north-south Ranges. It took only one year for surveyors to adopt the shorthand that soon passed into common conversation in the West, and continues, like the numbering system, to this day:

"T.4N, R.3E"—pronounced "Township 4 North [pause] Range 3 East"—in the third Range of Townships east of the Meridian, the fourth Township north of the Baseline. Or if you prefer, since after 1803 Townships run in Tiers as well as Ranges, you could translate that shorthand this way (although they don't actually say it this way in the West): "T.4N, R.3E"—"Tier 4 North, Range 3 East"—the Township at the intersection of the fourth Tier of Townships north of the Baseline and the third Range of Townships east of the Meridian.

In either case you will have specified the same Township. And if you cite the meridian from which that Township was projected—the Washington Meridian, in the case of southwest Mississippi—you will have specified a unique Township: no Township but one, in the entire

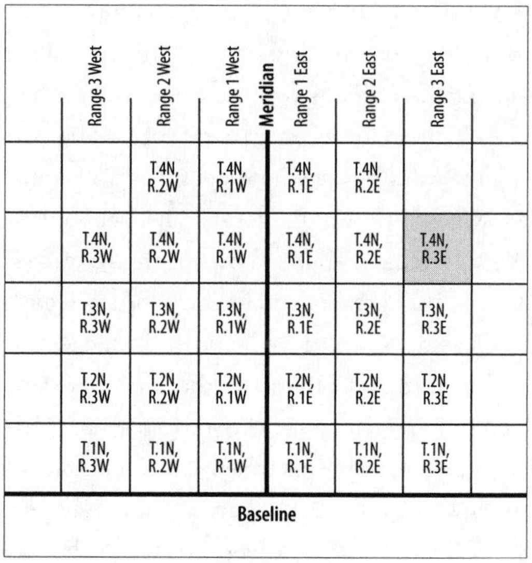

Range 3 West	Range 2 West	Range 1 West	Meridian	Range 1 East	Range 2 East	Range 3 East	
		T.4N, R.2W	T.4N, R.1W	T.4N, R.1E	T.4N, R.2E		
T.4N, R.3W	T.4N, R.2W	T.4N, R.1W	T.4N, R.1E	T.4N, R.2E	T.4N, R.3E		
T.3N, R.3W	T.3N, R.2W	T.3N, R.1W	T.3N, R.1E	T.3N, R.2E	T.3N, R.3E		
T.2N, R.3W	T.2N, R.2W	T.2N, R.1W	T.2N, R.1E	T.2N, R.2E	T.2N, R.3E		
T.1N, R.3W	T.1N, R.2W	T.1N, R.1W	T.1N, R.1E	T.1N, R.2E	T.1N, R.3E		
Baseline							

FIG 8.05

Public Domain, bears the name "T.4N, R.3E, Washington Mer." Once the Rectangular Survey really began to unroll across the continent, Americans gained an appreciation, and finally a fondness, for the ease of saying exactly which tract they meant—out of all the thousands and then millions eventually surveyed—with so few words.

Prescient innovations didn't just flow one way, out from Washington to the field. Surveyors like Isaac Briggs sent progress reports back to the capital, and in them they sometimes proposed innovations. Like contract surveyors before and after, Briggs complained that he couldn't do the work accurately on his $4-per-mile stipend. He disparaged his predecessor as well, reporting that the mile markers on Andrew Ellicott's 31st-parallel line weren't a true mile apart and thus couldn't be used to mark Section corners. (Was this the truth? Snippiness? An attempt to impress the boss? We'll never be certain.) But in the same report came a prescient proposal: he and an assistant would project not one but two Meridians up from the Baseline, starting from points "6 miles 12 perches apart" and carry them northward 42 miles—7 Townships—where they would be precisely 6 miles apart. You will remember that a perch is one-fourth of a chain, thus 25 links in length. So Briggs was proposing to start two due-north lines 6 miles and 3 chains apart and have them arrive together 42 miles later, 6 miles and 0 chains apart. What's going on here?

What's going on is the *convergence of lines of longitude*. If you pull your focus back from the single Township to the planetary view of figure 8.06, you can easily see that all lines of longitude converge; they are effectively parallel at the equator, but they converge faster and faster as they approach the poles of the earth. In the area of the "Lower 48" states, the difference is not so extreme, but it's still measurable. For a Township laid out against the 49th-parallel border with Canada, the two sides of a Township would converge 83.5 links in their 6-mile run

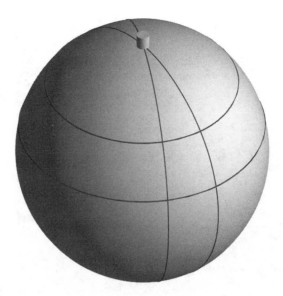

FIG 8.06

from south to north. For Briggs at the 31st parallel, his Township sidelines would converge 43.6 links in their 6-mile run to the north. Multiply 43.6 links by the 42 miles in 7 Townships, and you get 305.2 links, thus 3.052 chains. Compare this convergence with Briggs's plan to displace his two Baselines by 3 chains. Snippy, kvetching, or obsequious Briggs might have been, but the man knew his surveying technique!

Now don't for second suppose that it was not until 1803 that anyone noted the problem of the convergence of meridian lines. Many of the congressional authors of the early land apportionment laws were themselves surveyors or landowners familiar with surveying concepts. They knew about convergence, but they also knew that whatever errors in acreage might result from converging meridians, those errors would be trivial compared with the inaccuracies inherent in then-available surveying technology and in the skill of its usual practitioners. Jefferson's

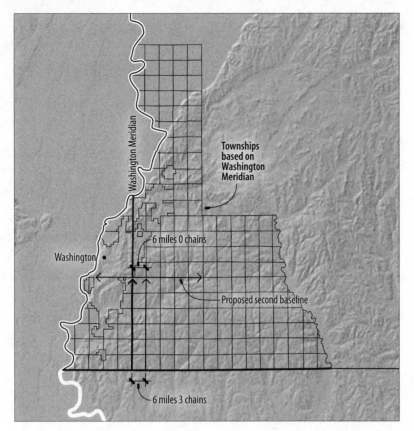

Washington Meridian

Townships
based on
Washington
Meridian

6 miles 0 chains

Washington •

Proposed second baseline

6 miles 3 chains

FIG 8.07

plan for linking state borders and land parceling even had a system for containing errors of convergence: the survey of each of his small rectangular states would "restart" at every other degree of latitude.

In his reply to Briggs's report, Albert Gallatin gave no support to the dual-meridian approach, and Briggs was forced to fall back on the single-meridian mandate contained in his original instructions. But which of the two would be anointed the Meridian that would control the survey of the territory?

Upon consideration Briggs chose not the meridian he had run but the one surveyed by his assistant, the one shown by a heavier line in figure 8.07. Only 6 miles to the west, that meridian, it turned out, happened to cut through terrain more amenable to survey—fewer streams, less dense forest—and so was likely a more accu-

rate, more reliable line from which to project a survey. In this choice Briggs followed the precept first set out by Israel Ludlow in the Symmes survey—that the location of survey-controlling lines was a choice best made not in Washington but by the surveyor in the field. If those initial controlling lines were to be as accurate, thus reliable, as possible, then their Point of Beginning should be whatever point afforded the longest prospect for unfettered surveying. Surveyors knew that a set of lines chosen for abstract reasons (such as those lines likely to issue from Washington) might result in controlling lines run through such rough territory as to impair those lines and thus to taint the entire survey projected from them. Better, in the minds of surveyors, to save Washington from its own folly by choosing an initial point on the basis of conditions observed in the field. (In chapter 10, we'll see exactly this circumstance in the choice of the controlling Meridian of the survey of Montana.)

Briggs did just that in choosing his assistant's "easy" meridian to anoint as the Washington Meridian controlling the survey of the southwest quadrant of the eventual state of Mississippi. (The Meridian, by the way, is not named for our first president, at least not directly. It was named for the town of Washington, Mississippi Territory, the settlement nearest the termination of Briggs's two 42-mile "test" Meridians.) Just one year later Briggs's principle of choosing initial points would be affirmed.[8]

One of the mythical images of American history is that of white settlement pushing into Indian territories in a smooth outward expansion, rather like that seen when water is poured onto one corner of a dry cellulose sponge. The truth is that many Indian tribes held their lands long after white settlement had swept around them, while other tribes ceded lands deep in the wilderness, far from the edge of "civilization." One such enclave was the Vincennes Tract.

Many pages ago we saw how, when Virginia ceded its western lands, one of its conditions for doing so was that the inhabitants of the Vincennes settlement would have their land titles honored. The pretext was that the French settlers there, after their 1779 "liberation" from the British by Virginia's George Rogers Clark, had sworn allegiance to Virginia—and by extension to the United States. The settlers had in fact been there since 1735, and in 1742 the French authorities had extracted a cession of land from the local Indians. Since 1787 the people of this "Vincennes Tract" had been petitioning Congress to validate their titles, and on May 1, 1802, Congress granted them their wish—adding one proviso.

The tract would be surveyed by the rectangular method prescribed by the 1800 Act except in those places where settlers had prior land titles (from France or Britain or Virginia, for example). In those cases the parcel boundaries described in the titles would, after adjudication, be honored; but outside those tracts the grid of the Rectangular Survey would flow continuously across the territory, its straight lines stopping against one side of a nonsquare parcel but starting up again on the opposite side.

This principle was retained (sometimes in the breach) throughout the settlement period, and it points up one commonly held misconception about the Rectangular Survey. The intention of the survey was never to "grid up" the country. The goal was that every acre of the nation's land would fall within a surveyed tract—to establish a *cadastre* for the nation. And strictly speaking, the Rectangular Survey was conceived as a method for surveying land belonging to the United States so that it might be sold to settlers: there was no need—indeed no right—to resurvey lands already owned.

In June of 1803 the boundaries of the tract were confirmed by treaty with the local Indian tribes, and in September those borders were surveyed and monumented—enclosing the paler area shown in figure 8.08 on the following page. The Vincennes Tract was a rectangle about 40 miles by 60 miles, rotated so its long axis ran west-northwest, to be perpendicular to the general run of the Wabash River just north and south of the Vincennes settlement. Most of the rectangle was on what is now the Indiana side of the river, but it also poked a dozen or so miles into present-day Illinois.

In July of 1803 President Jefferson named Jared Mansfield the new Surveyor General for the Territories Northwest of the Ohio River, and in the following March Congress placed the Surveyor General in charge not just of the lands inside the Greenville Treaty line but all the Indian-ceded lands between the Ohio and the Mississippi. So it fell to Mansfield to conduct the Vincennes survey.

There was one problem to be worked out first. It was assumed that eventually the Indians would cede the land between the Greenville Treaty line and the Vincennes Tract. In terms of

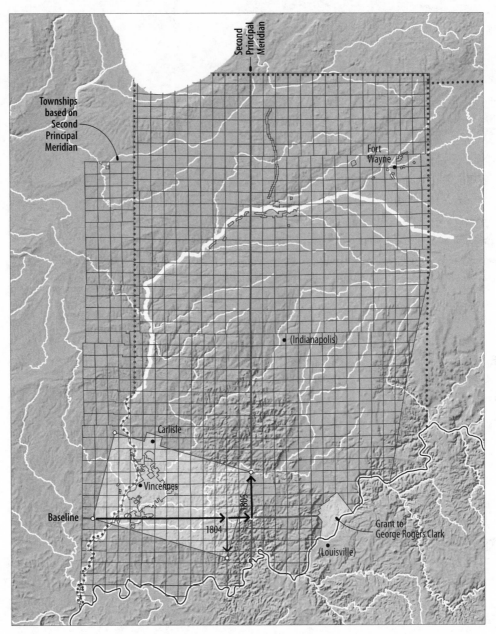

Second Principal Meridian

Townships based on Second Principal Meridian

Fort Wayne

(Indianapolis)

Carlisle

Vincennes

Baseline

1805

1804

Grant to George Rogers Clark

(Louisville)

FIG 8.08

the Rectangular Survey, the ideal would be for the grid of Townships already surveyed to continue unbroken across the lands to the west, to include the Vincennes Tract. But making the squares in the Vincennes Tract match up with the squares to the east would require marching overland, in precise 6-mile steps, right through Indian country—a task that would be difficult, time-consuming, probably dangerous, and in

Jared Mansfield's view, unnecessary. Why not simply start up a whole new 6-mile grid within the Vincennes Tract and then, when the Indian cessions allowed, extend that grid outward? Mansfield knew that no matter where a 6-mile grid was started in a territory whose boundaries come from some other principle, at some edge or other the grid will overshoot the boundary. But he also knew that he could get fewer chopped-off Town-

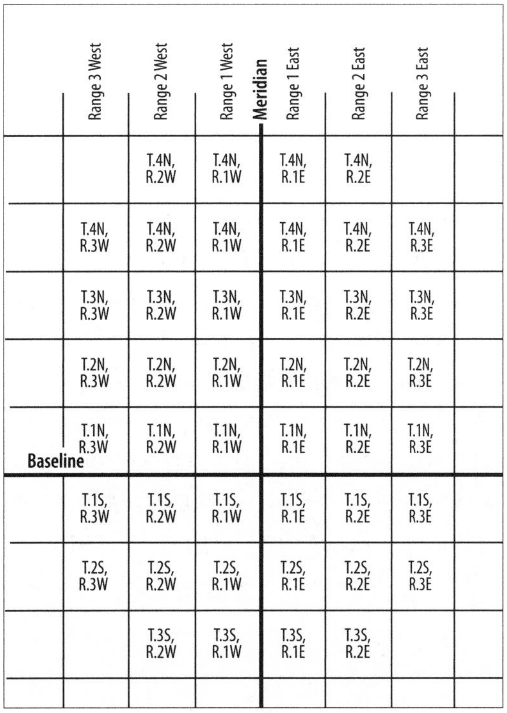

Range 3 West	Range 2 West	Range 1 West	Meridian / Range 1 East	Range 2 East	Range 3 East	
		T.4N, R.2W	T.4N, R.1W / T.4N, R.1E	T.4N, R.2E		
T.4N, R.3W	T.4N, R.2W	T.4N, R.1W	T.4N, R.1E	T.4N, R.2E	T.4N, R.3E	
T.3N, R.3W	T.3N, R.2W	T.3N, R.1W	T.3N, R.1E	T.3N, R.2E	T.3N, R.3E	
T.2N, R.3W	T.2N, R.2W	T.2N, R.1W	T.2N, R.1E	T.2N, R.2E	T.2N, R.3E	
T.1N, R.3W / **Baseline**	T.1N, R.2W	T.1N, R.1W	T.1N, R.1E	T.1N, R.2E	T.1N, R.3E	
T.1S, R.3W	T.1S, R.2W	T.1S, R.1W	T.1S, R.1E	T.1S, R.2E	T.1S, R.3E	
T.2S, R.3W	T.2S, R.2W	T.2S, R.1W	T.2S, R.1E	T.2S, R.2E	T.2S, R.3E	
		T.3S, R.2W	T.3S, R.1W / T.3S, R.1E	T.3S, R.2E		

FIG 8.09

ships in the Vincennes Tract itself if his grid's Meridian and Baseline ran through two of the tract's corners. So he issued his instructions, and here's what his Deputy Surveyors did during October of 1804.

First they went to the far southwestern corner of the tract to find the monument left by the previous surveyor. From that point they projected a Baseline eastward, blazing the course of the line and marking off miles and half miles but setting no permanent monuments. When they calculated that they were somewhere north of the tract's southeastern corner, they struck out to find that monument, from which they then projected a meridian line back north to the Baseline. Where the two lines crossed, they had a starting point for their survey, and with winter coming on, the surveyors left the field for the season.

The next season they came back to extend Townships from the Baseline; as the surveying season drew to a close, they carried the Baseline 2 Townships east—just outside the Vincennes tract—and then ran a meridian line north to the tract's corner. It was this Meridian, required as part of his Vincennes survey but nearer to the center of the Northwest Ordinance's "middle state," that Mansfield designated as the **Second Principal Meridian** (he was presuming that Israel Ludlow's meridian at the Great Miami would be extended as the **First Principal Meridian**).[9]

Even if "second," this was the first time a Principal Meridian and Baseline system had been placed where it could extend in all four cardinal directions. (It was also the first use of the term *Principal* Meridian.) And Mansfield jumped on this fact to complete the idea broached by Gallatin for the Washington Meridian. In lands surveyed

out from the Second Principal Meridian, Ranges would be numbered as in figure 8.09. The Township numbers would increase east and west away from the Principal Meridian, as before, but Townships would also be numbered away from the Baseline north and south. Not much of a leap of imagination, granted; but Mansfield made the leap, and his numbering system was followed ever after. (Later would come the name **Initial Point**, for the place where a Principal Meridian and a Baseline crossed.)

Also followed wherever possible was Mansfield's practice of letting the placement of a survey's initial point be governed by local conditions. When Gallatin received Mansfield's report, he was displeased with the Meridian's location. He had favored aligning it with the town of Vincennes, the border of the "middle state." But soon he was persuaded of the good sense of Mansfield's approach, and the Jeffersonian idea of tightly aligning the Rectangular Survey with state borders was finally laid to rest.

With the institution of Mansfield's twin systems for locating Initial Points and numbering Townships, we have almost all the essential features of the survey in its final form. The big issues that remained were the problem of the convergence of meridians and the precise methodology of subdividing Townships into the actual plots to be sold. We'll deal with those two issues chronologically, but first there are other principles to be added to the survey system.

It can be said that by the Act of February 11, 1805, Congress finally adopted the principle of apportioning the Public Domain not into the parcels that it wanted to sell but into the plots that settlers wanted to buy. It directed that the Townships surveyed under the 1796 "every-other-mile" process be resurveyed to mark the missing mile-apart lines in both directions, and that those new lines be monumented every half mile, to mark off Half Sections. For those Townships already surveyed into vertical Half Sections under the 1800 law, they too would be subdivided, by east-west lines at the midpoints, as specified in the 1804 law.[10]

What really matters, for our story of land subdivision, is that in this act Congress formally enshrined into law the surveying principle that "the monument, once set, shall be regarded as correct." The principle is based on an ancient precept of Anglo-American common law. When people farm the land, they develop patterns of occupancy and come to depend on them. Barring some extraordinary circumstance (a fraudulent sale or someone surreptitiously moving the fence posts), those patterns ought to continue. Congress's bold move was to apply that long-occupancy rule to farms only recently occupied, and indeed to farms not yet occupied.

To conform this ancient principle with rectangular occupancy, though, Congress had to invoke a concept from surveying, the *aliquot principle*. The term means essentially "division by halving," and it works in the manner diagrammed in figure 8.10.

Suppose the government has surveyed a 320-acre vertical Half Section and you want to buy only the southern 160 acres (allowable under

"320 acres" "320 acres"

Markers set by government surveyors

Half Section

"160 acres"

Private surveyor finds government markers and runs line

"160 acres"

Quarter Section

Equal | Equal

"80 acres" "80 acres"

Half-Quarter Section

Equal Equal

Centers calculated and marked by private surveyors

Equal "40 acres" Equal

Equal "40 acres" Equal

Quarter-Quarter Section

FIG 8.10

the 1804 Act). What you do *not* do is to have a surveyor carefully measure the western, southern, and eastern sides of the Half Section and precisely calculate the placement of an east-west line that would put exactly 160 acres on your side. Instead under the aliquot procedure, your surveyor would locate the half-mile markers set by government surveyors on the eastern and western sides of the Section and run a line between them. Your 160-acre Quarter Section would be the land lying south of the line between those monuments.

What if the two long sides are found to be less than 80 chains in length, or not to be parallel, or to differ in length? No matter: the aliquot principle still governs. And no matter where the halfway points might fall, if you are buying a Quarter Section that's not on the northern and western edges of the Township, your title will specify you as the owner of a full 160 acres of land.

This principle is important, because it would later allow the government to sell land in amounts smaller than a Quarter Section without then having to survey those smaller parcels. (Except under special circumstances, federal surveyors would never set their monuments on a finer grain than one-half mile apart.) In 1820 Congress authorized the sale of Half-Quarter Sections (80 acres), to be achieved by finding the midpoint of both the northern and the southern borders of a square Quarter Section and connecting them with a straight line. The Half-Quarter is thus achieved in the way the Half Section was: by cutting a square into two equal "tall" rectangles. Then in 1832, plots of a quarter of a Quarter Section were authorized (40 acres). And just as the tall-rectangle Half Section was cut "horizontally" through its middle to make the square Quarter Sections, so was the tall Half-Quarter rectangle halved into two square Quarter-Quarters. In both cases the end points of the halving lines were to be located on the principles directed and prescribed by the Act of 1805—that is, by the aliquot principle.

Out of this aliquot division would come, some years later, the legal description of private property that so mystifies easterners but is second nature to people raised amid the Rectangular Survey. To explain the system, let's go back to Township 4 North in Range 3 East of the Washington Meridian (even though this particular Township was never sold off in the way it is apportioned in figure 8.11).

We saw that this Township can be distinguished from all other Townships in the Public Domain by its description: T.4N, R.3E, Washington Meridian.

- *"Sec. 23, T.4N, R.3E, Washington Mer." is thus the unique-in-all-the-world name for Section no. 23 in this one Township.*
- *The eastern Half Section of Section 23 is "E½, Sec. 23, T.4N, R.3E, Washington Mer."*
- *The southeastern Quarter Section of Section 23 is "SE¼, Sec. 23, T.4N, R.3E, Washington Mer."*
- *The eastern Half-Quarter of that southeastern Quarter Section is "E½, SE¼, Sec. 23, T.4N, R.3E, Washington Mer."*
- *And the southern Quarter-Quarter of the southeastern Quarter Section is "SE¼, SE¼, Sec. 23, T.4N, R.3E, Washington Mer.," a 40-acre plot.*

Out of this would come another significant aspect of the aliquot principle: its later adoption by private landowners. We saw earlier how the 6-mile-square Townships left almost no mark on

the landscape of southern Vermont: the proprietors apportioned the land as they saw fit, and today the straight township lines can be seen only on political-boundary maps. In a flight over southeastern Ohio, where the basic unit was, at first, the mile-square Section, you can pick out traces of the original Section lines, but they're overwhelmed by other lines and patterns. But as you head into Indiana, the straight lines begin to emerge, and then to predominate. Finally, as you near Iowa, the rectangles overwhelm and control the landscape.

What changed as the Rectangular Survey moved west was not the survey itself but the use of the aliquot principle by private landowners, in their land sales to each other and in the apportionment of their own lands. For centuries American farmers have bought and sold their holdings. A family might say that they've "been on this land for seven generations," but you can be certain that the boundaries of the farmstead have changed, probably several times. And certainly the layout of the fields and pastures have changed. What makes eastern Ohio look, from the air, more like southern Vermont than Iowa is that Iowa farmers made their changes to the land based on the aliquot principle—halving and rehalving, combining and halving again.

What began as "the only way you can buy land from the government" became "the easiest way to sell land to each other" and finally "the most natural way to apportion my own land"—in short, a habit of mind, part and parcel (pun intended) of the culture of the place.

And we owe it, in some sense, to those aliquot acts Congress passed in 1805, 1820, and 1832. With the last of those acts, Congress pretty much

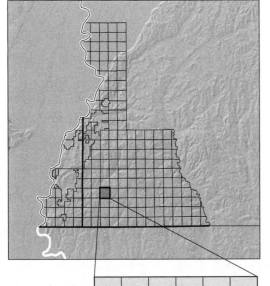

got out of the business of dictating how the Rectangular Survey would be conducted. It continued to pass laws about how settlers would acquire land, and how much they could acquire, but all those provisions were made in terms of the basic rectangular units finally arrived at in 1832. Once the details of the survey were worked out in the early nineteenth century, Congress handed off its management.

T.4N, R.3E, Washington Mer.

6	5	4	3	2	1
7	8	9	10	11	12
18	17	16	15	14	13
19	20	21	22	23	24
30	29	28	27	26	25
31	32	33	34	35	36

Sec. 23, T.4N, R.3E, Washington Mer.

E$^1/_2$, Sec. 23, T.4N, R.3E, Washington Mer.

SE$^1/_4$, Sec. 23, T.4N, R.3E, Washington Mer.

E$^1/_2$, SE$^1/_4$, Sec. 23, T.4N, R.3E, Washington Mer.

SE$^1/_4$, SE$^1/_4$, Sec. 23, T.4N, R.3E, Washington Mer.

FIG 8.11

You'll remember that in the first decade of the nineteenth century, the management of the survey was split between two officials, The Surveyor General for the Territories Northwest of the Ohio River and The Surveyor of the Lands of the United States South of the State of Tennessee—the two ostensibly coequal, despite the asymmetry of their titles. After the Louisiana Purchase, the Surveyor South of Tennessee was given responsibility for the Territory of Orleans (basis for the later state of Louisiana) in 1805, and the following year the rest of the purchase was placed under the Surveyor General for the Northwest.

By the 1810s both officials had multiple surveys going on simultaneously, and so the practice evolved of appointing Deputy Surveyors, one for each project. To oversee their deputies, both Surveyors had relocated from Washington to field offices deep in the survey territory.

Congress understandably decided that a bureau was needed in Washington to oversee all this far-flung activity. So on April 25, 1812, it established the General Land Office, its head to be called the Commissioner. Both the Surveyors had served under the direct supervision of the secretary of the Treasury, but now the Commissioner of the GLO would be inserted in the chain of command between them and the secretary.

Up until then, the two Surveyors had operated pretty much as free agents, and so the first Commissioners of the GLO were unable to exercise much direction over their subordinates. Plus, all the complexities of granting and adjudicating land titles had been transferred from other offices of the Treasury into the GLO: the first Commissioners had more than enough to occupy their

CHAPTER EIGHT

attention until those procedures could be made routine and thus entrusted to others.

What we'll see later is GLO Commissioners gradually asserting more control over how surveyors in the field handled the problems of longitudinal convergence and the subdivision of Townships. The story of the GLO itself, though, can be quickly sketched.

Over time the unworkability of two "super-Surveyors" overseeing the Rectangular Survey became apparent, especially after the Public Domain expanded all the way to the full Pacific coast in 1848. Gradually, big chunks of the territory of the Surveyor General for the Northwest were hived off onto Deputy Surveyors in charge of, for example, Illinois and Missouri, and then Michigan. With time and the erection of new territories and states, the Commissioners of the GLO evolved a system under which a subordinate Surveyor General would have charge of the conduct of the survey in one territory or state. The title *Deputy Surveyor* would then designate an individual who signed a contract to survey a designated area for a stipulated per-mile fee.

For a long time, the GLO issued instructions to its Surveyors General on a state-by-state basis. Only in 1855 would there come a truly national set of instructions. This *Manual of Instructions for Deputy Surveyors* (later called the *Manual of Surveying Instructions for the Survey of the Public Lands of the United States*) has been revised eight times and in its current 1973 edition still guides the work of the Rectangular Survey.

In 1836 Congress reorganized the GLO, giving it more control over the conduct of the survey. Then in 1849, after all the great land acquisitions of the 1840s, Congress felt the need for a Depart-ment of the Interior, which would specifically include the GLO. The next big change came in 1910 when the system of contracting with Deputy Surveyors was ended.

In our era of privatizing government services, the concept of hiring "entrepreneurial" private surveyors would be deemed good public policy. But in the Progressive Era, corruption in government contracts was seen as the greater problem, and the solution was a salaried civil service. After 1910, and continuing to this day, the surveyors of the Public Domain would be tested for their qualifications and then hired as government employees.

The final big change came in 1936, when the General Land Office was reorganized as the Bureau of Land Management.[11] By the 1930s much of the job of surveying land in the lower forty-eight states had been accomplished. Most of the lands still unsurveyed were either in national parks or forests or unlikely ever to find buyers: in neither of those cases was there a need to run Township and Section lines. Plus, by the 1930s the idea had arisen that government's proper function with regard to public land was not to sell it off but to conserve it—thus the shift from Land Office to Land Management. We'll look more closely at the evolution of this idea in the epilogue.

1855: A BETTER TECHNIQUE FOR CONVERGING MERIDIANS

We saw how Isaac Briggs recognized the problem of converging longitude lines when he began his two "test" meridians 6 miles and 3 chains apart, knowing that 42 miles to the north, the distance between them would converge to exactly 6 miles. His primary intention was not so much to handle the convergence of meridians. His concern was that he be able to start, at that 42-mile point, a new Baseline which would be more trustable than the 31st-parallel line marked out by Andrew Ellicott. In the method of that "new Baseline" lay the seed of a solution to the convergence problem.

It's obvious that when the eastern and western boundaries of a stack of Townships are projected northward from a Baseline, those "meridian" boundaries will converge until at some point the distance between them will have shrunk so much below 6 miles that to go any further would result in Sections that were unacceptably narrow. But if at that distance north from the first Baseline you the surveyor were to survey a whole new baseline, you could "restart" the convergence—measure out new, wider Township boundaries—and then project them northward until they too began to converge too much, at which point you could survey a third Baseline and start the process yet again.

This is an obvious corrective strategy, and before the advent of nationwide instructions, individual surveyors applied variants of the technique in ways they thought proper. Briggs's method was one: begin the Meridians "too far apart" and project them until they are "just right"—a solution that favored land buyers, giving them tracts that would always be ever so slightly in excess of the stated acreage. Less farmer friendly was the

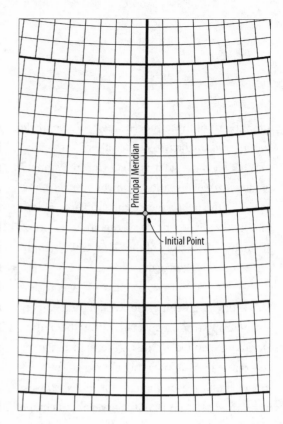

FIG 8.12

technique of starting the Meridians 6 miles apart on the Baseline—meaning that almost nobody would get full acreage. A compromise was to start the Meridians "too far apart" and carry them north until they were "too close together," which had the virtue of at least averaging losses and gains between the two Baselines.[12]

All these methods (and others) were tried by surveyors in the field, but in its 1855 edition of the *Manual of Instructions*, the GLO made one technique a national standard. Surveyors were to begin their Township meridians exactly 6 miles apart every time they "restarted" the Meridians at a new Baseline. Those new Baselines were to be called Standard Parallels, and surveyors were to establish them every 30 miles (5 Townships) when they were working south from the main Baseline, and every 24 miles (4 Townships) when working north from that Baseline—reflecting the faster convergence of meridians toward the north.[13] Figure 8.12 shows the pattern.

CHAPTER EIGHT

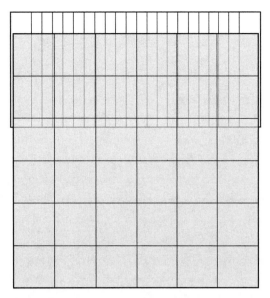

FIG 8.13

How much convergence would thus trigger a restarting of meridians? At the 45th parallel (through Minneapolis-St. Paul and northern Oregon), two meridians running north for 24 miles converge about 200 feet. At the 35th parallel (Albuquerque and Bakersfield) the convergence is about 150 feet.

It's hard to picture how much real difference 150 feet makes in a measurement of 6 miles, so try the analogy shown in figure 8.13. Imagine that the top third of a Township is shrunk to the size of a football field. Six miles is 31,680 feet, so a football field is roughly $\frac{1}{100}$ of that length; so with a Township 150 feet "too narrow" we are talking about a football field whose length is "off" by $\frac{1}{100}$ of 150 feet, or 1½ feet. That may sound like a tiny discrepancy, but remember that by that same reduction factor, a Section would be 50 feet on a side, and an acre would be about 2 feet square.

(This shrunken Township can also tell us something about the difficulty of surveying: side-line officials can "survey" the length of a football field with just 10 overendings of the first-down chain. But a surveyor has to measure 6 miles with a 66-foot chain, requiring 480 overendings. This is the equivalent of measuring a football field with a 7-inch string!)

But the 1855 edition of the *Manual* called for even more precision. Standard Parallels were not to be simply lines that resulted from surveying Townships. They were to be surveyed *before* the Township exteriors, to serve as a control lines for them. The manual set out an idealized procedure for accomplishing this.[14]

Surveying Township Exteriors

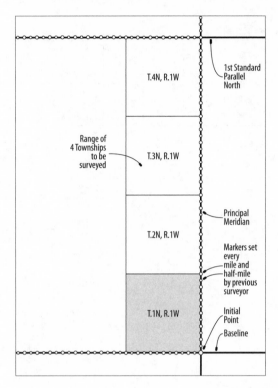

FIG 8.14

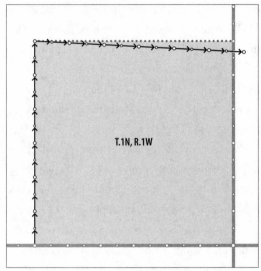

FIG 8.15

With the Baseline, Principal Meridian, and First Standard Parallel in place, you the surveyor are ready to lay out the Township exteriors according to the dictates of the *Manual of Instructions*. Your job is to lay out the exterior lines of a Range of four Townships, up from the Baseline to the First Standard Parallel North. You start at the Initial Point and trace the Baseline west and locate the 6-mile marker left by the Baseline surveying team. In figure 8.14, the full-mile markers set by the previous surveying team are shown as the bigger circles, the half-mile markers as the smaller circles.

In figure 8.15, you start at the southwestern corner of Township 1 and begin projecting a Meridian line to the north, setting mile and half-mile markers as you go. When you reach the 6-mile point of your line, you stop and turn east, back toward the Principal Meridian. On the Principal Meridian there already is a 6-mile marker, placed there by an earlier survey team. Your job is to *close your line* on that marker—have a line run straight from your 6-mile marker to that one.

In the Seven Ranges, the biggest errors occurred because the surveyors were neither required to close their east-west lines on lines surveyed previously, nor instructed on how to do so. The 1855 edition of the *Manual* both requires closure of the lines and sets out a procedure for accomplishing it—a method surveyors call **random and true**.

As you project the best eastward line you can manage, you don't blaze trees or set permanent markers. You leave only erasable signs to mark your path and set only temporary markers. When you have finished your 6-mile line and have placed your final temporary marker, you look around for the 6-mile monument set up on the Principal Meridian by the earlier survey team.

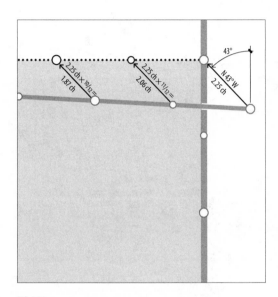

FIG 8.16

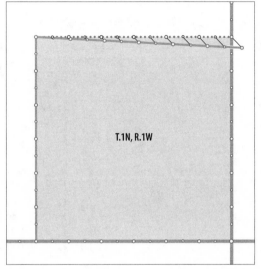

FIG 8.17

When you find that monument—in the close-up view of figure 8.16—you carefully measure the distance between it and your temporary marker, and with your surveyor's compass you calculate the bearing from your marker to it. You find your 6-mile line overshot the Principal Meridian's monument to the southeast: your temporary marker is 2 chains 25 links away from the monument, and the line to the monument bears 43° west of north (N43°W).

That distance and that bearing are all you need to shift your twelve temporary markers off your *random* line onto a **true** line. You move all the markers in the same direction, but decrease the length of each move proportionally, in the manner shown in figure 8.16.

Now pull back from the close-up to the whole-Township view of figure 8.17. You repeat the proportional-shifting process with all your temporary markers, and in so doing you *monument*—this time permanently—the *true* line that closes the corners of T.1N, R.1W.

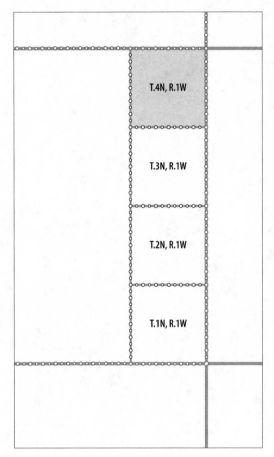

FIG 8.18

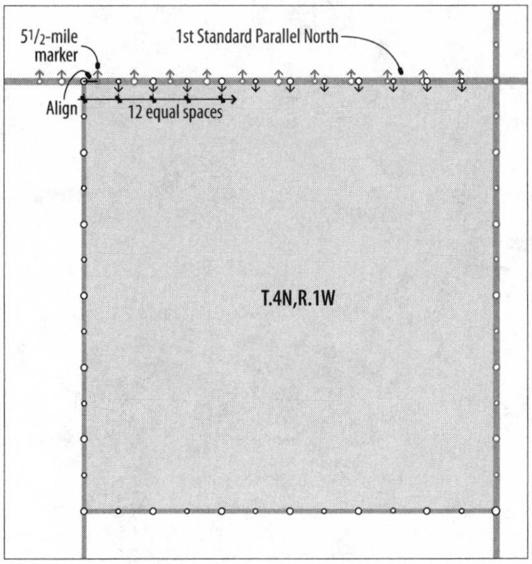

FIG 8.19

Now pull back again, to the full four-Townships view in figure 8.18. You have arrived back at the monument you set on your meridian line. Now you set off north again, setting mile and half-mile markers as before. After 6 miles, you repeat the "random-and-true" process for T.2N and repeat it again for T.3N. Township 4 North is the exception in the process because its northern boundary is on the First Standard Parallel, already surveyed and monumented every mile and half mile.

In close again, in figure 8.19: As you run T.4N's western border up to the Standard Parallel, you don't want to close your line on the Parallel's 6-mile monument: the markers on the Parallel are to be used to run the Township on the *north* side of that line. But even if you don't want to close your line on that monument, you still want your line to close on the Parallel.

So you go over to the 5½-mile marker on the Parallel and project a line back to the Parallel's 6-mile marker to find where the Parallel crosses your line. At that point you set the monument marking T.4N's northwestern corner—carefully differentiating it from the monument for T.5N's southwestern corner.

To place the mile and a half-mile monuments for along T.4N's northern border, you set out along the Parallel, placing temporary markers, including the final 6-mile marker. When you set that final marker, you carefully measure the distance from it to the monument already set on the Principal Meridian (due to convergence, you should have overshot it). Using that distance, you repeat the "¹¹⁄₁₂, ¹⁰⁄₁₂ . . ." proportioning calculation and move your temporary markers to their permanent positions.

Having "run the exteriors" of the four Townships in Range 1 West, you now go back to the Baseline, and repeat the process for Range 2 West—this time closing your east-west lines on the corners that your crew set.

You'd in fact repeat the process for all the Townships north of the Baseline, but south of the Baseline, you'd work differently.

You might think that you would head south from the Baseline, but in fact the mandated procedure was to go to the First Parallel South and project the Township borders toward the north. The reason is obvious when you imagine a surveyor in the field. If the compass man looks south to project his lines, he is sighting toward the sun. Sighting toward the north puts the sun in the pole man's eyes—bad for him, but by the same token, the sun makes his pole easier for the compass man to see through his sights.

In the Rectangular Survey, there's a reason for just about everything.

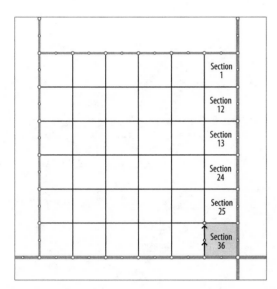

FIG 8.20

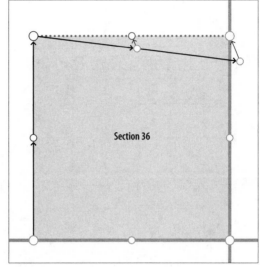

FIG 8.21

The method of subdividing Townships likewise grew out of field experience—and it too involved an "up, over, and back" procedure that provided handy checkpoints for surveyors. Here is a shortened version of the method laid out in the 1855 edition of the *Manual of Instructions*.

Figure 8.20 shows the first Township you surveyed, T.1N, R.1W, directly against the Principal Meridian and Baseline. You begin at the southeastern corner of the Township and head west on the southern boundary. At the first mile marker you set up your equipment and project a line 1 mile north (setting both half-mile and mile monuments on the line, of course).

Figure 8.21 zooms us in on Section 36. At the end of your line to the west, you project a random line eastward toward the mile marker already set at the northeastern corner of Section 36, setting a temporary marker at the half-mile point. When you have run a mile, you seek out that mile marker, measure the distance and bearing to it, and shift your markers to the true line. The process here is simple: you move the half-mile marker in the same direction, but just half as far.

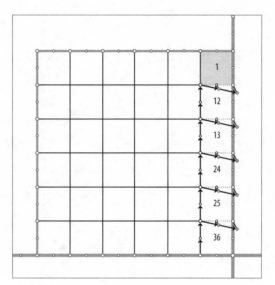

FIG 8.22

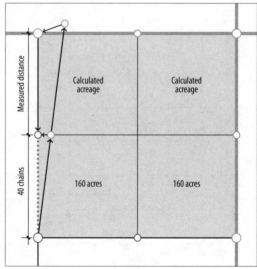

FIG 8.23

Figure 8.22 zooms back out to the whole Township. Once back at the end of your first north-south line, you project a second line 1 mile north, and repeat the over-and-back procedure. In fact, you repeat the that procedure four times, until you get to Section 1 at the northern edge of the Township. At this point the procedure varies. As a trained surveyor, you know that "excesses and deficiencies" are to be thrown into the most northerly and westerly Sections of the Township. Here is how you accomplish that.

Figure 8.23 shows Township 1: from its southwestern corner, you go north a half mile and place a temporary monument. Then you project a random line toward the marker previously set on the Township exterior. You use the same procedure as before to get your temporary markers onto the true north-south line except that, instead of setting the half-mile marker at the line's center point, you carefully locate the marker exactly 40 chains—a half mile—from the southern Section corner you set. You then carefully measure the distance from that half-mile marker back to the Township exterior. That distance will almost certainly be something other than 40 chains, but you carefully record it in your field notes.

When the time comes for you to draw up your plat of the Township, you divide Section 1 into 4 Quarter Sections. The two on the south—which you made exactly 40 chains tall— you record as containing 160 acres each. You then calculate the areas of the two Quarter Sections on the northern edge of the Township. Suppose the distance you measured was 38.60 chains: multiply that by 40 chains, and you get 1,544 square chains. So each of the two Quarter Sections at the northern edge of the Township contains—legally and for all time—154.4 acres.

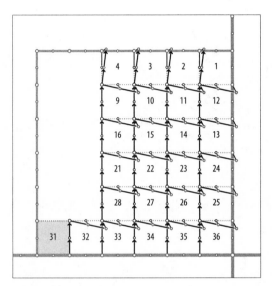

FIG 8.24

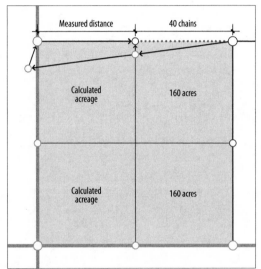

FIG 8.25

To continue your subdividing, you go back to the southern edge of the Township and repeat the "up, over, and back" procedure as shown in figure 8.24—the difference being that, this time, when you aim your random lines east, you'll be aiming at a marker *you yourself* set. At the base of the fifth stack, however, your procedure changes, so as to "throw excesses and deficiencies" into the westernmost Half Sections.

Figure 8.25 shows us Section 31. At the corner between Township 31 and 32, you run a line north 1 mile, set a monument, project a random line east, and calculate a true line back. Here the change occurs: you now run a random-and-true line to the *west*, toward the existing Township-exterior marker. As with the Sections on the northern edge of the Township, you make the first "half" of your line exactly 40 chains long, and carefully measure the remainder—which you will later use to calculate the areas of the westernmost Quarter Sections.

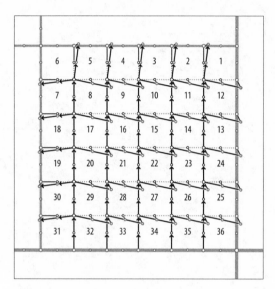

FIG 8.26

You continue this process northward, as shown in figure 8.26, until you reach Section 6. Here, only the southeastern Quarter Section gets its full 160 acres: the other three—abutting the northern and western borders—get calculated acreages.

1881 AND 1894: THE FINAL STANDARDS
FOR RECTANGULAR SURVEYING

By the 1880s, the pace of surveying had quickened considerably, many more surveyors were in the field, and the GLO felt a need for surer containment of errors. The 1881 edition of the *Manual* instituted what came to be known as the *quadrangle system* for handling convergence—the system in use today.

First, the *Manual* called for Standard Parallels every 24 miles, both north and south of the Baseline. To complement the Standard Parallels would be a system of Auxiliary Meridians (renamed Guide Meridians in 1890) every 24 miles east and west of the Principal Meridian—thus the name *quadrangle* for a square 4 Townships by 4 Townships. Figure 8.27 shows the new, tighter pattern, and figure 8.28 shows a Standard Parallel doing its job of correcting for the convergence of meridians in northern Montana.

State Surveyors General had been using Guide Meridians in a rough-and-ready manner for some time. When called upon to survey land not adjacent to areas previously surveyed, they would often project a Standard Parallel out into the area of new survey and then, at some convenient 6-mile marker, project a meridian northward. Later surveyors could then project their Townships' borders from these lines. The 1881 *Manual* merely systematized that technique. (The growth of the Montana survey, which you'll see in chapter 10, shows both the informal and systematized use of Guide Meridians.)

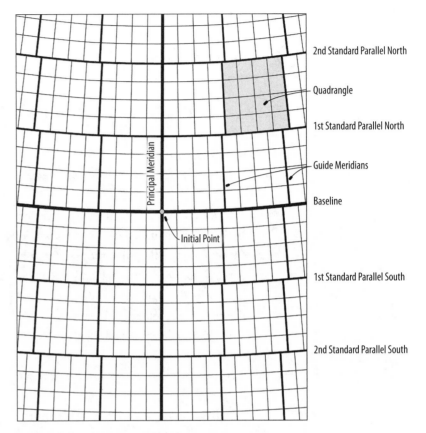

2nd Standard Parallel North

Quadrangle

1st Standard Parallel North

Guide Meridians

Baseline

Principal Meridian

Initial Point

1st Standard Parallel South

2nd Standard Parallel South

FIG 8.27

FIG 8.28

The GLO recognized how important these controlling Baselines and Meridians were. The 1894 edition of the *Manual of Instructions* called for four distinct classes of lines:

- *Baselines and Standard Parallels;*
- *Principal and Guide Meridians;*
- *Township exteriors; and*
- *Township subdivision lines and meander lines.*[15]

Not only would a higher per-mile rate be paid for the more important lines, but surveyors would have to demonstrate their qualifications to run such lines. (From the "Task at Hand" sections, you know why the GLO ranked Baselines and Parallels at the very top of the hierarchy—both for importance and degree of difficulty. "Meander lines," by the way, are how surveyors render the curves of watercourses as a string of measured facets.)

In subsequent years the GLO and then the Bureau of Land Management would mandate ever more precise instruments and call for more permanent corner monuments. Nonetheless, with the publication of the 1894 *Manual of Instructions*, the methods of the Rectangular Survey assumed the form they have today.

NINE

The Survey Is Extended across the Public Domain

WILLIAM AUSTIN BURT AND THE COMPASS
THAT SURVEYED THE WEST

In addition to specifying how to lay out Townships and Sections, the 1855 edition of the *Manual of Instructions for Deputy Surveyors* required one more thing of the nation's Deputy Surveyors—that in all their work they were to find north by means of an instrument called Burt's Improved Solar Compass.[1] If there were a Hall of Fame for American surveyors, the Babe Ruth of surveyors would be William Austin Burt, the inventor of the solar compass. Before Burt a surveyor might be able to calculate the true meridian by careful observations of the North Star; but the only way he could "refind" the meridian—during the day, when he would be doing his actual work—was to apply his very precise calculations to a very imprecise basis, the "north" indicated by his magnetic compass. Burt's solar compass allowed surveyors to find true north afresh, every time they needed to point their sights, because they could find true north merely by sighting on the sun.

Once Burt's solar compass became available, both surveyors and General Land Office officials recognized its usefulness. The 1855 *Manual of Instructions* required surveyors to use "the solar" to determine their magnetic compasses' declinations from true north. By the time of the 1890 edition of the *Manual*, deputies were told to dispense with their magnetic compasses altogether and rely exclusively on a solar compass to find the true meridian.

That date of 1890 is significant, because it was on the basis of the 1890 census that the historian Frederick Jackson Turner, in his speech-turned-essay "The Frontier in American History," advanced the notion that though the prospect of a frontier of empty lands had in large measure formed the American character, by 1890 population densities were such that a frontier no longer existed. Jackson's implicit thesis—that the "Frontier Era" began in 1607 and ended in 1890—became the presumed national ideology, a story of such force that, despite a century of attack by historians, to this day it is a factor in Americans' self-definition as a people.

The plain fact, though, is that while the frontier might have been closed by 1890, with farmers and ranchers scattered widely across the American West, only about half the Public Domain had by then been surveyed into rectangles to which settlers could assert secure rights of title. As you'll see in chapter 10, when we look at Montana as an exemplar, the Rectangular Survey was not a continuous mat of surveyed sections. By 1890 the survey had grown more in the manner of tendrils, reaching out from early settlements to "grid up" those places in which settlers had expressed an interest. Between those tendrils of settlement lay vast unsurveyed stretches whose value would become apparent (or realizable) only in the twentieth century.

Only by about 1940 could the West be said to have been "gridded": the spaces between the tendrils of settlement filled in and the landscape marked out in Townships and Sections. Much of that infill work was done by an adaptation of Burt's compass to the more precise surveyors' transit that came into use at the end of the nineteenth century.

So it does not go too far to say that Burt's solar compass and its descendants were the instruments that surveyed the American West. All of the grid lines that you see from an airplane window, in a great fan outward from Iowa, Wisconsin, and Michigan, were scribed on the land by surveyors using either solar compasses or transits directed by solar compasses.

Among surveyors, Burt is accorded the name William Austin Burt, to differentiate him from William, the eldest of his three surveying sons,[2] and to imbue his name with a grandiloquence commensurate with his importance. He is also called among surveyors Judge Burt, to honor his election as a magistrate in Buffalo.[3] He had migrated there from Massachusetts, the place of his birth in 1792. His forebears had been seamen, which might account in part for his interest in the movement of the sun: one of the most ancient aids to determining a ship's direction was to note the bearing of the sun as it rose and set at charted positions on the horizon of the sea. Burt might never have put that knowledge to practical use, however, had he not been the very *type* of the Yankee tinkerer. In Buffalo he made his living as a millwright; and later, in his thirties, he would invent one of the several prototypes that led to the realization of a practical typewriter.[4]

Despite his success in Buffalo, Burt, still in his twenties, succumbed to the wanderlust that so infected young men in Jacksonian America. He abandoned his prospects as a man of affairs and set out for the Northwest Country, determined to be a surveyor in the just-opening Michigan Territory. He stopped at the office of the Surveyor General in Chillicothe, Ohio, but was told that the Deputy Surveyor contracts for Michigan had all been let.[5] Undeterred, Burt settled himself north of Detroit and again found himself involved in local politics: it was the press of correspondence with constituents that impelled him to invent his typewriter.

During his time around Detroit, from his settlement there in 1822 through the early 1830s, he honed his skills as a surveyor, making the very good living a surveyor could expect in those days. Finally, in 1833, at the age of forty-one, Burt obtained a contract to serve as a Deputy Surveyor. His assignment was to help extend the grid of Townships up Michigan's Lower Peninsula toward Lake Superior.[6]

Once in the field, though, he found that the needle on his surveyor's compass was giving him very erratic indications of north. Not until 1844 would the first outcroppings of iron ore be found in Michigan's Upper Peninsula, but outliers of those deposits in the Lower Peninsula were having their effect.[7] No matter how carefully surveyors would find north by the stars at night and calculate the magnetic declination, when they moved their compasses any distance, they'd find their needles pointing in any number of directions. The cause of the problem was not a mystery: Burt himself found an iron ore outcropping.[8] But even with the problem identified, there was no way of rendering magnetic needles a reliable, portable indicator of direction.

Burt's innovation was to discard the magnetic compass and rely, for his true north, on a phenomenon that varied at least as much as magnetic attraction but had variations whose changes could be predicted with pinpoint accuracy, the movement of the sun through the sky. As a student of solar navigation, he must have sensed intuitively that it would be possible to compute the position of the sun in the sky for every moment of the day and for every parallel of latitude and meridian of longitude. But the printed tables of such calculations would be a heavy (and perishable) volume no surveyor would want to carry. He would also have known that the position of the sun could be calculated by multifactor formulas of spherical trigonometry, but he had seen the mathematical abilities of the typical surveyor—perhaps even recognized his own limitations in that regard— and knew that calculating the sun's position by formula would be as impractical as reading that position from published tables.

Burt realized that it might be possible to track the sun's motions through the sky by mechanical means—an instrument that could be adjusted with such precision that, when a set of sights was pointed at the sun, a mechanically linked pair of pointers would be aligned to true north.

Pointing the Solar Compass at the Sun

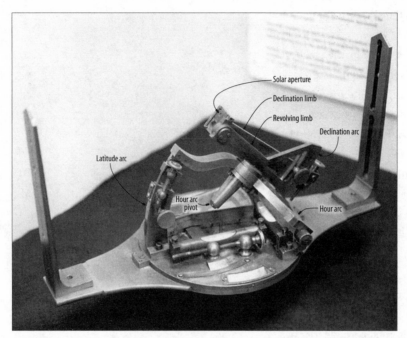

FIG 9.01

To understand how Burt's Improved Solar Compass works, let me first give you a tour of its features.[9] The function of all the mechanisms will become clear only at the end of this section, but bear with me. Burt's solar compass looks very much like a surveyor's compass, with the solar mechanism substituted for the big magnetic compass case. Figures 9.01 and 9.02 show you two views of a solar compass used in the survey of Michigan.

The first thing you should note is the *solar aperture* and its matching *solar reflector*. It's best that the surveyor not have to look directly at the sun, so in Burt's compass the sun's rays pass through an aperture that focuses the rays to a point on a reflective plate. The effect of such a solar sight is that a straight line is "drawn" from the reflector, through the aperture, and on to the center of the sun.

The rest of the solar compass consists of devices to rotate this solar sight through three axes of movement so as to align it with the sun. The aperture and reflector are at opposite ends of the *declination limb*. The limb is hinged to rotate upward and downward to account for the height of the sun in the sky on a particular day. The declination apparatus sits on an *hour-arc pivot*, allowing it to rotate from side to side to account for the time of day. The hour-arc pivot is in turn mounted on the *latitude limb*, which pivots upward and downward to account for the latitude where the surveyor stands.

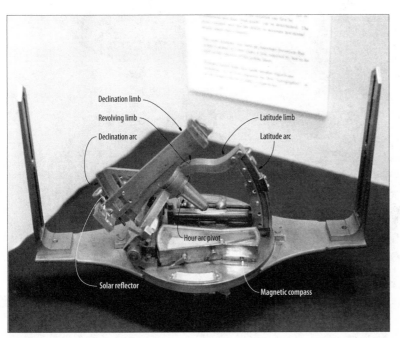

FIG 9.02

In the diagrams to follow, I'm going to be using the schematized version of Burt's compass shown in figure 9.03. I've drawn its parts more simply and thickly so that they'll be visible when reduced in size. You can analogize its components, one to one, with those in figure 9.02.

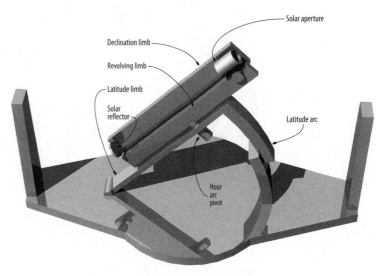

FIG 9.03

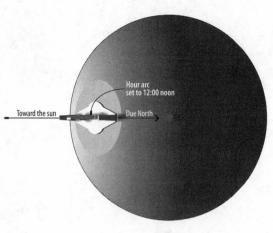

Hour arc
set to 12:00 noon

Toward the sun   Due North

FIG 9.05

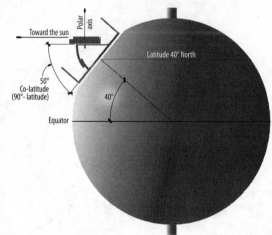

Polar
axis

Toward the sun

Latitude 40° North

50°
Co-latitude
(90°- latitude)

40°

Equator

FIG 9.06

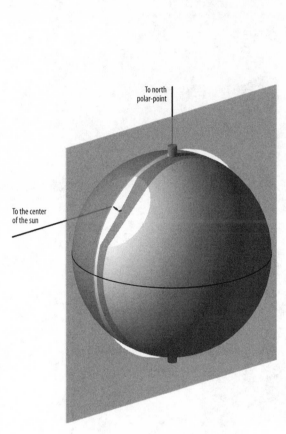

To north
polar-point

To the center
of the sun

FIG 9.04

First, cast your mind back
to the image shown again in
figure 9.04, and think again of
"solar noon" as a plane, constantly aligned with the sun,
under which points on the earth
rotate. We will again be standing on that horizon plane at
latitude 40° north.

The solar compass must first
be plumbed up on its tripod
so that its base is perpendicular to the center of the earth.
(I'm showing this condition in
figures 9.05 and 9.06 by setting
the base directly on our horizon
plane.) You then pivot the latitude limb upward through the
number of degrees equal to the
co-latitude of the place where
you stand. The **co-latitude** is
your latitude subtracted from
90°, so our co-latitude is 90°
minus 40°, or 50°.

The line through the solar
sight is now parallel to the
equator. With the parts of the
solar compass adjusted to this
position, when the compass
rotates under the solar noon
plane, the rays of the sun will
pass through the aperture
to the center of the reflector plate—and the sights of
the compass will be aligned
north–south.

If you look at figure 9.06 and
imagine the compass rotating
with the earth, you can see that
at any time other than noon,
the solar sights are *not* pointing at the sun's center, and so
the sun's rays don't come to a
focus at the center of the reflective plate.

That's how the solar sight
points at the sun at noon. How
do you adjust the solar compass
so that its sight points at the sun
at times other than noon?

The key is in the line marked
Polar Axis in the figure 9.05. It's
called the polar axis because
when the compass is properly
set up, the axis of the spindle
is parallel to the polar axis of
the earth.

The hour arc is calibrated in hours and minutes, and so you use it to pivot the limb and its solar sight to the "3:20 p.m." position, as shown in figure 9.08A. In effect you are rotating the solar sights "back" to compensate for the earth's rotation. With the solar sights in that position, the sun will again pass through the aperture to the center of the reflector—but only when the base of the compass is aligned north–south.

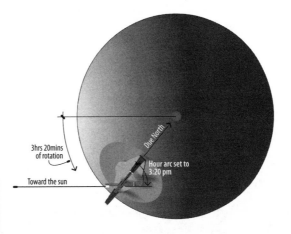
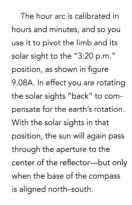

FIG 9.08A

Figure 9.08B shows the compass at 3:20 p.m., set up correctly for latitude and time, but the base and the sights are aligned in a direction *other than* north–south. The solar sights are now pointing to a place other than the center of the sun, so your compass's sighting vanes aren't "looking" north–south.

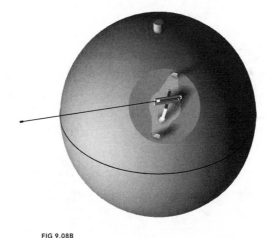

FIG 9.08B

FIG 9.07

Figure 9.07 shows the actual spindle. The limb carrying the solar sight can rotate around that spindle, and how much it rotates away from the "noon" position (shown in figure 9.06) is measured by the marks you can just make out on the hour arc.

Suppose that after finding north at noon, you do some surveying work, but then at 3:20 in the afternoon your work requires you to find north again. (You will, of course, have found the time of solar noon where you are working, and synchronized your timepiece. Because you want the true noon—not the "averaged" noon of Greenwich—you don't do an equation of time.)

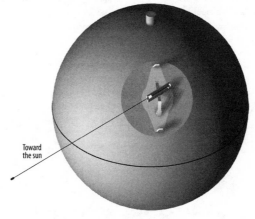

FIG 9.08C

But you know your solar compass is set up correctly, so you gently rotate the whole compass around, atop its tripod, until the sun's image finally falls at the center of the reflector plate, as it would in figure 9.08C. When that happens, you know that the sighting vanes of your compass are aligned north–south. You now clamp the compass in place and make your sighting.

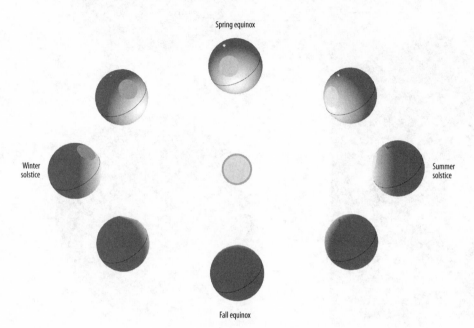

Spring equinox

Winter
solstice

Summer
solstice

Fall equinox

FIG 9.09

We've seen how the solar compass handles latitude, and how it handles time of day. There's one more factor it has to deal with, and that is the time of year. In the diagrams I've shown so far, a line from the sun's center passes through the equator, but that happens at only two moments in the entire solar year.

Those moments are, of course, the fall and spring equinoxes (around September 23 and March 21). Use the image in figure 9.09 to remind yourself how the earth moves in its orbit. The earth is tilted about 23.5 degrees in its orbit, with the result that the planet "leans in" at the summer solstice (June 21) and "leans back" at the winter solstice (December 22).

Figure 9.10 shows this more graphically: this is what the sun would "see" if it looked at the earth on each of those days when it's noon on our 40th-parallel horizon plane.

Spring equinox   Summer solstice   Fall eqinox   Winter solstice   Spring equinox

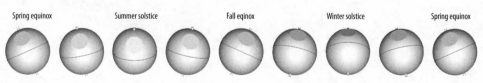

FIG 9.10

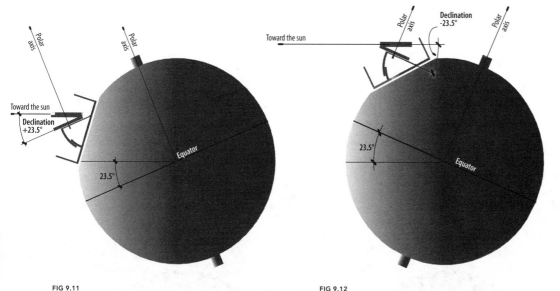

FIG 9.11

FIG 9.12

In figure 9.11 we see the earth at the summer solstice, tilted "forward" the full 23.5 degrees. On that day the sun will be 23.5 degrees "above" the equator—which surveyors express by saying that the **declination of the sun** is 23.5 degrees. The solar compass handles this with its **declination arc**: we tilt the limb carrying the solar sight upward by 23.5 degrees.

When the earth is "leaning back" (that half of the year between September 23 and March 21), the sun is "below" the equator and thus has a negative declination (see figure 9.12). Burt's compass has a clever feature for handling this (it's one of the innovations that made it Burt's *Improved* Solar Compass). The entire declination assembly can be spun 360 degrees so that its hinge can be

toward the sun or away from it. To make this work, the plates at the ends of the declination limb contain both an aperture and a reflector, one over the other, but they are flipped: the plate at the hinge end has an aperture on top and a reflector below, while the opposing plate has the aperture on the bottom, the reflector on top. Figure 9.13 shows these paired apertures and reflector plates.

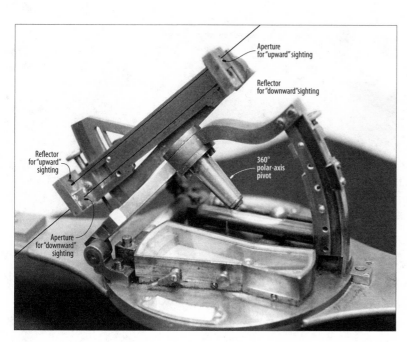

FIG 9.13

For negative declinations like the one shown in figure 9.12, you rotate the declination assembly so the hinge is toward the sun, and the sun's ray passes through the upper aperture to the upper reflector. For positive declinations, when the hinge is away from the sun, the sun's ray is caught by the bottom pair. Clever man, that Judge Burt.

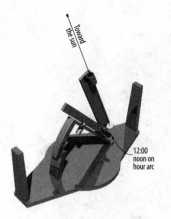

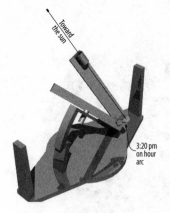

Toward the sun

12:00 noon on hour arc

FIG 9.14A

Toward the sun

3:20 pm on hour arc

FIG 9.15A

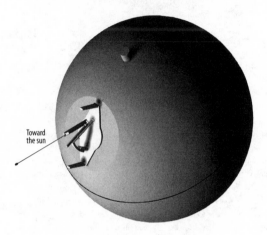

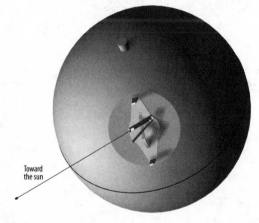

Toward the sun

FIG 9.14B

Toward the sun

FIG 9.15B

One last time. The schematic compasses in figures 9.14A and B are set up for noon at the summer solstice. It's noon on our horizon plane: the sights of the compass point at the center of the sun, and the vanes of the compass are aligned north–south.

In figures 9.15A and B, the schematic compass is still set up for the summer solstice, but the sighting limb has been swung around to the "3 hours, 20 minutes" position. It's 3:20 p.m. on the horizon plane: the sights of the compass again point at the center of the sun, and the vanes of the compass are again aligned north–south.

Burt's first compass looked very different from his "improved" model. He began tinkering with the idea in 1835, at the age of forty-three, and had America's leading instrument maker, William J. Young of Philadelphia, construct a working model to his specifications. Burt meanwhile submitted similar plans to the U.S. Patent Office, and in 1836 it granted him Patent no. 9428X, good for fourteen years, for the compass shown in figure 9.16.[10] The patent drawing, like Burt's original compass, has about it that air which the first versions of inventions often have: it is an almost direct translation into metal of the principles that govern the sun's movements.

It was a compass like this that Burt took with him to survey the area west of Milwaukee and then, through 1839, to survey much of Iowa. Those years in the field showed him how to make his compass more durable and more versatile (if perhaps less elegant), and by 1840 he had the instrument in its final, improved form.[11]

It's hard for us to think in these terms, but in its day Burt's solar compass was cutting-edge technology. For it to work accurately, it had be fabricated to the smallest tolerances then achievable. The brass parts had to be cast, then machined and tested, then machined again—or, if impossible to machine into alignment, melted down and recast. When Young's firm began to manufacture the improved compass, the process was laborious in the extreme. In 1850 Young had five men working on compass production, and they were able to complete only six working instruments that year. The compasses were naturally expensive: $130 each in 1854,[12] the equivalent of thousands today.

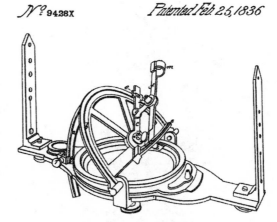

FIG 9.16   U.S. Government Printing Office.

Burt was to get $10.00 in royalties from Young for each compass sold, but he often waived his fee to make the price more affordable. Later in life, in need of money, he would come to regret his early generosity. In a story so often repeated, Burt—and later his three surveyor sons—petitioned both the Michigan legislature and Congress for financial recognition of the compass's centrality to the national surveying project, but to no avail.

Burt did get recognition from another quarter. In 1850, at the age of 58, he retired from surveying and the next year sailed for London to display his compass at the Great Exhibition there. Under the glass vaults of the Crystal Palace, Burt's Improved Solar Compass received great attention and was awarded one of the prizes for innovative technology. He returned home, had at least the comfort of seeing three sons well established as surveyors of the Public Domain, and died on August 18, 1858.[13]

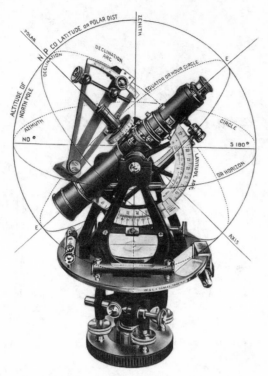

FIG 9.17   Used by permission of Gurley Precision Instruments.

I mentioned earlier that from 1855 onward, Deputy Surveyors were required to use the solar compass to run their lines. As the nineteenth century drew to a close, surveyors gradually shifted from the open sights of the surveyor's compass to the telescopic sights of the now-familiar surveyor's transit. In response, Burt's solar compass became a *solar attachment* to the transit, like the one shown in figure 9.17. It was this combination instrument that the 1890 edition of the *Manual of Instructions* mandated for Deputy Surveyors, and it was this instrument that surveyors used up until the middle of the twentieth century.

You will remember how corners were monumented in wooded country: a cairn of rocks or a post hewn from a tree trunk would be set in the ground at the corner, with nearby trees marked as "witnesses," and the lines between corners marked by a string of blazes axed on trees. All this was done, of course, so that a settler might find the corners, and thus the borders, of the parcel he wished to make his own.

By the time surveyors began working in Indiana, and even more so in Illinois, they were moving out of virgin forests into grassland. Here the trees occurred mostly in clusters, sometimes disappearing altogether except in river bottoms. The surveyor's only recourse for a marker was to dig up the soil and pile it in a mound over the corner. The 1855 edition of the *Manual of Instructions*, as we have seen, compiled insights from the previous "regional" manuals, and it gave precise directions on how to build a corner mound. Figure 9.18 combines some images from later editions of the *Manual*.

The mounds themselves were to be pyramid shaped, and quite substantial: 5 feet square at the base and 2½ feet tall at the apex. If possible, the surveyor was also to take 4 feet of a tree limb or small trunk and use an axe to remove the bark off the final foot of one end, to form four flat faces. The Township and Section numbers could then be indicated by hacking V-shaped notches into the corners between faces. The surveyor would set the post upright in the ground at the corner point, and then build the mound around it, leaving the carved portion sticking out of the mound's apex. The final step was to cover the mound with strips of sod, so the grass would reroot and prevent erosion, and then to dig a 6-inch ditch around the base of the mound.[14]

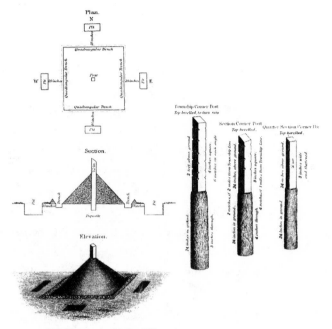

FIG 9.18   U.S. Government Printing Office.

The surveyors would naturally have to dig up enough earth to build a mound, and the *Manual* told them how they were to turn this necessary task to their advantage. They were not to dig up the earth indiscriminately but get it from compact rectangular pits dug parallel to the four sides of the pyramid mound and 18 inches away from it.

A pretty laborious process—one that we can imagine even the most conscientious surveyors shortcutting. Nonetheless, in later editions of the *Manual*, Washington bureaucrats made the process even more complex, specifying distinctive configurations of mound and pit for every conceivable condition—all this in plain contradiction to reports from the field that the mounds weren't proving very durable. If the sod strips failed to root in the mound, they'd simply slide off in the first rains, leaving the exposed mound to melt like a sugar cube.

Then there were the buffalo. If a herd in its passage didn't trample the mounds flat, or wrench off the sod covers by their grazing, there was the problem of scratching. Buffalo groom themselves whenever they can by rubbing their flanks up against anything they can find—a cottonwood in a river bottom, a steep bank, even a mud wallow. You can imagine how delightful a buffalo would find the sharp corners of a surveyor's post, set at just the height of his rib cage.

Curiously, the part of the surveyors' work that sometimes proved most durable was neither the mound nor the post but the pits. You doubtless know that the tan color so emblematic of the Great Plains indicates not that their prairie grasses are dead but that they are dormant. With any significant rainfall, the grasses will, almost overnight, become green and stay that way until the water is used up. In the pits dug by surveyors, water could collect, and the grasses that took root there could stay green longer, grow taller. Imagine looking out from the high seat of a covered wagon, over a waving sea of golden grass, and spotting in the distance four perpendicular tufts of green.

It was the fragility of earthen mounds that led the General Land Office authorities, even in 1855, to advise surveyors to make more enduring markers. One expedient was to dig a pit near the post and fill it with the charcoal and ashes from the previous night's campfire. Anyone later looking for the exact location of the corner, and getting a general idea of its whereabouts from anomalous grass or a hump in the terrain—or from measuring off from some other corner—could explore with a shovel until he hit a patch of black within the brown of the earth.

This notion of "zeroing in" on a corner's precise location from imprecise indications points up three important concepts in surveying: *perpetuation*, *recovery*, and *retracement*.

It was never intended that the markers put in the earth by government surveyors would be the monuments that would record the corners for all time. It was assumed instead that settlers would find it in their interest to continually *perpetuate* the locations of the corners through their own actions. In farming country, fence posts were the most common means of perpetuating corners and the lines the corners were intended to mark. If the corner established by a government surveyor marked the boundary between two farmers' holdings, it would be in the interest of both of them to set the first fence line by joint agreement. You can imagine an unbroken chain of perpetuation from, say, 1840: first a rough-hewn post in a wooden fence; then, after 1870, a post driven into the ground to hold strands of barbed wire; then in the 1920s a factory-made steel post to support more barbed wire; then finally, perhaps, a brass plate on a concrete plug set flush with the lawn of a suburban house.

Recovery is obviously the first necessary step in the chain of perpetuation. The instant our imagined settler found that deposit of charcoal and recovered the corner, you can be certain that he perpetuated that corner with something more durable and more visible. But if, after long searching, the corner simply couldn't be found, his next resort would have been *retracement*.

You'll remember that both surveyors' plats and field notes were archived, either at the federal Land Office or (later) at the appropriate state or county bureau of records. One of the jobs of a private surveyor in a settled area was that of claims arbitrator, and those records were part of the "evidence" the surveyor used to render his decision. Using the plats and the field notes, he would try, to the very best of his ability, to retrace the path the original government surveyor had taken, from a recovered corner to the missing corner. Here's the real reason for noting all those trees, those streambeds and ridges, the precise distances between them, and most especially the magnetic variation of the compass needle: all of those notations would aid the new surveyor in reenacting the work of his predecessor.

Our farmer could, of course, challenge the surveyor's location of the corner, but only on the basis that the surveyor hadn't reenacted the running of the line correctly: by the provisions of the Land Acts passed by Congress, a corner is, for all time, at the place where the original surveyor said he put it. The same acts tell the surveyor how to make his judgment in those cases where the plat and notes are missing or ambiguous—often by the aliquot principle that we saw earlier: establishing a corner by halving the distance between recovered corners.

Given the actions of buffalo and wildfires and erosion, and the surreptitious moving of corner markers, you can see how the procedures of the Rectangular Survey drastically simplified the litigation of land claims. Even so, the most unchallengeable basis for a claim would be the actual recovery of a government corner marker in its initial position. So in the 1930 edition of the *Manual of Instructions*, the GLO required its surveyors (civil servants since 1910, you'll remember) to mark their corners with government-specified iron pipe. The pipes were 3 feet long, and their

diameters were 3 inches for Township corners, 2 inches for Section corners, and 1 inch for Quarter-Section corners. They came with one end split lengthwise for a distance of 4 or 5 inches, the idea being that the surveyor could take however many pipes he'd need for the next few days' surveying, stick the split ends of the pipes in his campfire, and then, when hot, bend those halves outward to form flanges (the reason for doing the bending in the field being that the unflanged pipes would stack more compactly in a wagon or truck).

At a corner point, the surveyor would dig a hole wide enough to accommodate the flanges and about 2 feet deep. He'd set the pipe in the ground, plumb it up, and fill the hole back in, tamping it firmly. Then he'd take some dry-mix concrete from the truck, mix it with water, and fill the pipe with it. Finally, he would take one of the brass caps supplied by the district office—the ones with a warning against tampering with government property engraved around its circumference. With a mallet the surveyor would drive the cap home on top of the pipe, and then with a set of steel dies he'd stamp, in the center of the cap, the Township and/or Sections whose common corner the pipe marked.

Of all the markers used by government surveyors over the past two centuries, these pipes are about the only ones hikers today might come across. There's something undeniably thrilling about finding a government survey marker out in the middle of nowhere. Such a marker, when encountered in wilderness, has such an air of *human presence* about it that the thought of disturbing it seems a violation of something profound. Of course, survey markers *are* disturbed, often for no better reason than "for the hell of it." But in the settlement period, moving—or *removing*—a survey marker could have lasting consequences.

The classic tale of what happens to those who tamper with survey markers is O. E. Rolvaag's *Giants in the Earth*, the epic of Norwegian immigrants settling on the Great Plains.[15] Rolvaag (whose name is imperfectly rendered as "RUHRL-vogg"), himself a Norwegian, wrote the book in his native language. It was published in Norway in two parts, in 1924 and 1925, and the author's English translation came out in America in 1927, both versions to great acclaim.

In the story a group of Norwegian families, led by Per Hansa and his wife, Beret (the "t" is pronounced), move by wagon from Minnesota to the Dakota Territory. They set up their farmsteads on adjoining Quarter Sections, erect their first houses, and plow under enough grassy turf to get in their first crops. But at first they locate only the boundaries between their parcels, by means of the few corner markers they have in common. They've not yet found the stakes marking the outer corners of their holdings. And they've not yet gone to the land office to file their claims: that is a journey of several days, and the time can't be spared until after the harvest, if then.

One day, though, Per Hansa finds a stake. He leans down to inspect it and finds it inscribed "O'Hara." The land his group is farming has already been claimed!

Days later, after much soul-searching, Per Hansa pulls that stake, and others, out of the ground and hides them in his barn. Later he burns them as kindling—but not until after his wife has found the stakes and watched, unseen, as he chopped them up.

When Irish families—the O'Haras among them—arrive on the scene and attempt to occupy the land, the Norwegians stand their ground, all of them (except Per Hansa and Beret) convinced that these latecomers are claim jumping. There being no evidence, the Irish move on, set up their own community a little way farther on, and the crisis passes. For the rest of a very long book, the Norwegians endure the hardships and experience the joys of making new lives, but Per Hansa meets only with success upon success. This frightens Beret, for in Norwegian folklore, obliterating a boundary marker is not only a crime but a sin: Per Hansa's triumphs can only mean that the gods are setting him up for a fall. Convinced of this, she sinks from anxiety to real madness, and in the end Per Hansa . . .

I won't reveal the outcome to you, but do take from this synopsis of *Giants in the Earth* a sense of the centrality of the Rectangular Survey to the American settlement experience. What I want to do in this chapter's final section is help you imagine those life-changing survey stakes as part of a continentwide effort of two centuries' duration. It is estimated that government surveyors set something like 4 million corner markers in the contiguous United States. You have some sense of how they did their work. Let me now give you an image of the *extent* of that work.

You have seen the growth of the concept of starting a surveying campaign from a Baseline and Principal Meridian projected from an Initial Point. Eventually there were to be thirty-two such points in the contiguous states (in addition to the Point of Beginning in Ohio), with an additional five in Alaska, each controlling a separate grid of Townships and Ranges. You saw earlier how Israel Ludlow set up his meridian at the mouth of the Miami River in 1798. What I want to do now is return you to that First Principal Meridian, show you the Townships eventually surveyed from it, and then repeat the process for each of the thirty-two contiguous surveys, in chronological order. Then I will quickly summarize the newer Alaskan surveys.

What you will be seeing in the maps is the ultimate state of each survey: some of the Townships might have been laid out as much as a century after the marking of the Initial Point. A tiny number will have been surveyed or resurveyed in the past decade, for the Public Land Survey System is an ongoing process—never really completed, but still conducted following the principles laid down in 1785 and 1796.

To refresh your memory about surveying in Ohio, here are some of the major efforts you've read about thus far, from the northeastern of the state to the southwest:

- *Along Lake Erie, Connecticut's Western Reserve and Fire Lands were laid out in 5-mile-square townships.*
- *South of those lands, 6-mile-square Townships were laid out, most as extensions from the base established by the Geographer's Line at the Seven Ranges, with the exception of—*
- *The U.S. Military Reserve, which was apportioned into 5-mile-square townships to accommodate the land grants awarded in multiples of 100 acres to Continental soldiers.*
- *Along the Ohio River were the 6-mile-square Townships begun by the Ohio Company and later continued by surveyors under contract with the federal government.*
- *Finally, west of the Virginia Military Reserve came the 6-mile-square Townships in the "tilted" pattern, begun by Israel Ludlow and followed for the remainder of the land between the Little and Great Miami rivers.*

FIG 9.19

We saw earlier how Israel Ludlow had based his survey of land inside the Greenville Treaty line on a meridian line projected north from the mouth of the Great Miami River at the Ohio. His field notes from 1798 give us a hint of what many later surveyors would find as they pushed into "virgin" territory: on the first 2 miles of his meridian, Ludlow encountered no less than eight cabins with cornfields or "small improvements."[16] Ludlow carried his meridian to the Greenville Treaty line by the end of 1799, and by the following year most of the land inside the line had been apportioned into Townships.

On July 4, 1805, in the Treaty of Fort Industry, six Indian tribes ceded the land between the Cuyahoga River (part of the 1795 Greenville line) westward to a line 120 miles west of the Pennsylvania border. This north-south line would mark the far boundary of Connecticut's Western Reserve and Fire Lands, and in fact would be the mandated limit of white settlement for the next dozen years.[17]

In a pair of treaties in 1817 and 1818, a number of tribes ceded the land beyond the Western Reserve, along with most of Indiana, meaning that the survey of the new state of Ohio could be completed. When Indiana was admitted as a state in 1816, it became important to survey the Ohio-Indiana border, and the line was marked the following year. Significantly, the northern border Ohio had challenged—the line due east from the southern end of Lake Michigan—was also marked.[18] Making use of these facts, in 1819 the Surveyor General for the Northwest, Edwin Tiffin, set out his plan.

Run a Base line from the Southwest corner of the "Connecticut Western reserve" due West until it shall intersect the State line between the States of Ohio and Indiana, which is called and known by the first principal meridian . . .

Tiffin also noted how his plan would cope with converging meridians.

Beginning the Base line from the point here proposed (which is on or near the forty first degree of Latitude) it will pass nearly through the middle of the purchase; so that the width of the ranges in their southern and northern extremities, will not be materially increased or diminished.[19]

WASHINGTON MERIDIAN, 1803

We saw how Isaac Briggs took the problem of converging meridians more seriously, in his scheme to "restart" the survey at a second baseline six Townships north of the main one. Treasury Secretary Albert Gallatin rejected Briggs's idea, and in a letter he later wrote to Briggs, we can see the reasoning that guided Gallatin—and indeed many of the officials to follow him.

The principal object which Congress have in view is that the corners and boundaries of the sections & subdivisions of sections should be definitively fixed; and that the ascertainment of the precise contents of each is not considered as equally important. Indeed it is not so material either for the United States or for the individuals, that purchasers should actually hold a few acres more or less than their surveys may call for, as it is that they should know with precision, and so as to avoid any litigation, what are the certain boundaries of their tract.[20]

First Principal Meridian

Washington Meridian

FIG 9.20

SECOND PRINCIPAL MERIDIAN, 1804 AND 1806

We've already seen how Jared Mansfield used the already-surveyed Vincennes Tract to lay out his Principal Meridian and Baseline in Indiana. When those lines were first marked, they had no official name (as was the case with the First Principal Meridian). It was in 1806, in a letter to Treasury Secretary Gallatin, that Mansfield suggested the idea of numbering the Principal Meridian systems, with his Indiana Meridian to be called "the Second" and the meridian from the mouth of the Ohio (begun in 1805) to be "the Third."[21]

ST. STEPHENS MERIDIAN, 1805

In March of 1799 Andrew Ellicott was surveying the 31st-parallel border with Spanish Florida. Near Mobile Bay he decided to do a careful check on the latitude of his line, and so he went a little ways north, to a bluff with a clear view of the sky, and set up an "observatory" for taking precise sightings on stars. After two weeks of observations, he determined that his line had veered about 1.5 minutes too far north, so he set a prominent stone on the corrected latitude and continued his line.

We saw earlier that the first important Indian cession north of the Florida border was the Choctaw cession of 1801. The Washington Meridian would control the survey of the western part of that tract, but to speed things along, Surveyor South of Tennessee Isaac Briggs decided that a second Principal Meridian north from Ellicott's line was needed. In 1805 he chose as his Point of Beginning the stone Ellicott had set near Mobile Bay. As it turned out, the Meridian run north from the stone happened to pass close by Fort St. Stephens (at the northeastern corner of the Choctaw tract), and so that was the name given to this Meridian system.

With more Indian cessions, the Meridian was extended northward, and with Spain's surrender of Florida in 1819, it was pushed south.[22]

Second Principal Meridian

St. Stephens Meridian

FIG 9.21

THIRD PRINCIPAL MERIDIAN, 1805

The town of Kaskaskia sits on the Illinois side of the Mississippi River, about halfway between St. Louis and the mouth of the Ohio. It's a minor town now, but around 1800 it was an important trading post, and when in 1803 the Kaskaskia tribe ceded lands to the east, it seemed obvious that the town would be a center of settlement. Treasury Secretary Gallatin believed that the great regulating lines of a survey should originate at geographically prominent places, so in 1805 he directed Surveyor General Jared Mansfield to begin a new Principal Meridian at the confluence of the Ohio and Mississippi rivers. Gallatin also wanted the Townships in this newly opened region to continue the pattern of Townships laid out from the Second Principal Meridian, to the east. Unfortunately, the Kaskaskia tribe's cession didn't extend that far. How then to connect the two surveys?

Mansfield's solution was to send William Rector into Indiana, where he would find one of the Township lines already marked, and then project that line westward—through Indian country—into the Kaskaskia cession lands. Rector chose a line several Townships south of the Second Meridian's Baseline, and projected it all the way to the Mississippi. From there he floated down to the mouth of the Ohio and projected his Meridian line north, crossing his east-west line in November of 1805.

From these two lines Townships were laid out, primarily south to the Ohio, but not as far north as the Baseline of the Second Principal Meridian. Only in 1807, with more Indian cessions, could the survey be carried that far north and the Second Meridian's Baseline extended to become the Baseline for the Third Meridian system. Eventually the Third Principal Meridian was extended north to the Illinois River, and with statehood, to the new northern border of Illinois.[23]

LOUISIANA MERIDIAN, 1806

In July of 1805 Treasury Secretary Gallatin instructed Surveyor General Briggs to begin the survey of the new Territory of Orleans. Gallatin's plan was to pick up the 31st parallel, already surveyed to the eastern bank of the Mississippi, extend it westward to the Sabine River, and use that as the Baseline for Townships from the Red River (Oklahoma's southern border) to "the Seashore." Gallatin didn't specify a location for the Principal Meridian of the new survey, only that it should begin well to the west of "the inundations of the Mississippi River."[24]

Late in 1806 Briggs's survey team projected the 31st parallel across the Mississippi and set out on the 55 miles to the Sabine River, much of it through "Impassable Quagmire."[25] They chose to start their Principal Meridian at the thirtieth milepost they set. After the turn of the year the Principal Meridian was extended north and south from the Baseline.[26]

HUNTSVILLE MERIDIAN, 1807

In 1805 the Chickasaw tribe ceded a large tract of land in what is now central and western Tennessee, but a small triangle of the cession extended below Tennessee's 35th-parallel border into what is now Alabama. In March of 1807, Secretary Gallatin instructed his new Surveyor South of Tennessee, Seth Pease, to survey this new tract of land. Gallatin suggested that Pease have his deputy, Thomas Freeman, survey the 35th parallel across the top of the tract as a Baseline and then project a Principal Meridian south from that line.

In September of 1807 Freeman set out for Tennessee and set up an observation post to pinpoint latitude 35° north. That fall he extended the Base-

Third Principal Meridian

Louisiana Meridian   Huntsville Meridian

FIG 9.22

line across the tract and marked an Initial Point
for the Principal Meridian. The following year he
extended the Meridian south—past the town of
Huntsville—to the apex of the triangular tract,
and completed laying out the Townships. Later,
with more Indian cessions, the Huntsville Meridian would control all the Townships in the northern half of Alabama.[27]

In 1807 a number of Indian tribes surrendered the lands east of a line drawn north from Fort Defiance in northwest Ohio, and in 1815 this was still the boundary of Indian lands.[28] Surveying that line would not only mark the boundary of Indian lands, but would provide a Principal Meridian for Townships east of the line. The survey team set out from Fort Defiance in the spring of 1815, drove the line northward to a latitude they calculated would pass a little north of Detroit, and declared that to be their Initial Point. It was only in the following year that the Baseline was pushed east to Lake St. Clair;[29] the far end of that line became the now-famous Eight Mile Road, the northern boundary of Detroit.

You'll recall that Congress's enabling act for Ohio set the state's northern border at the "southerly bend of Lake Michigan," while the Ohio Constitution asked for a diagonal line from that point to the northern bank of the Maumee River. The Surveyor General for the Northwest, Edwin Tiffin, followed Congress's interpretation,[30] and so when the matter was finally settled with Michigan's statehood in 1837, Ohio inherited a sliver of Townships numbered from the Michigan Meridian.

Meanwhile, the Indian tribes didn't surrender land west of the Principal Meridian until 1824. But when surveyors laid out the Baseline west of the Meridian, they placed their line about 900 feet south of the Baseline running east. Thus, the Michigan Meridian system, uniquely, has two Initial Points.[31]

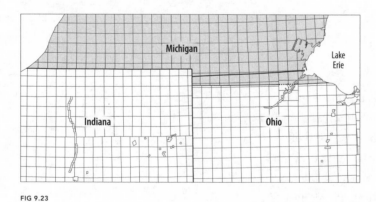

FIG 9.23

CHAPTER NINE

FIG 9.24

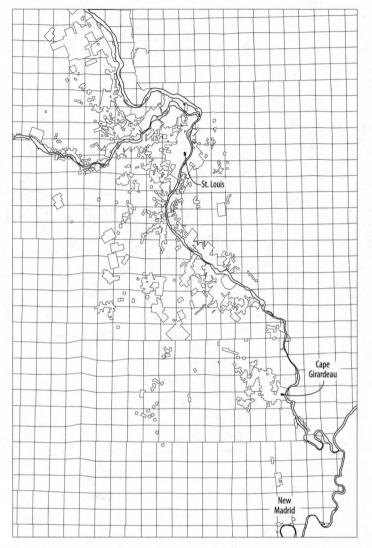

The proximate cause of the Fifth Principal Meridian was the need to lay out 2 million acres of lands as military service bounties. GLO Commissioner Josiah Meigs determined that sufficient acreage could be found in the triangle between the Arkansas and St. Francis rivers, so he directed that a new Baseline be drawn westward from the confluence of the St. Francis and the Mississippi, and a new Principal Meridian northward from the Mississippi's confluence with the Arkansas. Both lines were run in the fall of 1815. Little could Commissioner Meigs know that this Fifth Principal Meridian system, extended and extended again, would become the basis for a stack of 164 Township Tiers, all the way to Canada.[32]

FIG 9.25

Here is a detail of the Townships southeast of St. Louis. Remember that before the United States purchased Louisiana, both Spain and France had offered land to settlers. U.S. policy was that where the titles to such lands were clear, they should be honored. Thus the Township survey stopped up against such grants. The intention of the survey was to get every parcel of land quickly assigned to an owner—not to "grid up" the whole of the Public Domain.

Fifth Principal Meridian

FIG 9.26

The Initial Point of the Fourth Principal Meridian was established two days after that of the Fifth (November 12 versus 10, 1815), but the Fourth was authorized first, and so priority of numbering goes to it. As with the Fifth, the Fourth Principal Meridian system was set up to survey military bounty lands—in this case an additional 2 million acres north of the Illinois River. Surveyor General Tiffin had maps showing the Illinois River flowing southwest into the Mississippi, and so he decided that their confluence would be a good starting place for a new Meridian. In May of 1815 he issued these instructions, which, when later followed to the letter, were to cause a problem.

A true meridian line to be drawn North from the confluence of the Illinois & Mississippi rivers the exact distance of 72 miles, which line is to be denominated the 4th principal meridian, thence a Base line to be drawn due East from the 72nd mile post until it intersects the Illinois River.[33]

In October the surveyors drove their line north from the mouth of the Illinois, only to find the river veering west, across their path, thereby putting their Principal Meridian outside the territory allotted for bounties. They continued due north nonetheless, and gradually the river veered back toward their line, only to find that "the 72nd milepost"—where they were to set their Initial Point—fell right in the middle of the river. Nonetheless, they recorded the position. In November a second team carried the Baseline west to the Mississippi. The following spring a third team came to the Initial Point, with the assignment to begin laying out Townships. This team chose to reinterpret their instructions. They found the Baseline marked the previous year but moved the Initial Point a few chains west so that it would sit on dry land. From this new point they projected northward what would become the real Fourth Principal Meridian; the line from the mouth of the Illinois was eventually abandoned.[34]

Illinois was admitted as a state in 1818, with a northern border at latitude 42° 30', but it was not until 1831 that Congress authorized that border to be surveyed. GLO Commissioner Elijah Hayward decided to capitalize on this impending survey to extend the Fourth Principal Meridian system. The Fourth Meridian had been stopped where the Mississippi forms a bulge eastward into Illinois. Hayward's plan was to carry the longitude of the Fourth Meridian north across that bulge until it met the northern border of Illinois. The Meridian could then be extended north and a new set of Townships numbered from a new Baseline at the Illinois border.

Lucius Lyon got the contract, and in the fall of 1831, he was all over the region doing the work. First he extended the Third Principal Meridian north from the Illinois River, where it had been stopped. Then he went to the eastern bank of the Mississippi to find and mark latitude 42° 30'. Then he went south the carry the longitude of the Fourth Meridian northward. The Meridian and the Illinois border were brought together in December, and the work was suspended. The following February the work was resumed, at which time the Meridian was extended northward and a second Baseline pushed east.[35]

Fourth Principal Meridian

St. Helena Meridian

FIG 9.27

ST. HELENA MERIDIAN, 1819

In 1819 Spain at last ceded all of East and West Florida to the United States, and so that part of the new state of Louisiana between the Mississippi and Pearl rivers could be officially surveyed for sale. The plan of the survey was a simple one: use the already surveyed 31st parallel as a Baseline and extend the 1803 Washington Meridian southward as a Principal Meridian. This extended Meridian, as it turns out, passes through what is now Baton Rouge.[36]

CHOCTAW MERIDIAN, 1821

You'll remember that the Choctaw tribe had ceded lands along the 31st parallel, and that those cessions gave rise to the Washington and St. Helena Meridians. In 1820 the Choctaws were induced to surrender yet more land, a swath north and east of the Washington Meridian territory. The Washington Meridian had dead-ended at the Mississippi, and the GLO Commissioner determined that a new Principal Meridian system was needed. His instructions were that its Baseline should be the boundary of the old Choctaw cession, and from that line a new Meridian should be projected north into the new cession. Charles Lawson, the surveyor who did the work, knew that the old Choctaw boundary was unsuitable as a Baseline: it was not a line of latitude. Instead he chose the common northern corner of two Townships already surveyed from the Washington Meridian, declared that corner to be his Initial Point, and began projecting his Baseline west and then his Meridian north.

In 1830 the Choctaws were forced to give up all their remaining land and begin the trek to Indian Territory. That final cession, within the state of Mississippi, marks the extent of the lands surveyed under the Choctaw Meridian system.[37]

TALLAHASSEE MERIDIAN, 1824

Spain ceded East Florida to the United States in 1819, but it was not until 1823 that there were enough Indian cessions to justify starting a survey of the Florida Territory. The only part of Florida that seemed capable of settlement was the northern fringe (the idea of lying in the hot sun, almost nude, on an oceanfront sand dune would have seemed incomprehensible), and so it was

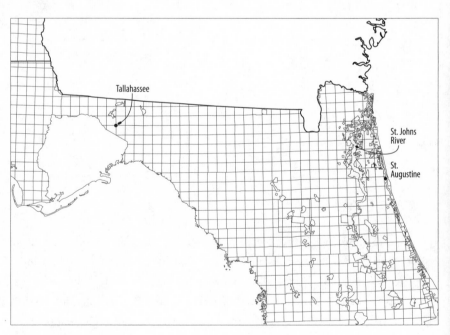

FIG 9.28

Here you can see early Spanish land grants along the St. Johns River. The unsurveyed area near Tallahassee contains the Apalachicola National Forest.

Choctaw Meridian   Tallahassee Meridian

FIG 9.29

natural to establish the surveys there. In 1823 the decision was made to place the territorial capital at Tallahassee, and along with that came the determination to locate the Initial Point of the territorial survey at a prominent place within the new capital city. State officials pointed out to Surveyor General Robert Butler the location of the proposed state capitol building, and he measured a quarter mile east and then a quarter mile south and established his Initial Point there.[38]

CHICKASAW MERIDIAN, 1833

In 1818 the Chickasaw tribe ceded all its land within the state of Tennessee, retaining a claim to its lands to the south. The state's southern border—and thus the outer boundary of the land the state could sell—had not yet been surveyed, and so the line was quickly run the following year. In 1830 Tennessee thought it might benefit from a more careful determination of its 35th-parallel southern border, and indeed that survey produced a new boundary 2 to 3 miles south of the 1819 line. Officials in Mississippi quite naturally refused to accept this diminution of their state, and there matters stood.

In 1832 the Chickasaws were compelled to give up their remaining lands in Mississippi and join the Choctaws and Cherokees in Indian Territory. The treaty stipulated that all the revenues from the sale of their lands (minus the cost of surveying them) would go to the tribe. Since it was a settled matter that their 1818 cession had ended at the Tennessee state line surveyed in 1819, that line was the true northern boundary of the lands the tribe now gave up. So it was determined that the Chickasaw lands be surveyed southward from a Baseline at that northern boundary line. Deputy Surveyor John Thomson located the 1819 line in August of 1833 and established his Principal Meridian at a point he determined to be about midway along the Mississippi border. The Baseline was run and monumented, along with the Meridian, and land sales began.

In 1837, though, Mississippi and Tennessee appointed a joint commission to decide their 35-degree border once and for all. The line they settled on was very near the one Tennessee had run in 1830—that is, a couple of miles *south* of the 1819 line that was the northern border of the Chickasaw cession. The result is another unique anomaly: the Baseline for the Chickasaw Meridian system—a line for apportioning the Public Domain—lies *outside* the Public Domain, a couple of miles inside Tennessee.[39]

WILLAMETTE MERIDIAN, 1851

The United States took possession of the Oregon Country in 1846, but it took two years for Congress to set up a territorial government, and three more years to authorize a survey of the territory—which had been attracting settlers for almost two decades. The valley of the Willamette River was where most emigrants set up their farmsteads, and the town of Portland, at the mouth of the Willamette on the Columbia River, had by 1851 become the principal settlement. GLO Commissioner Justin Butterfield directed that the Initial Point of the Oregon survey be just west of Portland, so that the Principal Meridian would extend south through most of the Willamette valley. The main consideration in locating the Baseline was that it be south of the southernmost bend of the Columbia, so as not to have to cross and recross the river (in the manner forced upon the Fourth and Fifth Principal Meridians). Butterfield's intention was that the this new Principal Meridian govern the survey of all lands west of the Cascade Mountains (about 80 miles from the coast in that part of Oregon), and so he instructed his surveyors to extend the Meridian from the Initial Point south to the border of Oregon at the 42nd parallel and north as far as Puget Sound.

Willamette Meridian

Chickasaw Meridian

FIG 9.30

Chief surveyor John Preston and his crew set out from the East Coast early in March of 1851, reached Panama at the end of the month, crossed the isthmus by mule train, caught a steamer first to San Francisco and then on to the mouth of the Columbia River. After scouting the territory, Preston set a white-cedar post at his Initial Point on June 4. The crew then split up and began extending lines east, west, and south. They reached the Pacific later in June, the Cascades in July, and the foot of the Willamette Valley in August. Later the Willamette Meridian would be extended to California and Seattle, and the Baseline pushed over the Cascades to the Snake River.[40]

By the time California became a state in 1850, the hills around Sacramento were filling up with mineral claims, and farmers were setting out grain fields in the flatter valley lands. A survey was needed—and quickly—to tie the claim borders into the Rectangular Survey net, and to give farmers rectangular plots they could gain title to. GLO Commissioner Butterfield, in Washington, looked at his maps and saw that

Mount Diablo is visible from a great distance, being represented as 3,677 feet high, and lying, it would seem from the Maps, in a direction favorable to a meridian line which would run through the Valley of the Sacramento [River].[41]

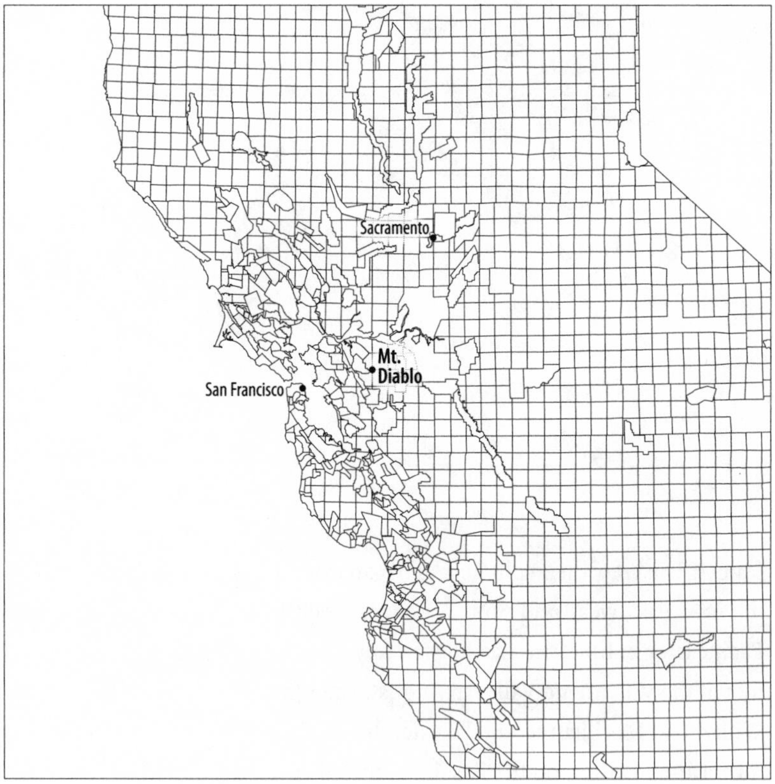

FIG 9.31

The Mt. Diablo survey had to work around not only Spanish and Mexican land grants, but early claims made under American jurisdiction.

His plan was to use Mt. Diablo as the basis for surveying all of California, the idea being to go eastward from the summit some distance on a Baseline, then project a Principal Meridian south to govern the survey of the Central Valley, and then perhaps another, offset Meridian for the areas south of the Tehachapis centered on Los Angeles. With instructions to that effect, Surveyor General Samuel King and his chief assistant, Leander Ransom, set off for San Francisco via Panama.

Mt. Diablo lies about 20 miles east of Berkeley and is quite a rugged peak—so rugged that when Ransom and his crew reached the summit on July 18, they found that the slopes down from the peak, in the four cardinal directions, were much too steep for the crew to project measured and monumented lines. So the surveyors erected a cairn of stones buttressing a flagpole, came down from the summit, and began to survey a rough square around the peak, on flatter ground about 12 miles outward from it. From this square they were able to project their Baselines and Principal Meridians.[42]

(The Mt. Diablo Meridian system would be extended to the Central Valley but not to Southern California. Instead it was pushed eastward, over the Sierra Nevada, to control the survey of Nevada.)

Mt. Diablo Meridian

FIG 9.32

SAN BERNARDINO MERIDIAN, 1852

Once actually on the scene in California, Surveyor General King soon realized the folly of trying to survey the whole of the state from one Principal Meridian system. He wrote back to Washington, convinced the GLO Commissioner, and in August of 1852 sent Ransom south to scout out the territory. Officials in Washington had suggested establishing the Initial Point near the town of Los Angeles, but King knew that a Meridian sent south from there would quickly dead-end in the Pacific (somewhere near Long Beach, in fact). He instructed Ransom to go far enough inland that a Meridian pushed south might extend all the way to California's border with Mexico. King suggested Mt. San Bernardino (northeast of the present city).

Ransom reconnoitered the area, decided that Mt. San Bernardino was the best Initial Point, and in October assigned Henry Washington the task of climbing the mountain and starting the Principal Meridian and Baseline. Washington found that Mt. San Bernardino is both taller and more rugged than Mt. Diablo (over 10,000 feet tall—8,000 feet above the valley floor). Like Ransom, the best he could so was to erect a cairn and a 25-foot wooden pole as a marker for the Initial Point, then climb down to more congenial terrain. There, by sighting back on his pole, he was able to start the Baseline and Principal Meridian.[43]

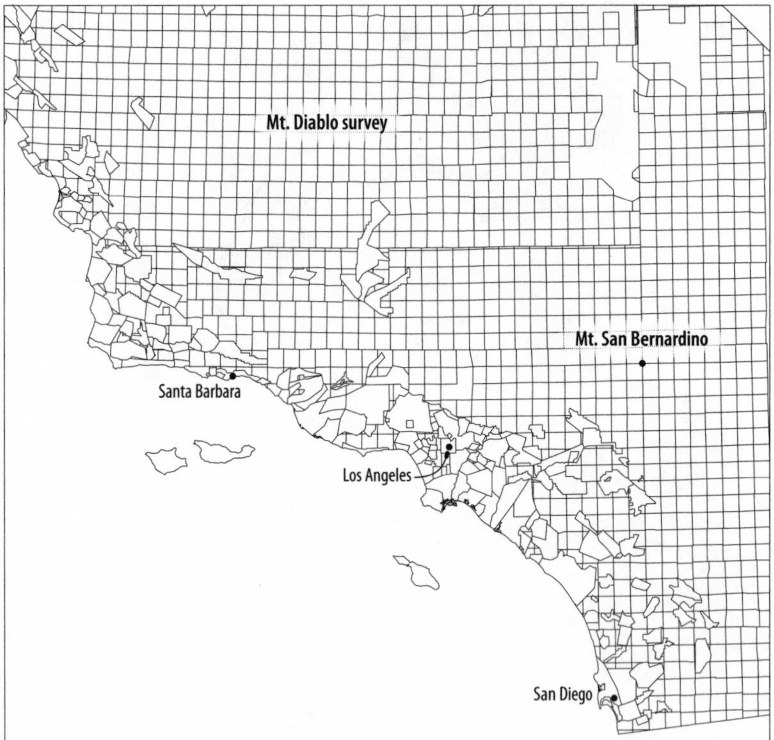

HUMBOLDT MERIDIAN, 1853

By 1851 the area in northern California around Humboldt Bay and the town of Eureka was being opened up, and settlers there asked to have a survey so they could gain title to their lands. In July of 1853 Surveyor General King turned to Henry Washington to establish the Humboldt Meridian system, leaving to his judgment the location of the Initial Point. After scouting the area, Washington chose a mountain peak directly east of Cape Mendocino: Mt. Pierce, named for the president. Again the peak was too steep to allow extending lines directly from the Initial Point, so California's third Meridian system extends outward from—this time—a triangle down the slope from the summit.[44]

FIG 9.33

Many of the existing land grants in Southern California were made by Spain and especially Mexico for large agricultural holdings.

CHAPTER NINE

Humboldt
Meridian

San Bernardino Meridian

FIG 9.34

In 1854 Congress set up both the Kansas and Nebraska territories, and that August GLO Commissioner John Wilson decided that the 40th-parallel border between the two territories would be an ideal Baseline for surveying them. As for the location of a new Principal Meridian—

For reasons of expediency, because of the apprehensions of hostile interruptions from the Indians, it is not deemed proper and prudent to survey a base line further to the west than one hundred and eight miles distant from the Missouri river.[45]

John P. Johnson was chosen to conduct the survey, and on November 16, 1854, he set up his instruments on the bank of the Missouri to locate latitude 40° north. A few short weeks later, on December 5, he had run that latitude the mandated 108 miles west. There he set up a sandstone post to mark the location of the Sixth Principal Meridian. Eventually the Sixth Meridian system would be extended well beyond Kansas and Nebraska.[46]

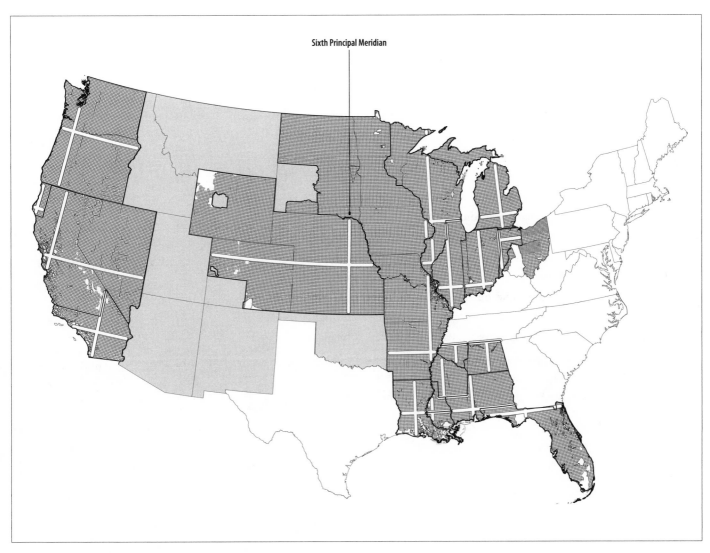

Sixth Principal Meridian

FIG 9.35   The unsurveyed area in northwestern Wyoming is, of course,
Yellowstone National Park, set aside for the nation in 1872.

New Mexico Territory, you will recall, was set up after Texas ceded its western extremes in 1850. In August of 1854 William Pelham was made Surveyor General of New Mexico, charged with starting up the survey of the territory. The location of the Initial Point was left to his discretion, with only the suggestion "that the Prime Meridian should run near the suburbs of Santa Fe"[47] and the Baseline, to the south of the major settlements.

Pelham arrived in Santa Fe on December 28 and began a reconnaissance southward through the valley of the Rio Grande. He indeed set his Initial Point south of Santa Fe, indeed south of Albuquerque, on "a hill about six miles below the mouth of the Puerco River."[48] By the following spring, Pelham's assistants had run the Principal Meridian 96 miles north to the Santa Fe "suburbs" and south all the way to Mexico, and had pushed the Baseline 21 miles to the west.[49]

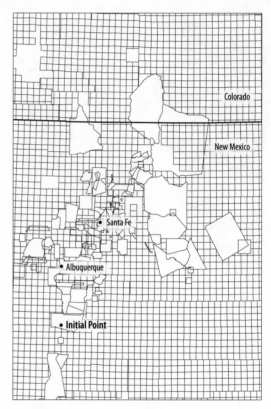

FIG 9.36

You can see here the small Spanish and Mexican land grants around Santa Fe and Albuquerque—and the truly huge grants to the north and east.

New Mexico Meridian

FIG 9.37

Utah Territory was created in 1850, and Congress authorized the beginning of a survey in 1855. But before then, the Mormon settlers had laid out their capital at Salt Lake City. On August 3, 1847, the chief surveyor for the colony, Orson Pratt, laid out the block where the Temple was to stand, with the southeastern corner of the Temple Block declared the center of the city. From that point all the streets (2 chains wide: 132 feet) were laid out in the cardinal directions, surrounding blocks that were 10 chains square.

David Burr was appointed Surveyor General of Utah, and he arrived in Salt Lake City in July of 1855. By the first week in August, he had chosen his Initial Point (presumably in consultation with Brigham Young): "the Southeast corner of the Temple Block in Great Salt Lake City." The Baseline and Principal Meridian were quickly extended through the new city and out into the countryside.[50]

Salt Lake Meridian

FIG 9.38

GILA AND SALT RIVER MERIDIAN, 1865

The act creating Arizona Territory also authorized a survey to be commenced. In 1864 John A. Clark, Surveyor General of New Mexico, was put in charge of both territories. In September the GLO authorized Clark to find an Initial Point for the Arizona Territory survey, and in January of 1865 he set out from his office in Santa Fe.

He was in no especial hurry, visiting friends along the way and stopping to admire the famous mission church of San Xavier Del Bac near Tucson. From Tucson he followed the Santa Cruz River northwest to the Gila River. The Gila, you'll remember, was part of the U.S.-Mexican border set by the Treaty of Guadalupe Hidalgo. When Clark followed the Gila to its confluence with the Salt River (near present-day Phoenix), he found "on the south side of the Gila . . . a conical hill, about one hundred and fifty feet in height." On its peak was a circular monument 8 feet high left over from the 1851 survey of the border. It was a marker seemingly ready-made to be an Initial Point, and Clark seized the opportunity. The following December (surveying season in Arizona) he contracted for the Principal Meridian and Baseline to be extended from the boundary monument.[51]

BOISE MERIDIAN, 1867

Congress created the Idaho Territory in 1863, and that same year a government was set up in Lewiston, located right against the border with Washington. In 1866 Congress authorized the appointment of a Surveyor General but specified that his office had to be in "Boise City," almost 300 miles south of Lewiston on the northern edge of the great Snake River plain. The new Surveyor General, Lafayette Cartee, opened his office in November of that year, and by the following spring he had chosen his Initial Point, "a small butte situated about twenty miles from Boise City, five miles from Snake river" (in the words of the April 9, 1867, *Idaho Statesman*).

We can wonder whether Cartee knew that by placing his Initial Point on that hill, the Principal Meridian projected north from it would stay within Idaho for its entire length, all the way to Canada. What was obvious was that, since the Snake River plain runs north-south near Boise, by starting south and west of the town, his new Meridian and Baseline could both extend for some miles before running into the mountains that enclosed the flatlands.[52]

Boise Meridian

Gila and Salt River
Meridian

FIG 9.39   The unsurveyed areas in Idaho are national forests. Within the
unsurveyed area in northeastern Arizona are the Hopi and Navajo
reservations.

PRINCIPAL MERIDIAN (MONTANA), 1867

In the next chapter we'll look much more closely at the Meridian system under which Montana was surveyed. Note here, though, that this is the one Meridian system that was never given a proper name. It is simply *Principal Meridian.*

NAVAJO MERIDIAN, 1869

By an 1868 treaty, the Navajo people were confined to a reservation straddling the northern end of the Arizona–New Mexico border. Small as it was, the reservation at least encompassed Canyon de Chelly, a flat-bottomed gorge a thousand feet deep, occupied by the Navajo since ancient times (the canyon is pronounced "de SHAY-ee," and is very much worth a visit).

It was U.S. policy at that time to survey reservations into rectangular farming parcels. The Navajo lands were hardly suited to conventional farming, but in 1869 a contract was let which resulted in a Baseline at the reservation's southern border and a Principal Meridian along its eastern edge.

Some Townships and Sections were projected from these lines, but few were taken up by the Navajo, and in 1936 the Commissioner of the GLO declared the surveys in New Mexico cancelled. By this time the Navajo Reservation had been extended far to the west and south, much of it left unsurveyed (except for the outer boundary) because held in common by the tribe.[53]

INDIAN MERIDIAN, 1879

With so many tribes shoehorned unwillingly into Indian Territory, it was probably inevitable that disputes would break out, especially over land. One of the more severe conflicts involved the Choctaws and the Chickasaws, and led to a formal treaty between the tribes, signed in 1855. The tribes' lands had been between the Canadian and Red rivers: the treaty divided their lands at the 98th meridian. (The Canadian flows across the "pan" part of Oklahoma, curving downward past Oklahoma City, then upward; and you will remember that the 100th meridian is the eastern edge of the Texas Panhandle.) The Chickasaws' lands were to extend essentially from the 98th meridian east to a north-south line whose description in the treaty meant that the Army's Fort Arbuckle would sit nearly astride the line. Surveyors marked these borders in 1858.

In 1866 the tribes signed a new treaty, under which the lands west of the 98th meridian were sold to the U.S. government for $300,000. The treaty also specified that, if they wished, the Chickasaws could have their lands east of that line surveyed at government expense.

The Chickasaws exercised that option, and government surveyors took to the field in 1870. Their instructions were that the line running past Fort Arbuckle would become a new Principal Meridian, and that they should project a Baseline from somewhere near the fort—west through the Chickasaw lands and past the 98th meridian into the land now available for sale to white settlers.

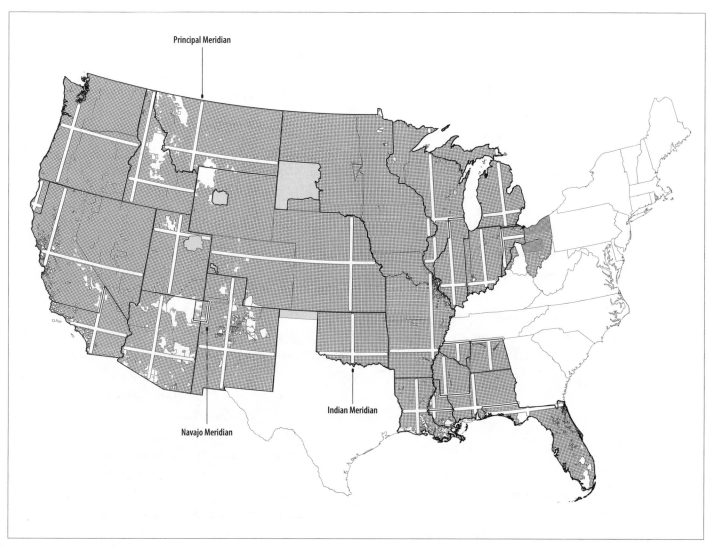

Principal Meridian

Indian Meridian

Navajo Meridian

FIG 9.40   Glacier National Park is inside the unsurveyed area in northwestern Montana.

In October of 1870, the survey crew, running the Principal Meridian from the north, reached Fort Arbuckle (by now abandoned by the Army). Within sight of the fort they established their Initial Point and marked it with an sandstone pillar 18 inches square and 54 inches tall. Of all the markers set at Initial Points by the original surveyors, the monument for the Indian Meridian is one of the few still standing.[54]

WIND RIVER MERIDIAN, 1875

The Indian wars of the 1860s and 1870s are a complex and bloody story, but one crucial moment occurred in 1868 when treaties were signed with several Rocky Mountain tribes. One tribe, the Eastern Shoshone, was assigned a reservation in what is now western Wyoming. Unfortunately for the Shoshone, gold and even oil was discovered in the southern part of their reservation (near the South Pass of the Rockies), and so in 1872 they were induced to sell off those lands for $25,000—to be paid out over five years—plus a $500-per-annum "salary" for their chieftain.

The agreement also called for the remainder of the reservation to be surveyed upward from a Baseline near its southern edge (the boundaries of the reservation being described not "rectangularly" but primarily in terms of natural features). James Miller was given the contract, and he set up a sandstone marker at his Initial Point in July of 1875, surveying the southern part of the tract. Others returned to advance the survey in 1885, 1887, 1890, and 1898.[55]

UINTAH MERIDIAN, 1875

In October of 1861, President Lincoln wrote to his secretary of the Interior,

In the absence of an authorized survey (the valley and surrounding country being as yet unoccupied by settlements of our citizens), I respectfully recommend that you order the entire valley of the Uintah River within Utah Territory, extending to both sides of said river to the crest of the first range of contiguous mountains on each side, to be reserved to the United States and set apart as an Indian reservation.[56]

Only in 1875 did Congress appropriate funds for the survey, and the contract was given to Charles DuBois. His team began at the eastern end of the valley, and set an Initial Point at the end of August. By the end of September, the snows began, and DuBois retired from the field with the Principal Meridian and Baseline barely advanced. Later surveyors resumed the work, finishing it in 1903.[57]

BLACK HILLS MERIDIAN, 1877

One of the Indian treaties of 1868 established the Great Sioux Reservation, which comprised all of what is now South Dakota west of the Missouri River. The reservation's borders gave most of the sacred Black Hills to the Indians—but not quite all: the western boundary was described, in the white man's way, as being the 104th meridian.

In 1875 gold was discovered in the Black Hills, and the U.S. government made an offer to purchase the region. The Sioux refused, but prospectors invaded the hills nonetheless. The Indians resisted, and the eventual result was the tragic campaign against Crazy Horse and Sitting Bull. In 1877 the Sioux were forced into a new treaty. Their reservation lands were extended slightly into North Dakota, but the western border was moved from the 104th to the 103rd meridian, east of the Black Hills.[58]

By 1877 the Surveyor General of Dakota Territory reported to his boss in Washington,

In addition I may say that the population of the [Black Hills region] is now variously estimated at from 15 to 25,000 (and by some as high as 40,000) persons—engaged in mining, agriculture and general business.[59]

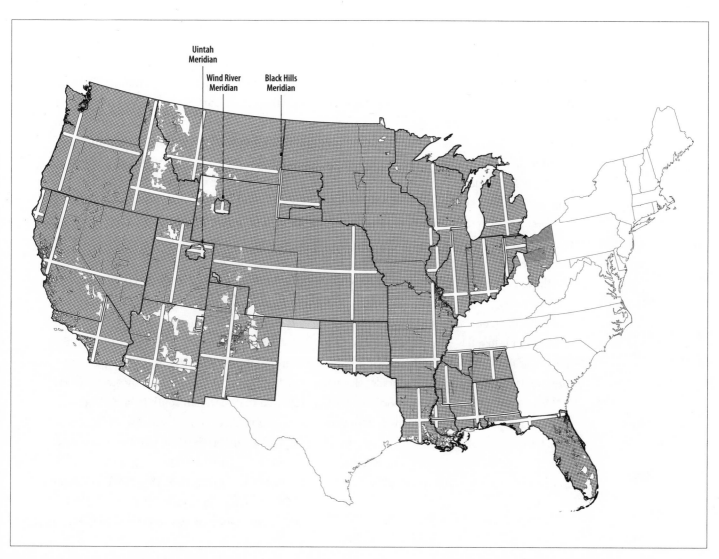

Uintah
Meridian

Wind River
Meridian

Black Hills
Meridian

FIG 9.41

GLO Commissioner James Williamson knew that the boundary between Dakota and Wyoming was being surveyed at that very moment (the survey that resulted in the tiny jog in South Dakota's western border), and he decided to use that line as the Principal Meridian for a new survey. From that Meridian a new Baseline could be projected east from one of the mile markers being set by the territorial boundary team, the choice of marker left to the surveyor on the scene.

Charles Scott took to the field in August of 1878. The boundary surveyors had numbered their mileposts upward from the northwestern corner of Nebraska, and Scott chose the one numbered 69, "a Pine Post 6 inches square in a mound of stone." He began extending a Baseline that season and returned to the work the following two years. In 1890 the Baseline and the Townships accompanying it were pushed eastward into what had been reservation land.[60]

One of the 1868 treaties gave the Ute tribes most of what is now Colorado west of the 107th meridian. In a repeating pattern, gold was found in the San Juan Mountains, in the southern third of the Utes' grant, and in an 1873 treaty they were made to surrender much of their land to the prospectors.

Enter (in 1878) Nathan C. Meeker, head of the White River Indian Agency and follower of Charles Fourier's communalist ideals. When Meeker tried to place the Utes on collective farms, they rebelled, conflict broke out, and Meeker telegraphed for the military. The result was the sad and familiar pattern of overreaction and massacre.

In 1880 the Uncompahgre branch of the Utes were forced into a few square miles at the confluence of the Colorado and Gunniston rivers (present-day Grand Junction, Colorado). Daniel G. Major surveyed the tract in 1880 and 1881, marking not just Quarter Sections but 40-acre Quarter-Quarter Sections for the Indians to farm. Researchers have found no records of any Uncompahgre Utes taking up the offer.[61]

Here is why the western end of the Oklahoma Panhandle doesn't quite align with the Texas–New Mexico border. That border was first surveyed in 1859 by John H. Clark, under contract with the U.S. Department of the Interior to mark the boundaries of its public lands outside Texas's new 1850 borders. Clark began his work at El Paso and ran the 32nd-parallel line east to what he calculated was the 103rd meridian. Then he carried the meridian north about 70 miles, where he ran out of water.

Resupplied, Clark returned, not to his line but up to the southwestern corner of Kansas Territory. In 1859, Kansas Territory had not yet been truncated by the formation of the Colorado Territory (see figure 5.25), and its southwestern corner fell on the 103rd meridian. Clark had been the "astronomer" for the 1857 survey of the territory's southern boundary, so he knew the line well. He followed that meridian south to latitude 36° 30' and then pushed a line about 2° south. There he stopped again, leaving a gap in the Texas–New Mexico boundary line. (That's the provenance of the line "Surveyed by Texas 1859" in figure 5.36.)

Other surveys followed in the region, but the one that matters to us was authorized by the GLO

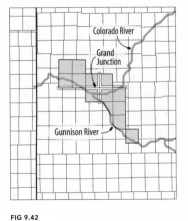

FIG 9.42

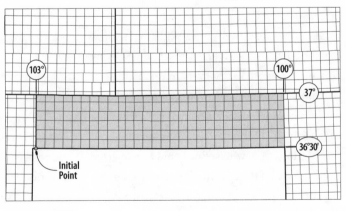

FIG 9.43

CHAPTER NINE

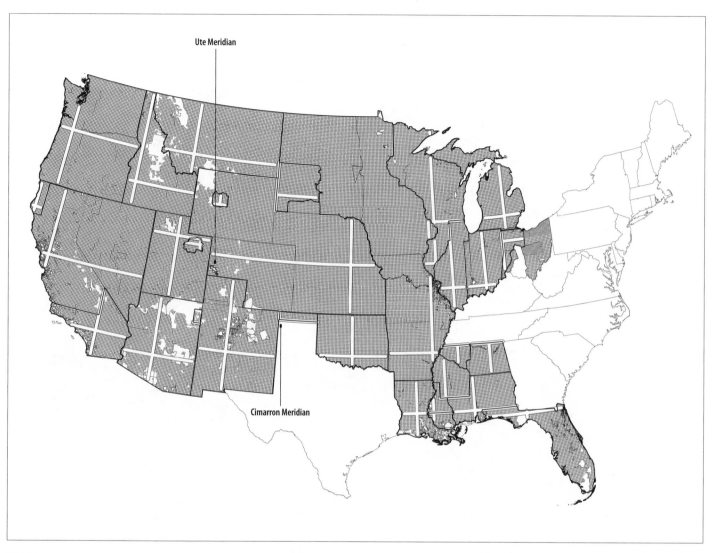

Ute Meridian

Cimarron Meridian

FIG 9.44

in 1881: establishing the Cimarron Meridian at longitude 103° and projecting a Baseline at latitude 36° 30'. Richard Chaney and William Smith were told to go to the town of Las Animas (in southeastern Colorado), set up a celestial observatory, and use time signals transmitted by telegraph to find the observatory's precise longitude. This they did (their stone pier south of the railway depot was at 102° 08' 41"), and then they carefully chained the distance, east and south, overland to

103° west and 37° north. Next they ran their new Principal Meridian south to latitude 36° 30', then pushed a Baseline east to longitude 100°.

But Chaney and Smith's Initial Point was a little over 2 miles east of the monument Clark had set at Texas's northwestern corner, and about 200 feet north. Litigation ensued, ending only in 1911 with a joint resolution of Congress declaring that the monuments set by Clark marked the true borders of Texas.[62]

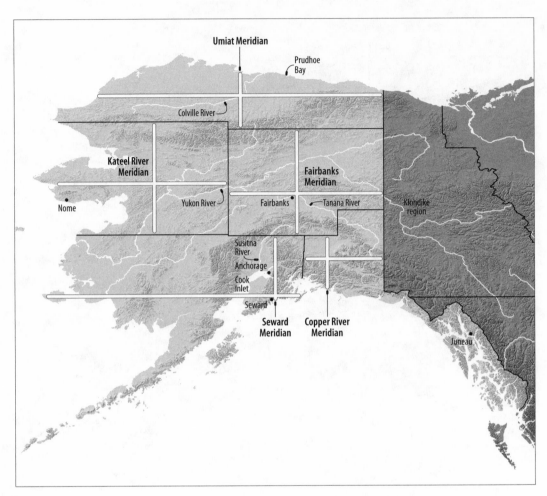

FIG 9.45

FIG 9.46   The lonely Kateel River Initial Point. Neither a Principal
Meridian nor a Baseline extends from this 3-inch steel
pipe with a brass cap.

Courtesy of C. Albert White.

Although purchased from Russia in 1867, Alaska didn't get a proper civil government until 1884. And only when gold was discovered in the Klondike region in 1896 did prospectors make an earnest push into Alaska's interior (the easiest routes into and out of the Klondike lay through Alaska). Nuggets of copper were found along one of the rivers leading inland, and that river was given the name of the mineral; but in 1898 real veins of copper were found, and a true boom ensued. The following year Congress decided it would be a good idea to extend the Rectangular Survey to Alaska so that the staked-out mining claims might be tied to a systematic survey.

A survey was authorized, with Washington asking that the Initial Point be set near "what is known locally as Stuck Mountain," about 12 miles south of the boomtown of Copper Center. On April 17, 1905, Alfred Lewis found a suitable stone up the side of Stuck Mountain, set a copper plug in a drilled hole, and established the Initial Point of the Copper River Meridian system.[63]

In 1901 gold was discovered in the valley of the Tanana River, and the town of Fairbanks sprang up on its parallel tributary, the Chena. Again pressure mounted for a survey. In the first decade of the twentieth century, the U.S. Coast and Geodetic Survey was directing its efforts toward Alaska, and it was thought prudent that the Geodetic Survey—not the General Land Office—should establish the Initial Point for the Fairbanks region. On August 9, 1910, a concrete marker was placed "on a prominent hill to the northeast of Fairbanks,"[64] to serve both as the Initial Point for a rectangular land survey and as a station point for a precise mapping campaign.

Problem was, the Geodetic Survey people took it upon themselves to begin laying out Townships—work that was definitely the province of the GLO. Hackles were raised, umbrage was taken, faults real and trivial were pointed out: Section lines were "drawn with colored inks"; the lettering was "oftentimes not clear or sharp in outline." Eventually, the work was corrected and carried forward, this time by GLO surveyors, who were by now working not under contracts but as government employees.[65]

By 1911 the Alaska Central Railroad was pushing north from the seaport of Seward, and there was hope for settlement in the Susitna River valley north of Cook Inlet. In April of that year, the GLO issued instructions for a third Initial Point. A month later the surveyor in charge, John Walker, chose to go to the marker set by the Coast and Geodetic Survey at the head of Resurrection Bay, the inlet on whose western shore Seward sat. He moved 10 chains north of the marker, and then 128 chains east before setting an iron post in the ground; but because the latitude and longitude of the Geodetic Survey marker had been carefully calculated, he would know the location of his Initial Point with equal precision.[66]

In 1912 Alaska was declared a territory of the United States, and in 1946 the residents voted for statehood. Delegates wrote a state constitution in 1955, and voters approved it the following year. With statehood in the offing, it was decided in 1956 to institute two new Initial Points to govern the survey in the far north and northwest of Alaska. These, though, would be Initial Points different from all that had come before.

Surveying instruments were advancing to the point where, no matter where a prospector or landowner wanted to stake a mineral claim or clear a farmstead, a surveyor could determine the longitude and latitude of either position with good precision. If he knew the precise latitude and longitude of the Initial Point, simple calculation could tell him where the Township and Section lines would run in his vicinity: no need to project actual lines across the earth from a Baseline or Principal Meridian.

Under such a system, a good Initial Point need not be a prominent geologic feature that could be sighted from miles away. All that was needed was a point roughly in the center of the surveying district whose latitude and longitude were known with precision. Such points already existed, in the form of station points set by the Coast and Geodetic Survey; so the Bureau of Land Management (the new name for the General Land Office when it reorganized in 1936) simply designated two of those stations as Initial Points.

On October 17, 1956, the station JAY 1953, set up that year near the Kateel River, was designated the Initial Point of the Kateel River Meridian system. On December 31 the station UMIAT 1953, on a bluff north of the Colville River, was declared the Initial Point of the Umiat Meridian system.[67]

Pretty prosaic, but it's all that's needed to get the job of surveying done in the modern age.

The annexation of Hawaii is one of the less savory episodes of American history. In 1893 Queen Liliuokalani proposed a new constitution for her islands, which would have given more power to her monarchy. Foreign interests (mainly American) objected, the queen abdicated, and a new government was formed. That government requested annexation, Congress debated the matter in June and July of 1898, and Hawaii became a territory of the United States on August 12 of that year.

The treaty of annexation stated that U.S. public land laws would not be applied in the Hawaiian Islands, and so the Public Land Survey System is not used there. In truth, it would have made little sense to mark rectangles up and down the slopes of volcanic mountains on small islands. More to the point, one prime aim of the annexation movement was to preserve the existing patterns of landownership.

Canada's Rectangular Survey

In the years after the Civil War, while America's attention was focused southward on the violent Reconstruction of the rebel states, across the border British North America was beginning its quiet, slow-motion march toward independence. The precipitating event was surrender by the Hudson's Bay Company of the immense territory it had held since 1670. You'll remember that the company had been granted all the lands that drained into Hudson Bay (with the exception of lands already granted in the east)—which meant, essentially, everything west of Ontario all the way to the Continental Divide in the Rockies. Under Confederation, British Columbia would keep control of its unappropriated lands, but the Hudson's Bay Company lands would be administered by the new central government.

Just as was the case with the Continental Congress in 1784, a brand-new government suddenly found itself in possession of a huge public domain. Just as the Congress had done, Canada looked to its own experience and traditions to devise a plan for apportioning that domain into tracts for settlement. Part of Canada's "experience" during its history had been to watch its neighbor to the south mark rectangles on the ground all across the continent. But during that time Canada had also been watching its mother country Britain as it worked out the fine points of the enclosure movement begun at the close of the Middle Ages. As new patterns of land tenure were worked out, Britons were able to force the landowners to allow passage across the newly laid-out fields—paths that were later formalized as public rights-of-way.

Canada's Survey of Dominion Lands combines these two traditions. The survey uses the 36-Section Township of the U.S. system but surrounds the Sections with rights-of-way that remain in public ownership. When first instituted, the system called for "a road allowance of one chain and fifty links on every side of a section."[68] Authorities later decided that a 99-foot easement every mile in both directions was a bit excessive. They narrowed the right-of-way to 1 chain and had it surround vertical pairs of Sections, as shown in figure 9.47.

Canada looked to the multiple Points of Beginning used by Americans and thought it could do better. The entire survey begins at a Principal Meridian that passes about 12 miles west of Winnipeg (about 98° west of Greenwich). The survey then "restarts" at meridians spaced every 4° east and west, as you can see in figure 9.48. All of the meridians begin at the 49th parallel and head north as far as settlement might require.

To American eyes, the Dominion Survey seems quintessentially Canadian: reasonable, rationalized, communitarian, maybe a little boring. It's a stereotype that Canadians embrace with a kind of rueful pride. It says something about both countries that each cherishes a distinctive national myth about frontier settlement. The American myth is that emigrants drove Conestoga wagons across the Plains and alighted in an unmarked wilderness. Canadians, by contrast, rode west on the national railway and got off at a town already equipped with a national bank, a national land office, and a local headquarters of the Royal Canadian Mounted Police. Myths equally untrue, and equally flattering, each to each.

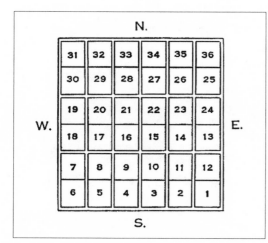

FIG 9.47

This chart from the Dominion Survey's 1918 edition of the *Manual of Instructions* shows how Canada's boustrophedonic township-numbering system echoes the south-to-north trend of the survey itself.

Canada Department of the Interior.

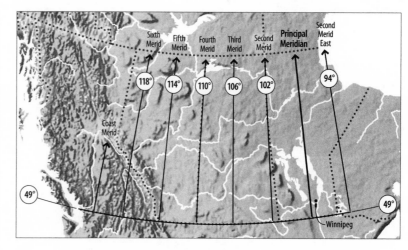

FIG 9.48

Although British Columbia held on to its public lands, a Coast Meridian was established to help guide the surveys of the railway route over the Canadian Rockies.

TEN

The Spread of the Survey across Montana

CHOOSING THE SURVEY'S INITIAL POINT

We saw in chapter 5 how Montana's "natural" boundary at the Continental Divide was over-ridden in 1864 to encompass the mining areas around Butte and the fertile valleys like the Bit-terroot. In 1860 there were fewer than 100 white people in what would become Montana Territory, and only a few head of cattle; but by 1870 there would be 20,000 nonnatives and 48,000 cattle, including some Longhorns trailed all the way up from Texas.[1] So on March 2, 1867, Congress authorized the appointment of a Surveyor General for Montana to head up a survey of the ter-ritory, starting that very year. President Andrew Johnson's choice was Solomon Meredith of Cam-bridge City, Indiana. An imposing figure at 6'6", Meredith was a prominent Indiana politician and Civil War hero, but his surveying experience consisted of what he had picked up from direct-ing his troops over terrain during the war. So to do the real work in Montana, he chose his home-town friend Benjamin Marsh, a trained surveyor who before coming to Cambridge City had spent more than a decade supervising the construction of railroads in the South and West.[2]

By May 9, Surveyor General Meredith had his instructions from GLO Commissioner Joseph Wilson. Buried deep in a document of twenty-nine "items" is this statement:

It is desired that "Beaver Head Rock," a remark-able land mark overhanging the river of that name be selected as the Initial point of the surveys, unless a more prominent and suitable point exists. . . . The stage road from Bannock to Virginia City passes by the "Rock" which is reported to be about midway between those two places, and the Rock is said to be visible for fifty miles up and down the stream and hence, eminently suitable for the Initial point of the public surveys in Montana.[3]

Virginia City and Bannack (as it is actually spelled) were two mining centers, with Vir-ginia City being also the territorial capital. Wil-son might have also noticed, from his maps, that a triangle with its points at Virginia City, Bannack, and Butte (the biggest mining cen-ter of all) has Beaverhead Rock near its center. And certainly somewhere in Wilson's consider-ations must have been the fact that the rock had figured prominently in the account of the ex-pedition to Oregon of Meriwether Lewis and William Clark.

In August of 1805, the expedition was moving up the Missouri in what is now western Mon-tana. There they discovered that the great river is formed by the confluence of three smaller streams, which Lewis and Clark named for their patrons back in Washington: from west to east, the Jefferson, Madison, and Gallatin rivers.[4] The Jefferson seemed to offer the best west-ward route, so they proceeded upstream on that river, and then up its tributary now known as the Beaverhead River. A few miles up that stream, Sacagawea, their Indian guide, saw a view like

that in figure 10.01 and suddenly recognized the area as her home place, from which she had been kidnapped five years earlier. On August 10, Clark wrote in his journal (recording directions with reference to his boat: "Stard."—starboard or right, and "Lard."—larboard or left):

Some rain this morning at Sun rise and Cloudy we proceeded on passed a remarkable Clift point on the Stard. Side about 150 feet high, this Clift the Indians call the Beavers head, opposit at 300 yards is a low clift of 50 feet which is a Spur from the Mountain on the Lard. about 4 miles, the river is verry Crooked, at 4 oClock a hard rain from the S W accompanied with hail Continued half an hour, all wet, the men Sheltered themselves from the hail with bushes.[5]

The rock is indeed a "remarkable" landmark. It looks exactly like what its name implies—a huge beaver surrounded by flatlands that look for all the world like a lake he's swimming through. Whatever symbolic or historical resonance it might have possessed, Beaverhead Rock's prime attribute, as a place to begin the survey, was that it lay near some of the principal mining centers of the territory, and the government was anxious to record the many mining claims on its maps.

Bear in mind that prospectors didn't make their claims in the aliquot, part-Section way that farmers and ranchers did, primarily because most miners had no intention of ever owning land. What they were staking claims to were the mineral rights to any ores in the land. Staking, by the way, is the operative term: a prospector would literally drive a wooden stake at each of the corners of his claim, the result being that most claims took the shape of long rectangles or trapezoids oriented not to the points of the compass but toward the direction the claimant hoped the vein of ore ran. The intention of instituting the survey in the mining areas was to dimensionally tie all those angled boundaries to the grid lines of Townships and Sections.

Mineral Rights

It seems incredible to us today, but until very recently, with the advent of environmental legislation, a prospector simply staked his claim and tore up the earth to look for ore. If he found valuable minerals, he'd cart them away and pay the federal government a tiny royalty. If he had no success, he'd simply walk away from the claim. The operative term of art in such an operation is **overburden**.

Overburden is the soil (and trees and grass, and even fields and houses) that's lying over the minerals to which the miner has a right because of his claim. The miner had no obligation to put the overburden back once he moved it. Worst of all, ownership of land didn't necessarily guarantee the property holder the mineral rights: if a miner obtained the mineral rights to the ore beneath someone's farmstead, all the farmer's fields and pastures, even his barn and house, became *overburden* and could be treated as such.

In the areas around the mining claims, on land not wanted by the prospectors, farms were being laid out. With all the prospectors digging at their claims, farmers knew there would be a ready market for their crops. Extending the Rectangular Survey would give these farmers secure title. To tie both mines and farms to the survey, it was necessary to begin the gridding-up of Montana, and so Meredith and Marsh set out from Indiana, arriving in Helena on June 22, 1867. The secretary of the Interior had been authorized to name the location of Montana's land office; he chose Helena even though Virginia City was the territorial capital. Possession of a land office was a great plum for any up-and-coming frontier town: Helena would become the capital in 1875[6] (and you do remember, from grade-school geography, that it's pronounced "HELL-uh-na"?).

Meredith's first task was to rent space for his office. The GLO Commissioner had insisted that it be fireproof, to protect the drafted Township plats that would be stored there. Meredith found such a place but encountered problems. So that you can hear the voice of an Indiana politico suddenly finding himself in a Wild West mining town, let me share with you the entirety of his first letter back to Washington.

Surveyor General office
Helena, Montana
July 29th, 1867

Hon Jos S. Wilson
Com. Gen. Land Office
Washington, D.C.

Sir:
Enclosed I send you a drawing of the Surveyor General office. The rents are very high in this place. The rent of the 2 rooms is $1500 per annum. My instructions required me to rent in a safe building isolated from others but no suitable one so located could be obtained as nearly all the houses are temporarily constructed of pine lumber adjoining or near each other. The rooms I have taken are in the only safe fire proof building I could obtain the walls being solid granite and the roof covered with 18 inches of earth.

The Register and Receiver, for the same reason have procured rooms in the back part of the building as indicated on the enclosed plan.

I have been doing business here for some time and obtaining information from reliable sources on the topography &c. of the Territory. I leave in the morning with Prof. Marsh to examine Beaverhead Rock, the point indicated in your instructions as the most suitable for the Initial point of the surveys.

When made I will report to you at once the result of our examination

Very respectfully
Your Obt. Servt.
S. Meredith
Surveyor General[7]

With his letter consigned to the postal service, Meredith set off from Helena the next day with Marsh, Walter deLacy (a local civil engineer who knew the terrain), and a crew of two chainmen and three axmen. We can get an image of the party from no less an authority than William Austin Burt. In 1854 Burt published A *Key to the Solar Compass and Surveyor's Companion,* which not only tells in great detail how to operate the Improved Solar Compass, but how to outfit a surveying team. Burt describes specifically the "outfit for a surveying company of 6 men for 4 months in the public surveys."8 Our Montana party was composed of eight men—diminished to six when Surveyor General Meredith and engineer deLacy returned to Helena—who were to be in the field for 3½ months, so they may very well have equipped themselves in a manner similar to this:

SUPPLIES OF PROVISION.
The following quality and kinds, or a substitute for
    them, is generally required.
8 barrels of flour.
2½ [barrels] of clear pork.
8 bushels of beans.
2 [bushels] of dried apples.
120 lbs. of good dry sugar.
70 lbs. ground coffee, or a substitute for it.
10 lbs. of saleratus [baking powder], or its
    substitute.
1 lb. of ground pepper .
1 small bag of table salt.
25 lbs. of rice.
4 lbs. of Castile soap.

CAMP FURNITURE.
1 large tent for the surveying company.
1 small tent for the packmen.
6 Mackinaw blankets.
3 common blankets to spread underneath them.
2 dozen boxes of matches. (best kind.)
1 good chopping axe.
4 tin pails, made to fit into each other.
14 tin basins.
1 set of knives and forks. (Small size.)
1 butcher, or meat knife.
7 spoons.
3 light frying pans.
2 half round cans, made to fit inside of the pails,—
    for lard and saleratus.
2 tin pepper boxes, with covers to fit closely over
    the sieve.
6 "soldiers' drinking cups," also needles, awls,
    thread, twine, small cord, &c.
2 mixing cloths, made of heavy cotton drilling, one
    yard square each.
4 papers of 3 oz. tacks for nailing boots.

FOR PACKING, ETC.
1 or 2 good horses, or mules, as circumstances
    require; one pack saddle; a bell and spancil [a
    rope fetter] for each.
20 stout bags, that hold one and a half
    bushels each.
4 linen bags, for pork.
6 small bags, for beans, dried apples, knives and
    forks, &c.
8 India Rubber bags for sugar and coffee. (Should
    be lined.)
2 strong drilling cloths, two or two and one half
    yards square, to do up the camp equipage into
    packs; also, strap and cords, to secure the packs
    to the horse and saddle.

1 solar compass.

1 case of drawing instruments.

1 measuring chain.

1 standard chain.

11 tally pins.

1 tape measure.

1 Telescope 16 or 18 inches in length.

2 marking tools.

2 pocket compasses.

2 marking axes, weighing three and a half pounds each.

1 hatchet, and two whet stones.

2 three-cornered files, for sharpening axes, &c.

2 small round files for sharpening marking tools.

Also, field books, mapping and writing paper, ink, pens, pencils, India rubber, mouth glue, and a small valise (or box) to carry them in.⁹

Burt then inserts some remarks on how to judge good-quality cooking equipment, and then helpfully appends a recipe for "camp bread" which sounds suspiciously like the formulation my grandmother used for biscuits.

Flour is mixed for bread on a cloth of cotton drilling, of about one yard square. It is done as follows:—

Spread the cloth on a blanket, folded and laid on the ground; pour enough flour upon it for a mixing, and make a hollow in it; then pour in some lard from the can, and add saleratus and salt dissolved in warm water, stirring the flour with a spoon to a proper consistency for kneading with the hand, taking care not to reach the bottom of the flour so as to wet the cloth.—Bake the loaves in the frying pans before the fire, and when done, fold the cloth, and lay it aside for future use.¹⁰

Following this are some discussions about tents and how best to carry supplies on horse- or mule-back. In rough country, supplies would sometimes have to be carried by the men themselves, and Burt describes two methods: one in which the strap attached to the pack passes over one shoulder and across the chest, in the manner of a bicycle messenger's bag; and a second in which two straps run over the shoulders and an additional strap passes across the forehead. Burt remarks, in passing,

Packs which are carried by a man, to supply a surveying company in the field, usually weigh from seventy-five to a hundred and twenty pounds each.¹¹

To complete your image, here is what Burt recommends for

SURVEYORS' WEARING APPAREL

The common wool hat is best for any season of the year, especially in timbered land.

Trowsers should be made large, and strong cloth.

A light coat, or frock, should be provided, well supplied with water proof pockets, to keep books and papers dry in wet weather, and a light India rubber, or waterproof cape should also be provided to keep the compass dry, when travelling in wet weather.

Flannel for under clothes, is preferable to cotton, for all seasons and kinds of weather.

Boots may be made of good kip [kid?] skin, and rather larger than for ordinary use; the fronts of the legs should be cut narrower, and the backs wider, than is usual to cut them. A thick single sole projecting about one quarter of an inch from under the upper leather, and well nailed over the bottoms with sparables, or tacks, are the most durable. The nails keep the feet from slipping, and the broad sole protects the upper leather from wearing against brushes, grass, and &c. A large silk handkerchief, of any colour but red, to tie over the ears and neck, is a good protection from flies and musquitoes.[12]

(Can we infer from this that American buffalo, like Spanish bulls, were apt to charge a red target?)

Reaching Beaverhead Rock, the team mounted the beaver's back and climbed up on top of its head to view the surrounding country. There they discovered what Lewis and Clark could have told them: although prominent in the local landscape, Beaverhead Rock is in reality an anomalous spur off the mountains enclosing one of the flat-bottomed valleys so characteristic of that part of the Rockies. And because the Beaverhead valley tends southwest-northeast, any lines projected in the cardinal directions from the rock would very soon run into those enclosing mountains, as you can see from figures 10.02 and 10.03.

Of course, a Meridian running north would also head straight for Butte and thus, if projected, could quickly "grid up" hundreds of claims fanning out from the town, which was sited on what would come to be called The Richest Hill on Earth. But the members of Meredith's team were thinking like surveyors, not deed recorders. Besides being tremendously laborious, a Principal Meridian and Baseline run through mountains would be hard to keep accurate, and any inaccuracies in those controlling lines would only promulgate as the survey progressed outward from them.

So, with the concurrence of his associates,

Surveyor General Meredith took the bold step of rejecting the wishes of his boss in Washington. The team clambered down the rock; cut overland to the east; moved southeast up the Beaverhead River's tributary, the Ruby River; then east up the Ruby's tributary, Alder Gulch, to Virginia City. From there they struck east until they arrived at the valley of the Madison River, flat-bottomed like the valley of the Beaverhead. They followed the Madison valley north some dozen miles to a point where a curious topographic circumstance occurs, one which probably determined the location of Montana's Principal Meridian.[13]

At that point the Madison valley encounters the Tobacco Root Mountains coming in from the west and the Madison range coming in from the east. Between them a saddle of land rises up from the valley floor, with the Madison River veering east to drive through the toe of the Madison range in the steep ravine of Bear Trap Canyon. Rather than follow the Madison through the ravine, the party mounted the saddle and came to an important realization. When they reached the crest of the saddle—that place where the "sagging" east-west contour of the land between the mountains meets the "bulging" contour of the valley bottom rising north-south over the gap—they could look south and see almost the whole of the Madison valley.

That seeming propitious, the party broke out their instruments and ran a test Meridian north about 18 miles, at which point the surveyors found themselves on top of a prominent rise of land. From that crest they could see that they were about 3 miles south of the Jefferson River, and they could look east and see that the land ran relatively flat for many miles before encoun-

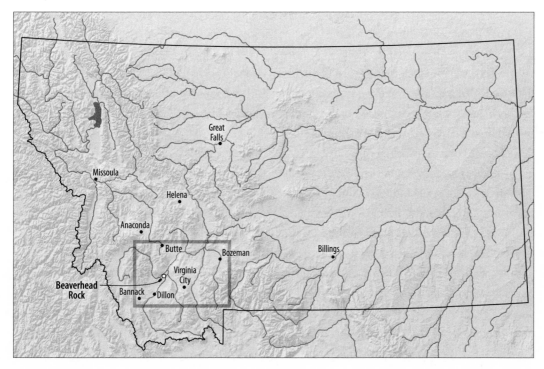

FIG 10.02

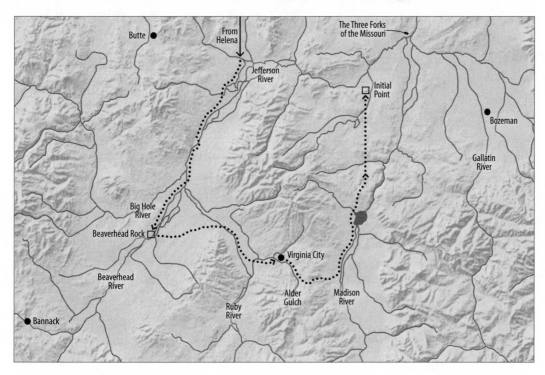

FIG 10.03

tering what they knew to be the Gallatin Mountains.[14] They were, in fact, in the midst of one of the broadest of the valleys in the northern Rockies, formed by the alluvial soils pulled down by the three tributaries of the Missouri as they flowed toward one another. The combined valleys of the Jefferson, Madison, Gallatin, and the resultant Missouri form a rough diamond of terrain between mountains, extending some 60 miles east to west and only a little less north to south. One of the richest agricultural areas in the state, it is little wonder that Montanans chose to locate their agricultural and mechanical college—Montana State University—in Bozeman, at the eastern end of the beautiful Gallatin valley.

None of the team could foresee this future, of course, but to their eyes this hill, found almost by accident, seemed an ideal spot for the Initial Point; but to be certain, the men split up and reconnoitered the surrounding terrain for alternatives. Finding none, they reassembled atop the hill and declared the place they had found to be the Initial Point for the Rectangular Survey of Montana. To mark the spot exactly, they incised a cardinal-points cross into a place where the hill's limestone core broke through the surface, and then placed witness cairns of stone around the spot so it could be seen from a distance.

We don't know the exact date when the team carved their X and made the Initial Point official, but it was some small number of days before August 19, 1867, because it was on that date that Benjamin Marsh, with his two chainmen and three axmen, plumbed up his solar compass over that cross and began marking the Baseline of the survey to the east.

Certain that the work was in good hands, Surveyor General Meredith (and, it appears, engineer deLacy) set off back to Helena. Once secure in his fireproof office, Meredith, as required, made a fair copy of the Contract no. 1 he had signed with Marsh—on August 1, on their way to Beaverhead Rock—and mailed it off to GLO Commissioner Wilson on September 27. Wilson got the letter on October 14, and that very day fired off his response—in effect:

"You wrote to me last July 29 that you and Marsh were leaving to examine Beaverhead Rock, a place indicated in my instructions as most suitable for the Initial Point, and yet only three days later, before you could have gotten there, you signed this contract with Marsh. That contract is hereby suspended and furthermore,"

You are therefore hereby required to report to this office at once as to the Initial Point, whether it was established by you at Beaver Head Rock as indicated in our instructions and if not, its locality and reasons which influenced your action in departing from them, and await further instructions.[15]

This correspondence was occurring while Marsh was in the field extending the Principal Meridian and Baseline away from the Initial Point. Marsh finished his work on October 12, then returned to Helena to draft the plats of his work and copy out his field notes. Meredith approved them on October 25, and then—days later—the letter arrived from his boss in Washington rescinding the contract he had made with not only his good friend but the man he had been counting on to do the real work of the survey.

Meredith's long letter in response, dated November 2, betrays a man thoroughly chastened (and perhaps in something of a panic). He bolsters his words with an appended report from his local expert, Walter deLacy. All this seems finally to have mollified Commissioner Wilson, but in light of Beaverhead Rock's proximity to mining interests, we can only imagine his reaction to the last sentence in deLacy's report:

I would also remark that the Meridian and Base Lines as now located will accommodate a larger scope of the agricultural & animal interests that could be done in any other part of the country.[16]

Not for the first (or the last) time would people on the ground in the West foresee the future more clearly than officials back in Washington.

I want to show you one of the documents prepared by Deputy Surveyor Benjamin Marsh to record his surveying around the Initial Point in the summer and fall of 1867.

The document is on drafting vellum, now mounted on a cloth backing, and measures just under 9 inches by 70 inches. It resides at the Bureau of Land Management headquarters in Billings, and the staff there were kind enough to let me handle this record of the very beginning of the Rectangular Survey of Montana.

Like all such surveying records of its era, it was drafted using very fine steel "crow-quill" pens dipped into an inkwell and then carefully wiped before committing ink to vellum. Besides employing a variety of line thicknesses to record his findings, Marsh used several different styles of lettering; and to the delight of any drafter like myself, his faint pencil guidelines are visible at the top, bottom, and sometimes middle of each line of letters.

The drawing records the Principal Meridian, from the Baseline at the Initial Point, up to the "top" of Township 10 North. In figures 10.4 and 10.5, I'm joining two images that cover the bottom 22 inches of the roll, just up into T.3N. Just north of the Baseline, you can see the hachures indicating the slope down from the hill on which the surveyors had established their Initial Point. Just past 3 Sections north of the Baseline, you can see the Jefferson River flowing across the Meridian, and around the southern edge of T.2N are more hachures marking "Road from Jefferson Ford to Helena."

In the title block, you can see Marsh citing his "Contract No. 1" with its August 1 date that got his boss in such trouble with Washington. The drawing is dated "Helena, Montana Territory December 7, 1867," and we can imagine Marsh toiling away in one of those fireproof rooms, under that sod roof, as fall gave way to winter. Finally, in the lower left-hand corner is the beautiful signature of Marsh's friend from Indiana, Solomon Meredith, Surveyor General of Montana Territory.

There would have been other drawings that mapped the extension of the Principal Meridian southward, and the Baseline to the east and west. These, though, would have been subsidiary to the Township plats, one per sheet, that constituted the prime requirement of Marsh's contract. In the summer and fall of 1867, Marsh and his crew were able to survey fifteen Townships, reaching north as far as Helena.

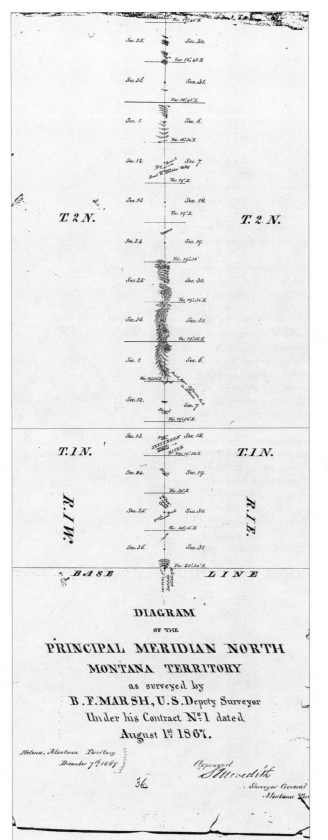

FIG 10.04–10.5

Bureau of Land Management,
United States Department of
the Interior.

A Note on the Maps

In the spring of 1999, the Bureau of Land Management moved its Montana headquarters from an office block in downtown Billings to a new facility out on the interstate, purpose-built to incorporate heightened security. But the records compiled by all the Surveyors General since 1867 moved with them, and it is from those records that I compiled these maps.

The basic source was a series of folio-sized volumes containing hand-lettered copies of all the contracts let to Deputy Surveyors for work in Montana. The contracts were copied as they were issued, with a notation after each as to how (or whether) the work had been carried out. In almost all cases, the work was completed within the calendar year, so the maps make the assumption that all work contracted in, say, 1867 was completed in 1867.

The contracts detail whether the Deputy was to run Meridians or Parallels, or Township exteriors, or the lines dividing Townships into Sections. On the maps I have marked a given Township as Surveyed if a contract from that year called for its exterior lines to be run. As you move into the maps for the 1880s, a few curious "unrun" Townships appear. These blanks might be due to errors in my transcription; or it might be the case that some long-ago Surveyor General acceded to the fact that when surveyors run the exterior lines of the four Townships surrounding a "blank," the exteriors of that Township were inevitably marked as well—and no contract need be let until its interior lines needed to be run.

The names of natural features—which might not coincide with local usage—are taken from the DeLorme Company's *Montana Atlas & Gazetteer*. To help you get your bearings—and a sense of the scale of the landscape—I've taken the Meridian line you just saw and mapped it onto the base terrain we'll be using for the rest of this chapter.

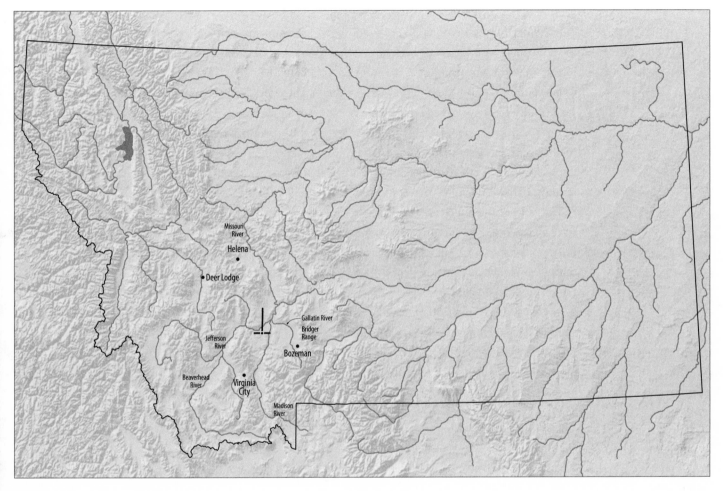

FIG 10.06

From their Initial Point, the surveyors pushed outward in all four directions. To the south they retraced their test Meridian back to the saddle between the Madison and Gallatin ranges, then extended it down through the valley of the Madison River. They pushed their Baseline east across the floor of the Gallatin valley to where the Bridger range rises up. To the west they took the Baseline across the foothills of the Madison range into the valley of the Jefferson River.

The bulk of the surveyors' work was to the north. They extended the Principal Meridian a full 60 miles across the Missouri, stopping at the foot of the Big Belt Mountains. Along the way, at the mandated 4-Township intervals, they began their First and Second Standard Parallels North. The First Parallel stopped in the west against Bull Mountain; and on the east it reached as far as the Horseshoe Hills and Sixteenmile Creek. The Second Parallel also stopped against hills: the Big Belt Mountains to the east and the Boulder Mountains to the west.

In marking Township exteriors, the surveyors stuck to a principle they would follow throughout the Montana survey: lay out Townships in the flatter lands first; cover hilly lands later, if at all. They ran a Range of Townships on the east side of the Principal Meridian, the flatter side nearest the Missouri. Then, once they had gotten past the Elkhorn Mountains, they could push Townships west, across the bench of flatlands east of Helena and on to the town itself.

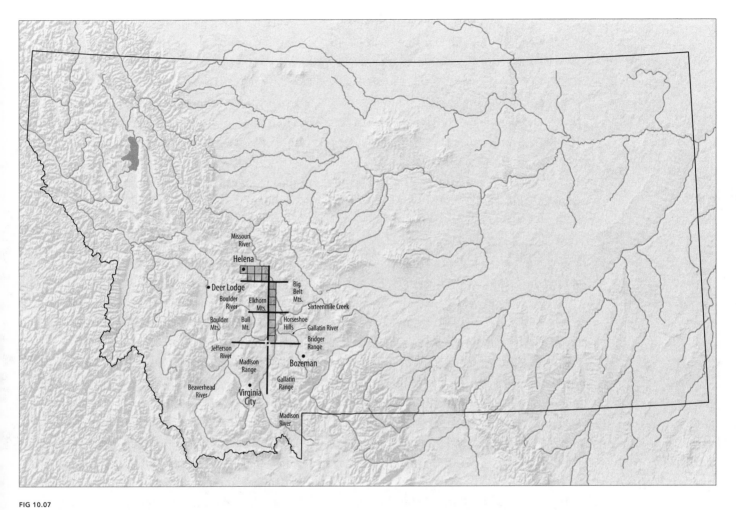

FIG 10.07

In the vicinity of the Principal Meridian, surveyors pushed out Townships in all directions. To the east, they virtually covered the floor of the Gallatin valley—east to the Bridger range, south into the Gallatin range—right up to the town site of Bozeman. To the north they laid out a second Range in the valley of the Missouri, and a third Tier north of Helena. To lay out this third Tier, the surveyors extended the Principal Meridian 6 additional miles north, stopping it in the heights of the Big Belt Mountains. Significantly, the Principal Meridian would remain stopped at this point for decades as surveyors worked around the obstacle of the mountains.

To the south, surveyors laid out Townships across the Madison-Gallatin saddle, and into the Madison valley. In the pattern mandated in the 1855 edition of the *Manual of Instructions*, they began their First Standard Parallel South at a point 5 Townships south of the Baseline, dead-ending it in the Gallatin range on the east, and pushing past Virginia City to reach the Beaverhead River in the west, very near Beaverhead Rock.

In this part of the Rockies, the Continental Divide makes a rather dramatic switchback, due to the course of two rivers. The Big Hole River, almost skirting Montana's western border, drains into the Missouri-Mississippi system; yet the Clark Fork River, whose tributaries extend as far east as Butte, flows eventually into the Columbia. (Along the narrow valley of the Clark Fork ran the main route through Spokane to Oregon.) In 1868 the decision was made to bring the survey to the valleys of these two rivers, the one to the west draining to the Atlantic, the one to the east flowing to the Pacific.

To get to the Big Hole, surveyors pushed the Baseline west, up over the Highland Mountains then down, then up again into the Pioneer Mountains. At a peak in the Pioneers, they decided not to advance the Baseline farther west but to strike out north—starting a Guide Meridian—down to the valley floor of the Big Hole. Once at the river, they followed its course to the southwest. At the proper calculated latitude, they restarted the Baseline.

Along the way the surveyors had started a Guide Meridian that would pass north down the middle of the valley of the Clark Fork. Following that Meridian north, at a point east of the mining town of Anaconda, they calculated that they were at the latitude of the First Standard Parallel North, so they projected that parallel east, to join up—at the crest of Bull Mountain—with the line projected west the previous year.

In the valley of the Clark Fork, the team spread out Townships, engulfing the river town of Deer Lodge. North of the town they calculated that they were at the latitude of the Second Standard Parallel North, and so they pushed a line east toward it (the two ends of the line would join four years later).

CHAPTER TEN

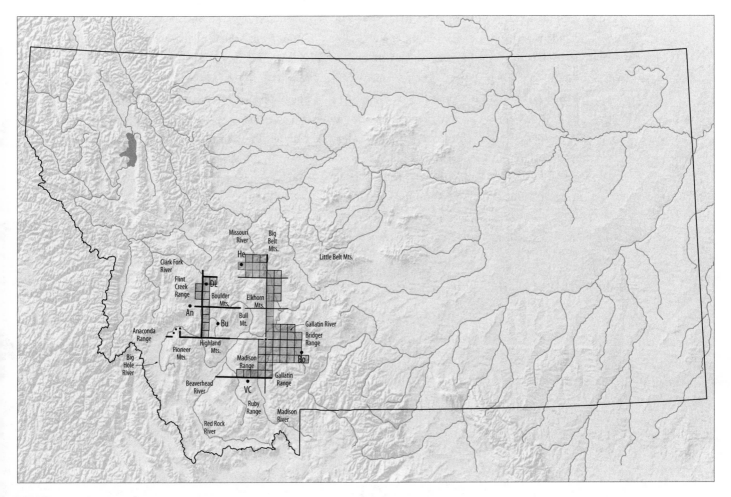

FIG 10.08

Although surveyors had begun to lay down the regulating lines for a survey of the Big Hole River's valley, work in that area was postponed for three years. Instead, to the east, a new Guide Meridian was run up the valley of the Beaverhead River, and a chain of Townships laid out in the flatter lands around the Madison range. As part of this effort, the Principal Meridian was extended south to the territorial border, and a Second Standard Parallel South begun 5 Townships south of the First, as well as a new Guide Meridian up the valley of the Ruby River.

To the north, surveyors had abandoned the effort to push the Principal Meridian over the Big Belt Mountains, opting instead to project a Guide Meridian north from a Township corner. They positioned the Meridian so that it would pass up the valley of the Missouri, but then they extended it across the Dearborn River and on to the Sun River. In breaking through the mountains, the surveyors had come down onto the westernmost reaches of the Great Plains—where the survey could truly spread out. By lucky happenstance, a Fifth Standard Parallel North (120 miles north of the Initial Point) would pass down the Sun River. They pushed that line eastward, laying out two Tiers of Townships in the flatlands north of the river. At a point they calculated to be north of where the Principal Meridian had stopped, they shot a meridian line south to (eventually) meet it.

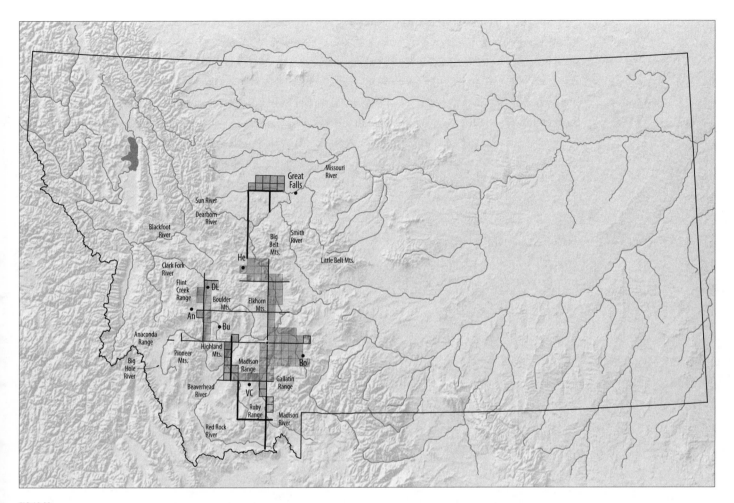

FIG 10.09

In the south, the Surveyor General chose not to follow up on the guidelines pushed across the Ruby range, but instead to lay out Townships along the Beaverhead River, toward the mining center of Bannack. Where the Beaverhead meets the Big Hole River, the combined stream takes on the name *Jefferson*, and a mat of Townships was spread out on both sides of the Jefferson River. As the Jefferson heads toward the Three Forks, the Boulder River joins it from the north, and a stack of Townships was pushed up the Boulder—with a jog to the west to avoid the heights of the Elkhorn Mountains.

To the east, the Second Standard Parallel North was extended east to the Smith River, in preparation for Townships to be laid out there in the following year.

The most dramatic advance, though, came to the west. After laying out two new Townships near Deer Lodge, surveyors there had run out of flat land where the Clark Fork dove into a narrow valley. But everyone knew that farther on, the river flowed across the northern end of the Bitterroot Valley, and so the surveyors meandered the Clark Fork—mapped its curves into measured straight-line segments, you remember—to its confluence with the Bitterroot River. There they calculated the latitude of a Third Standard Parallel North and laid out that line east and west, just north of the valley, across the foothills of the Sapphire Mountains. From that parallel they pushed a Guide Meridian north to Missoula, at the confluence of the Clark Fork and Bitterroot, laying out three Townships around the town. They also projected that Guide Meridian southward, into the valley proper, in preparation for Townships they would lay out two years later.

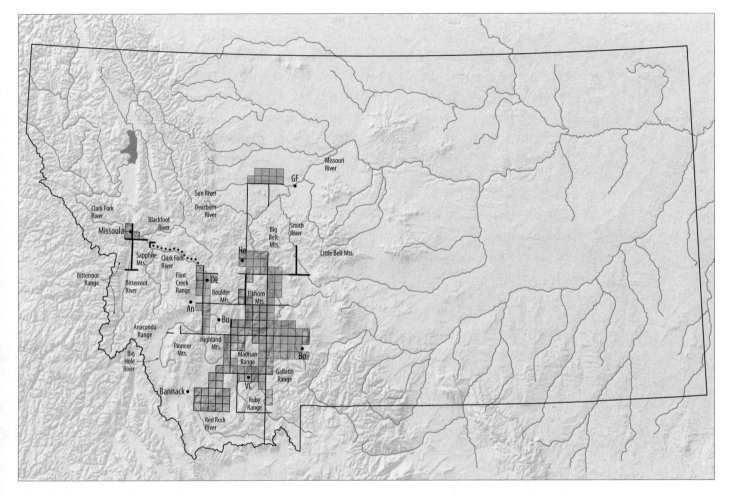

FIG 10.10

Rather than complete the Bitterroot valley survey, the survey teams to the west expanded the Townships around Deer Lodge. In preparation for later surveying down the Blackfoot River, they extended their Guide Meridian north, then struck east with a Third Standard Parallel North, stopping the line against the mountains. (The Blackfoot is the river that "runs through it" in Norman Maclean's classic story.)

To the south, more Townships were added along the Beaverhead River; and to the east, the Smith River got a stack of Townships, split to accommodate the Little Belt Mountains.

The big "push" of 1871 was in the north, where Townships were spread west—along the Sun and Dearborn rivers, up to the foothills of the Lewis and Clark range—and eastward along the Missouri. The surveyors were now at the great Falls of the Missouri, and the town named for them; the extension of the Fifth Standard Parallel North crossed the river near the town site. The crews could lay out 6-mile-square Townships in a stairstep fashion, to follow the northeast-flowing Missouri; but the 24-mile interval between Standard Parallels entailed bigger jogs. They took their Fifth Parallel east, well past the Missouri, then projected a Guide Meridian 24 miles north to a Sixth Standard Parallel North—which, happily, would run along the Teton River, another likely site for settlement.

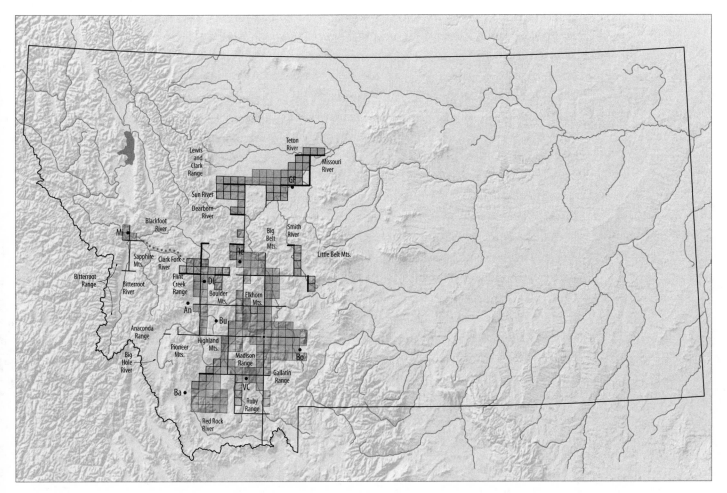

FIG 10.11

Rather than extend the Townships at the mouth of the Teton, it was decided that two Tiers of Townships were to be laid out near the river's headwaters, up against the Lewis and Clark Range. So a Guide Meridian was driven north, the latitude of a Sixth Standard Parallel calculated, and the Townships projected from that new Baseline.

A few miles south, "stairstep" surveying continued northwestward along the Blackfoot River. Farther south, surveyors returned to the guidelines they had marked near the Big Hole River and used them to project Townships around the flanks of the Pioneer Mountains, setting out a Guide Meridian as soon as the general run of the mountains allowed for a long north-south line. And at the far southwestern corner of the territory, the Second Standard Parallel South was pushed into the Tendoy Mountains, to serve as a Baseline for Townships in the hills southwest of the silver-mining town of Bannack.

To the west, surveyors returned to the Bitterroot to complete the survey of its valley, jogging their Guide Meridian westward to accommodate the shape of the valley floor. The Bitterroot valley, flat-bottomed but edged by alpine hills, is one of the signature landscapes of Montana: it's no wonder that Montanans chose to site one of their land-grant universities there.

Equally iconic is Flathead Lake, the largest expanse of fresh water in the West, at either end of which are broad, fertile valleys. The southern valley had been reserved for the Flathead Indian tribe, but the northern valley was open for settlement. So a team of surveyors took off overland to the Little Bitterroot River at the edge of the Flathead lands. They followed that river north to where their calculations showed them the Sixth Standard Parallel North would fall. There they stopped and projected that line west into the mountains and east to the shore of Flathead Lake. Then they continued northward, then east, and finally entered the northern valley. There they again calculated their latitude and laid out a Seventh Standard Parallel North, across the breadth of the flatlands, projecting a clutch of seven Townships. Their work was to be the last time government surveyors would traverse the area until the coming of the Great Northern Railroad in the 1880s.

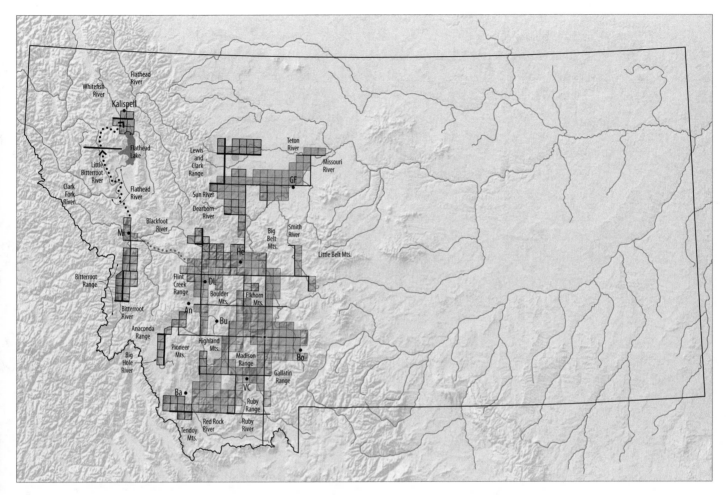

FIG 10.12

In 1873 surveyors pushed the Second Standard Parallel North over the pass between the Anaconda range and the Flint Creek range to establish Townships in the valley of Flint Creek. In this region too, this would be the last major surveying campaign until the coming of the railroads. For the next decade and beyond, the surveyors would focus their efforts along the three rivers that flowed east across the Plains: the Missouri, the Musselshell, and the Yellowstone.

Up by the Missouri, surveyors extended the Tiers of Townships along the Teton River, connecting their Sixth Standard Parallel to the line marked two years earlier. And they pushed the Smith River Townships south to the edge of the Big Belt Mountains.

The survey of the Musselshell began in earnest with the eastward extension of the Second Standard Parallel. From that line, Townships were projected south to the river and north until they hit the Little Belt Mountains.

Near Bozeman, the Baseline was carried east over the Bridger range, and the Townships projected from it just reached the banks of the Yellowstone.

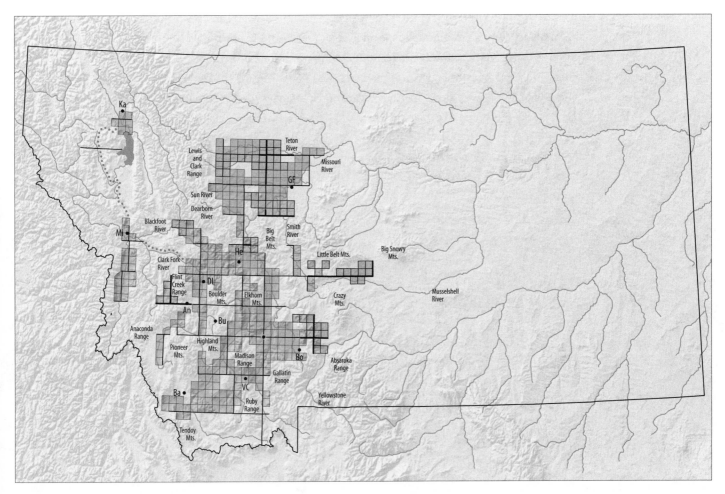

FIG 10.13

The Yellowstone River survey pressed on, extending a Guide Meridian southward to the calculated latitude of the First Standard Parallel South and then beyond. As with the Townships themselves, the Guide Meridian sidestepped westward to skirt the heights of the Absaroka range.

The Second Parallel North was pushed 30 miles east, and Townships were projected north from it until stopped at the Big Snowy Mountains.

To the north, the Principal Meridian was extended to a Seventh Parallel North, to serve as a Baseline for Townships along the Marias River.

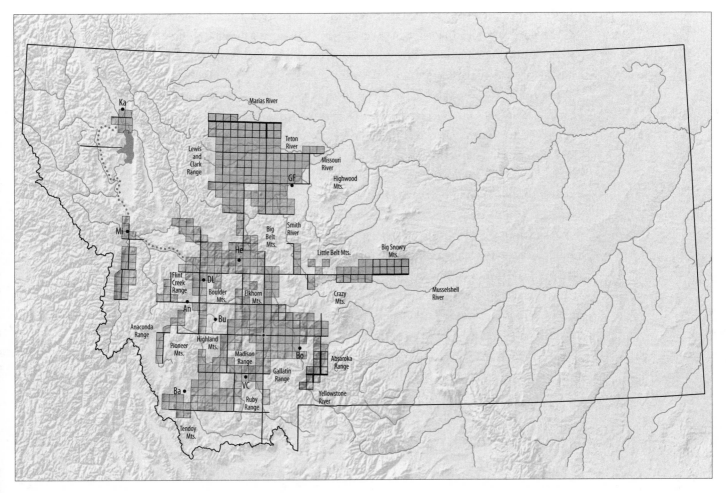

FIG 10.14

To the north, the Principal Meridian reached the Marias River, and Townships were laid out along it. In addition, a block of six Townships was added north of the Teton.

The real action, though, was to the east. In moving eastward, the Second Parallel North passed by the flatlands between the Little Belt and Big Snowy mountains. North of that saddle lay the Judith basin, a region watered by a rack of streams flowing northeast into the Missouri. So a Guide Meridian was struck north (offset, as required, at the fourth Township), up to another calculated Standard Parallel, this time the Fourth. From these guidelines Townships were extended east to the mountains enclosing the basin, and north along the streams parallel to the Judith River.

There was reason, in 1875, for extending the survey north into the Judith basin rather than east along the Musselshell and Yellowstone.

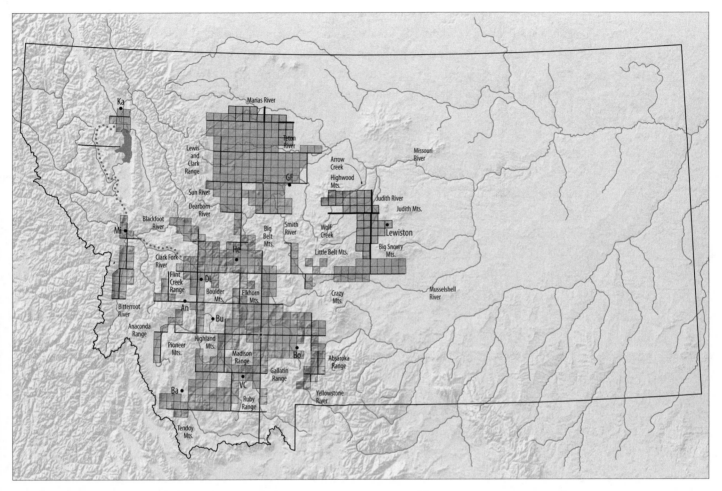

FIG 10.15

Recall that the Black Hills of South Dakota lie just east and south of Montana's southeastern corner, and that from the 1860s onward, there had been unrest among several Indian tribes over prospectors invading the tribes' sacred lands. The great wedge of Montana south of the Yellowstone River had been acknowledged as an Indian reserve for quite a while, but with the white influx into the Black Hills, many Sioux had migrated north into the area. At the end of 1875, President Grant decided that the Sioux must be forced back onto the truncated reservation set aside for them east of the Black Hills. From this decision came the last of the Indian wars, and with it, the battle near the Little Bighorn River on June 25, 1876.[17]

With the defeat of General George Armstrong Custer's Seventh Cavalry, the determination of Grant and Congress redoubled, more troops were sent, and by the spring of 1877 most of the Indian bands had surrendered. One result was a great reduction in the Indian reserve south of the Yellowstone. The Crow people got a tract along Montana's southern border, centered on an Agency outpost near the battle site, with a smaller Northern Cheyenne reservation to the east. The northern borders of both reservations stopped well south of the Yellowstone River.

In 1876 surveyors extended Townships southward along the Musselshell, east on the Yellowstone, and then north along Sweet Grass Creek.

Farther west, a new Guide Meridian was driven north over the Anaconda range, and Townships were extended up the Big Hole River, where a third segment of the First Standard Parallel South was calculated and begun.

Given that we now think of Butte, Montana, as sitting on The Richest Hill on Earth, it's perhaps surprising that the survey reached it, and nearby Anaconda, only in 1876. Butte began as a gold-rush town in the 1860s, but a few years later the gold played out and the town became a backwater. Early miners had found silver also, but the ore required expensive smelting. Only with the arrival of big outside money could silver mining take off, and in the 1870s it did, with Butte as the center of mining and Anaconda the site of smelting operations.

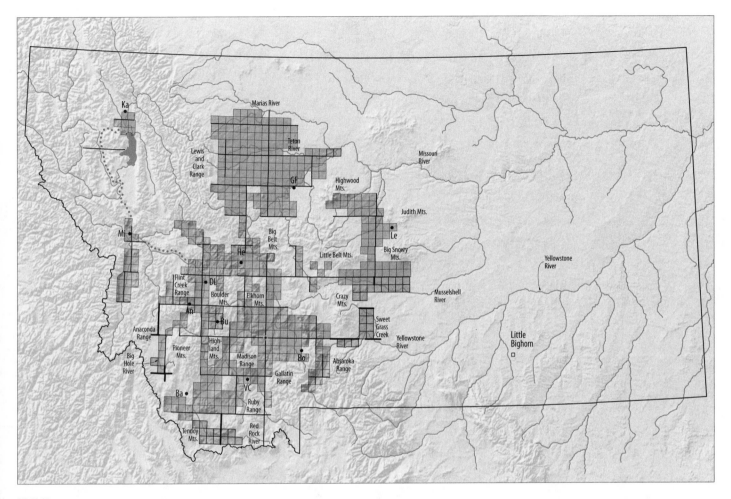

FIG 10.16

1877

The defeat by the Indians at Little Bighorn had an immediate effect on the course of the Montana survey. With the Crow and Northern Cheyenne tribes confined south of the Yellowstone, the survey's Baseline was extended 50 miles to the river's banks. The First Parallel North was extended a hundred miles east until it too touched the Yellowstone. Then a Guide Meridian was sent north, and a Second Standard Parallel sent a further hundred miles east. All three lines would serve as an armature for a major surveying effort to come.

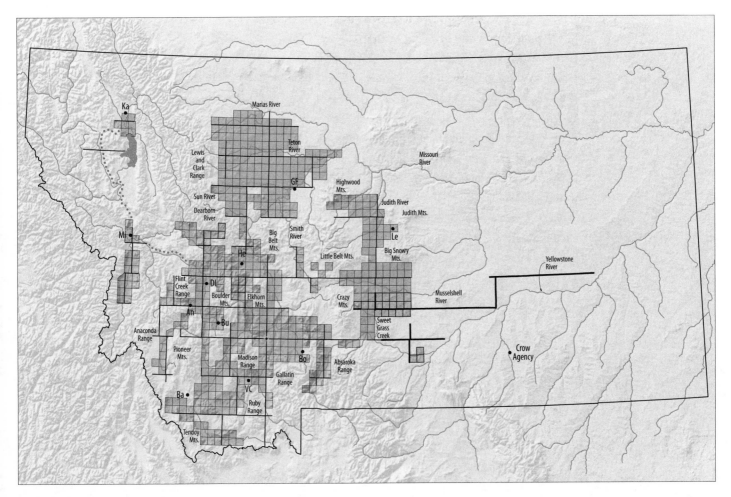

FIG 10.17

But rather than extend Townships directly from these guidelines, the surveyors at first stuck closely to the banks of the Yellowstone, stairstepping Townships to follow the river's course. They clustered their first Townships at places where streams flowed into the Yellowstone: those confluences would be likely town sites, and settlement would naturally spread up those streams. It was also presumed, by now, that a railroad would soon be built across Montana, and the likeliest route would be along the Yellowstone— another reason for laying out Townships close to the riverbanks.

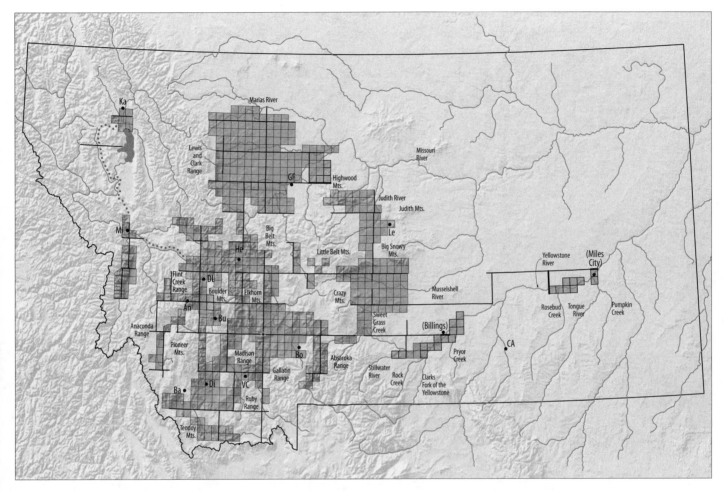

FIG 10.18

With the Indians confined to their reservations, white settlement could move up the many streams that flowed north into the Yellowstone. So the surveyors jumped south, calculated the latitude of the First Standard Parallel North, and prepared to lay out Townships along Rosebud and Pumpkin creeks and along the Tongue River.

Farther north, a crew pushed the Third Parallel North westward into the Little Belt Mountains. Note how the surveyors were now working more systematically: laying out their Guide Meridians and Standard Parallels far in advance of Townships, with the more-skilled surveyors marking the guidelines, the less-skilled "filling in" between.

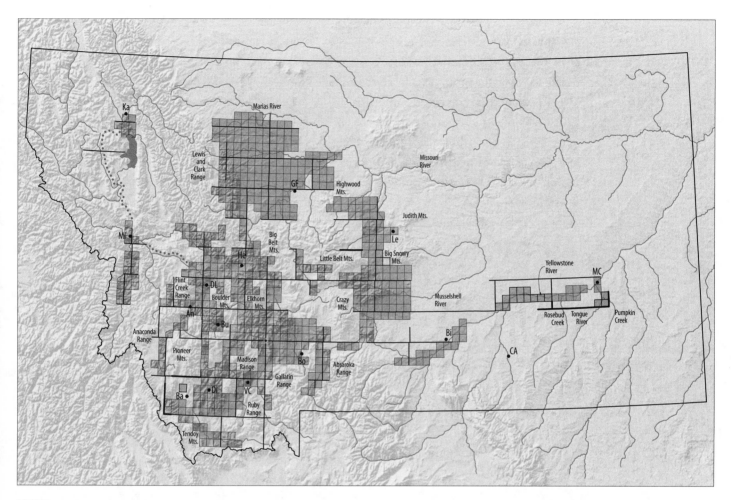

FIG 10.19

Sure enough, in 1880 that Third Standard Parallel into the Little Belt Mountains got its block of 12 Townships. And from that block, surveyors could extend the Fourth Parallel and a Guide Meridian around the Highwood Mountains.

Just to the east, survey teams extended the Second, Third, and Fourth parallels eastward. Significantly, as they carried those latitude lines over mountains, they didn't stop to monument the lines with half-mile Section markers, since settlement would be unlikely in rugged terrain.

Farther east, a team projected another Guide Meridian northward across the Yellowstone, calculated the seventh segment of the Third Parallel North, and laid out a block of Townships.

The big news, though, was coming up from the south. In 1880 the Utah & Northern Railroad was moving up the Red Rock and Big Hole River valleys, and by the end of the year, the new town of Dillon had been founded.[18] From here on out, the course of the survey would be determined as much by railroad lines as by riverbeds.

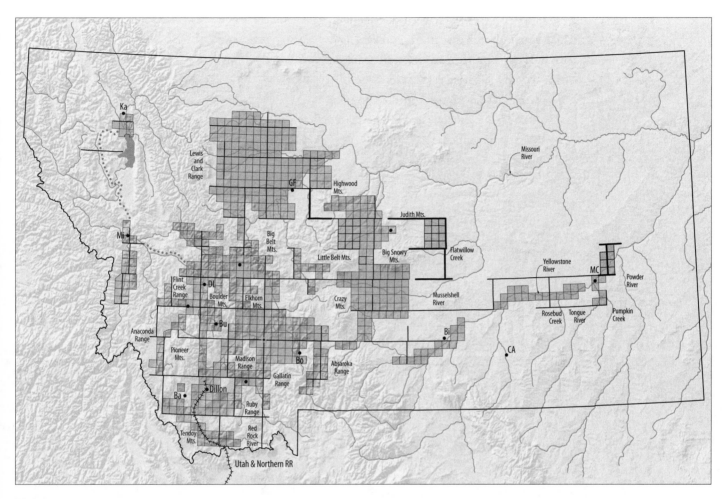

FIG 10.20

By 1881 the Utah & Northern had reached Butte,[19] which meant that all that silver (and later, copper) could now be carried south toward Salt Lake City and from there head east or west.

The same year saw a great spurt of surveying eastward on the Yellowstone and its tributaries. You will see why in a moment.

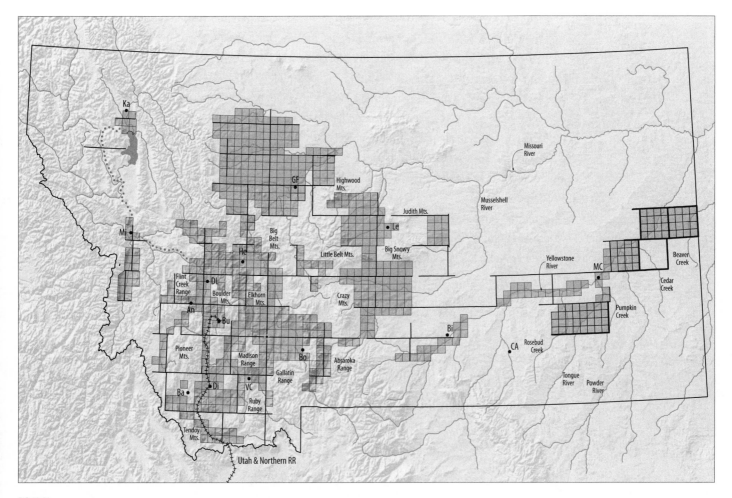

FIG 10.21

In 1882 the Northern Pacific Railroad was laying track up the Yellowstone from the new town of Sidney, at the territorial line. The Montana survey was quickly extended to meet it.[20]

In a second great "push," the three Standard Parallels begun the previous year were extended to their goal, the Musselshell River. Look closely at these new Townships and at those to the east along the Yellowstone, and remember that it was the 1881 edition of the *Manual of Instructions* that formalized the idea of a quadrangle of 16 Townships enclosed by Guide Meridians and Standard Parallels.

Look also to the west: in along the Clark Fork River came the tracks of the Northern Pacific, heading east from Spokane. And note the Fourth Parallel North, and two Townships, flung out to meet them.

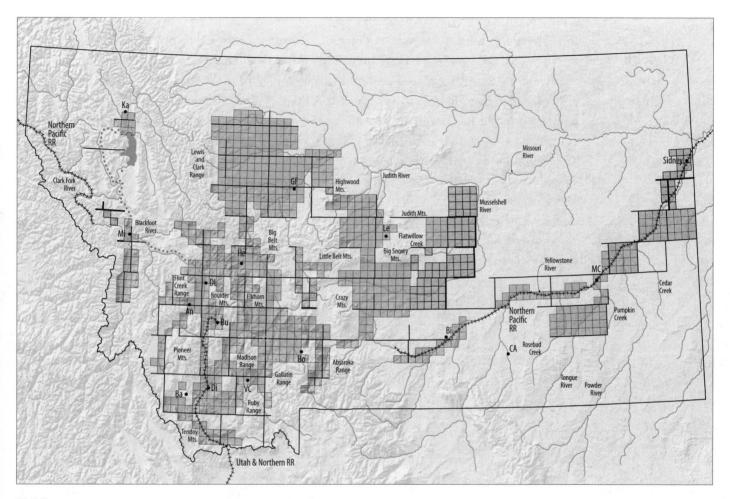

FIG 10.22

Montana had its own Golden Spike moment when the two branches of the Northern Pacific met up at Gold Creek on September 8, 1883.[21]

Surveying work that year consisted mostly of "infill." Nine more Townships were pushed toward the Judith River, with an isolated six Townships to the east, just north of the Judith Mountains. North of Bozeman a brace of seven Townships were pushed up against the Crazy Mountains. South of Bozeman was a small extension along the Yellowstone River, up against the border of the National Park. And to the east, there was more coverage along Pumpkin Creek.

The big campaign that year came at the request of the Crow Indians. They asked that the survey be extended into their reservation, and so surveyors moved up the Bighorn and Little Bighorn Rivers, working from a Guide Meridian that sidestepped at the latitude of the Baseline and then again (5 Townships south) at the First Standard Parallel South. Note that the battlefield site falls in the Township just southeast of the Crow Agency. What must the surveyors have felt as they ran their lines?

CHAPTER TEN

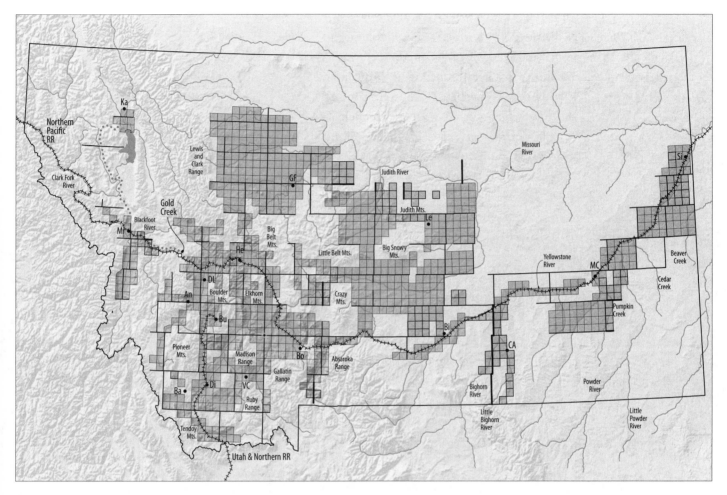

FIG 10.23

The southeastern corner of Montana was east of the Crow and Northern Cheyenne reservations, so that is where the survey went next. You'll note that apart from a few unexplained jogs, the surveyors adhered to the 4-by-4-Township quadrangle system—with the exception of the 4-by-5 quadrangles caused by the "5 Townships to the south" rule from the 1855 *Manual of Instructions.*

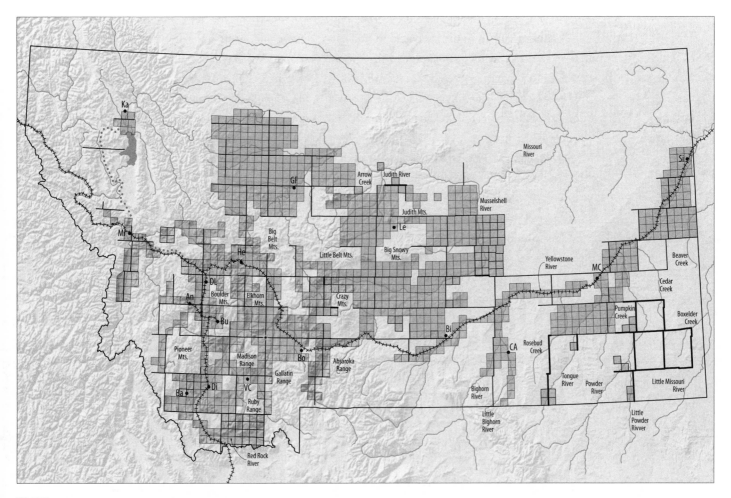

FIG 10.24

From the nascent grid of quadrangle guidelines spreading south from the Yellowstone, you can foresee the growth of the survey in the 1880s. The railroad you see stretched across the north of Montana just as surely predicts the survey's expansion in the 1890s through the turn of the twentieth century.

In 1887 the St. Paul, Minneapolis & Manitoba Railroad pushed west from Minot in northern Dakota Territory along the Missouri and then the Milk River to the new rail center of Havre (Montanans pronounce it HAV-er), and from there southwest to Great Falls. The same year the Montana Central reached Great Falls from Helena. In 1890 the two railroads were consolidated as the Great Northern, which soon pushed a line over the Rockies, through a pass mapped only in 1889, to reach Puget Sound in 1893.[22]

The Great Northern and Northern Pacific would later reorganize, merge, be split up, and adopt new names; but all the railroads in Montana had a common goal of promoting settlement along their tracks. To that end they founded towns all along their lines, as close as 10 miles apart, as trading centers for the farmers to come. Settlers were enticed to the Plains by advertisements, discounted land, and the new, scientific practice of "dry farming." We'll look more closely at this mode of farming—and the heartache it brought—in the epilogue. Suffice it to say here that by the eve of World War I, much of eastern Montana was planted to crops like wheat, all the farmsteads apportioned out of a Rectangular Survey that now spread almost unbroken across the state.

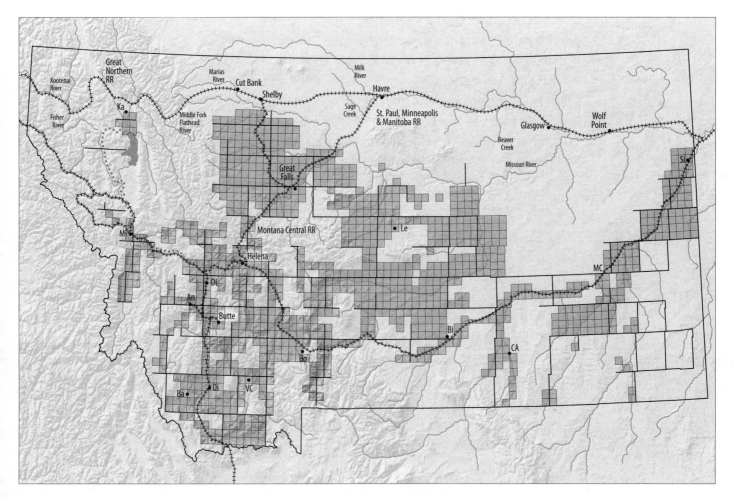

Great
Northern
RR

Kootenai
River

Fisher
River

Ka

Mi

Middle Fork
Flathead
River

Marias
River

Cut Bank

Shelby

Great
Falls

Montana Central RR

Helena

DL

An

Butte

Bo

Ba

Di

VC

Milk
River

Sage
Creek

Havre

St. Paul, Minneapolis
& Manitoba RR

Glasgow

Beaver
Creek

Missouri River

Le

Bi

CA

Wolf
Point

Si

MC

FIG 10.25

Once the idea arose of withdrawing land from possible development (by sequestering it in national forests or parks), it became apparent that there was no need to apportion those lands into Townships and Sections, and so no contracts were let to survey them. But apart from those specifically designated areas, the Rectangular Survey spread across the whole face of Montana.

Naturally, since the survey was started up from so many points, a map of the survey shows inevitable "misfits" where different surveys butted into each other as they expanded. Nonetheless, even with some Townships misshapen, within those Township lines lies the possibility the Rectangular Survey was set up to achieve: a true *cadastre* of the state—the entire surface of Montana apportioned into bordered parcels, and a unique owner assigned to each. By tracing surveyors' steps from their notes, the course of any line marked by government surveyors can be located again. By following aliquot principles, any subdivision of those government lines can likewise be found, even if obscured by time, even if lost to memory.

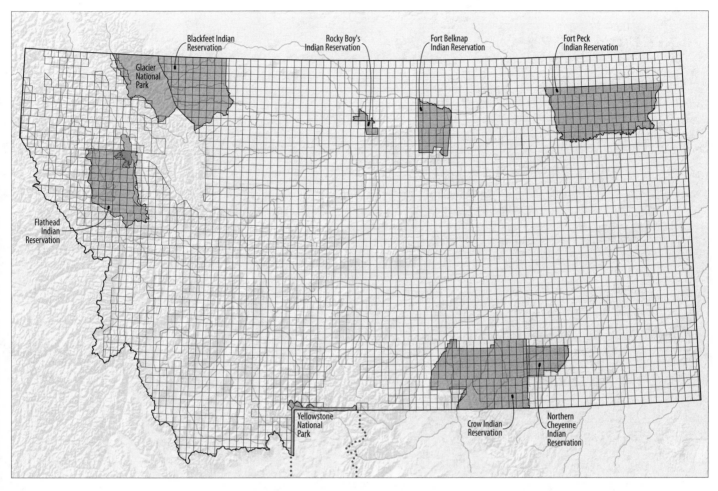

FIG 10.26

Other Ways to Apportion a Public Domain

TEXAS AS A COUNTEREXAMPLE

The great chunk missing from the U.S. Public Domain is, of course, the state of Texas (see figure E.01). Because Texas retained control of its public lands, in the nineteenth century it was faced, with the same two issues that the U.S. Congress had faced at the end of the eighteenth:

- *how to divide up the land into governable entities—in Texas's case, into counties; and*
- *how to convey the land out of government ownership and into private hands.*

Both the Congress and the Texas legislature (itself called the Congress during the Texas Republic) looked to their own traditions and experiences when addressing these issues. Texas had come to its independence through a route quite different from that taken by the United States, and the new republic and then the new state enacted policies quite different from those hammered out at the federal level. The differences say a lot about Texas, but—more important for us here—they hint at some consequences that might have resulted if the United States had chosen to follow different policies when it formed its states and apportioned its own Public Domain.

THE TEXAS LAND GRANT TRADITION

Texas, you'll recall, had been admitted to the Union in 1845 and assumed its present, iconic shape in 1850. To understand how this state kept possession of the unsold lands within its borders, you need to keep in mind the nature of state creation under the U.S. Constitution.

The law that creates a state—an *organic act*—is unique among legislative actions, because once Congress passes an organic act, it cannot repeal the act, or even modify it, without explicit consent from the state thus created. As a consequence, organic acts entail intense negotiation, and one of the topics for negotiation in all statehood acts was the disposition of lands owned by the federal government. Starting with the admission of Ohio, organic acts would grant to each new state a parcel of land in each yet-unclaimed Township: Ohio got one Section per Township, as did the other states east of the Mississippi. As

FIG E.01

The total area of the U.S. Public Domain in the contiguous states, if unappropriated, would have been about 1.44 billion acres, or about 2.25 million square miles.

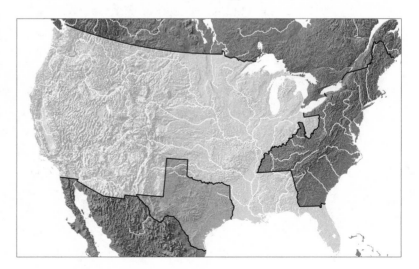

statehood pushed into the West, where land was not so uniformly arable, newer states argued that the number should be increased; and so it was, at first to two Sections and finally to four Sections per Township.

Texas came to the negotiating table with a stronger hand. It had an elected government that had functioned, however fitfully or provocatively, for over a decade; but more crucially for the negotiations, that government had been performing a defining task for a sovereign state, apportioning the lands within its borders. Texas was quite willing to continue performing this essential function, and the federal government, now faced with administering land all the way to the Pacific Ocean, relieved itself of that burden.

In the late 1840s, then, Texas was in much the same position as the new United States had been in the late 1780s: a brand-new government suddenly possessing a huge public domain and needing a system to apportion it; and as had the Continental Congress, the Texas legislature looked to its traditions for guidance. We saw how Congress was influenced by almost two centuries of colonial experience. Texas had a shorter history to draw upon, but its institutional memory of that history was more intense: members of the founding generation, and their immediate descendants, were still active in government. The method of land apportionment that had made those people prosperous was the **empresario system**, under which a huge tract of land would be granted to an individual, who would then apportion the land to actual settlers. In Texas there was neither the New England tradition of surveying tracts prior to settlement nor the southern practice of granting warrants of specific acreage to individual farmers. The "Texas way" (inherited from Spain and then Mexico) was for the government to parcel out the public domain in huge chunks to a few trusted owners, and then pretty much get itself out of the land business.

As had been the case with the township proprietors of New England, the grants to *empresarios* came with certain conditions about how they would apportion their land. The unit in the Spanish colonies for measuring small distances was the *vara*, the "Spanish yard," which was eventually codified under Texas law as $33\frac{1}{3}$ inches. The unit for measuring larger distances was the *legua*, or league, equal to 5,000 varas (in English measure, about $2\frac{5}{8}$ statute miles). From these two units of distance came the two basic measures of land area.

The *labor* was 1 million square *varas* (about 177 acres), and as the name implies, this was the quantity of land considered sufficient to sustain one farming family. A square tract of land 1,000 *varas* on a side would thus contain one *labor*, but as with the English acre, a Spanish *labor* could take any shape, as long as its area totaled 1 million square *varas*.[1] (Interestingly, a distance of 1,000 *varas* is a little over a half mile—2,778 feet versus 2,640 feet—so a Spanish *labor* with square sides would have been just a little bigger than an American Quarter Section—177 acres versus 160 acres. By entirely different routes, two very different governments arrived at similar definitions of "a farm.")

The unit for measuring very large tracts of land was the square league, equal to 25 million square *varas*. The square league was thus the area of land contained within a square 1 *legua* (5,000 *varas*) on a side, but like the *labor* and the acre, a league of land could take any shape. Nonetheless, if your league of land was roughly a square, it could be said among your fellow Texans that your tract had "a league to each wind."

After the new Mexican government took authority in 1821, some *empresarios* were instructed to give each of their settler families a tract comprising one *labor*. If, however, the family wanted to raise cattle in addition to farming, they could have a league of land on which to graze their stock. Upon learning this, many farming families decided to become cattle ranchers, and one of the most common land apportionments made by the *empresarios* was for "a league and a *labor*."[2]

By the time of its independence in 1835, much of what we now call Texas had been placed in the hands of *empresarios*, as you can see in figure E.02. Not all of the grants had been acted upon, and some would still be in dispute at the time of statehood, but their size and relative success gave Texans a ready template for future land apportionment. There was one great difference, though, between these original grants and those that would be made by Texas. Spain and later Mexico had made their allotments on the presumption that without these grants, settlement would not take place: they felt that they could ask no more from the *empresarios* than that they guarantee settlement on the grants. The Texans, from a very different culture, knew that there would be no problem getting people to settle on public lands, and they could therefore extract a price from grantees. The new Republic of Texas was cash-poor but land-rich (as had been the new United States), and from early independence through statehood, its people decided to

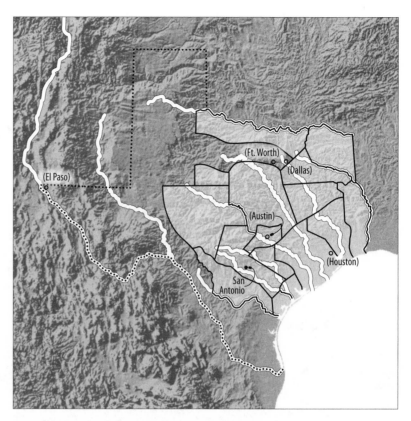

FIG E.02   Redrawn from Gournay, *Texas Boundaries*.

use that land wealth to buy for themselves public improvements: roads, railroads, universities, and a grand state capitol. Two of these grants illustrate the process and help explain why the landscape of western Texas is "squared up," even though Texas never formally adopted the U.S. Rectangular Survey.[3]

From its inception, Texas recognized the need to connect the far outpost of El Paso with the rest of the state. The U.S. Congress authorized a number of transcontinental railroads during and after the Civil War, and one of them—the Memphis, El Paso and Pacific—would have met Texas's desire for a route to the state's far west. By the end of 1874 track had been laid to just west of Dallas (as shown in figure E.03), but then Congress withdrew federal backing and construction came to a halt. Congress later reversed itself and restored financial support, but Texas itself made a grant to a reconstituted Texas and Pacific Railroad: for each mile of track built, the state would grant the railroad 16 Sections of public land. The Texas legislature followed the pattern used by Congress in its railroad grants in that the Sections were to be allocated in a checkerboard fashion, the Texas and Pacific getting the "black" sections, the state retaining the "red" ones. Since a Section is 1 mile per side, the checkerboard of the railroad's holdings would extend 16 miles to either side of the right of way—and even farther outward if the checkerboard encountered any lands already held by private interests. It would be up to the railroad to send surveyors out onto the land, mark its Sections on the ground (one aspect of the checkerboard pattern being that, in monumenting the boundaries of *its* Sections, the railroad would necessarily mark the boundaries of *the state's* Sections), and prepare plats of the survey.

By 1883 the Texas and Pacific, building westward, had linked up with the Southern Pacific, pushing east from California, and the state had its link to El Paso. It also had a swath of land, in places as much as 40 miles wide, surveyed by the railroad into Sections virtually indistinguishable from those being marked on the ground, in the rest of the West, by government surveyors using the Public Land Survey System.[4]

In 1879 the Texas legislature set about building a new state capitol. It held a competition for the design, and set aside 3 million acres in the Panhandle to pay for construction. Three years before, the legislature had erected new counties in the region, and the designated acreage was to come out of the ten of them outlined in figure E.03. Unlike the checkerboard railroad grants, the capitol grant was to be one contiguous, unbroken tract, directly against the New Mexico border—the only exceptions being any platted town sites or lands already in private ownership.[5] There were about 7 million acres in the ten counties, so the legislature could be reasonably sure that 3 million unclaimed acres could be found within them—but *which* 3 million acres?

The easiest method would be to survey the whole region into squares of a designated acreage and mark each square as Taken or Available. The grantee could then simply go to a map and check off Available squares until they totaled 3 million acres, then draw a border around the whole. (The process was, of course, much more complex in actuality.) The Panhandle was then wide-open rangeland, and the legislature directed that the each of the squares of the survey be a Spanish league (a square about 2⅝ miles on a side, you'll remember), not the more fine-grained mile-square Section mandated for railroad grants.[6]

In 1881 the architect Elijah Myers won the competition for the design of the capitol (he had earlier designed the Michigan state capitol and would later design the capitol of Colorado), and the contract for construction was awarded to a syndicate of Chicago builders.[7] The syndicate, now in possession of millions of uninhabited acres, chose not to offer the land to settlers, in the manner of the *empresarios*, but set about turning their tract into one huge cattle-raising operation. They secured about $5 million from British investors (among them the Earl of Aberdeen, a literal "cattle baron"), and thus was born the famous XIT Ranch.

Both the XIT and the capitol are the stuff of Texas legend. Over 35,000 calves branded in one year![8] A soaring edifice of pink granite, taller than the U.S. capitol and costing almost $4 million![9] But in apportioning part of the Panhandle into squares "one league to each wind," we can see how far the idea of surveying in rectangles had penetrated the minds of Americans—even Americans who were determined, in many matters, to avoid doing things the same way as the federal government.

We can see this rectangular mindset most clearly in the ways Texans apportioned their state into counties. Figure E.04 shows that in the area of the *empresario* grants, most county boundaries were drawn to follow the courses of the major rivers, the natural conduits of trade and communication. As settlement pushed toward the Rio Grande and what is now Oklahoma, straight-line boundaries were used, even where the rivers ran diagonally. By 1876 the rectangular idea had so taken hold that the legislature established 54 counties, all of them rectangular and most being roughly 30 miles square,[10] a pattern that would be followed until the state was completely filled out with 245 counties. (The great exception in this was the land "west of the Pecos," legend-

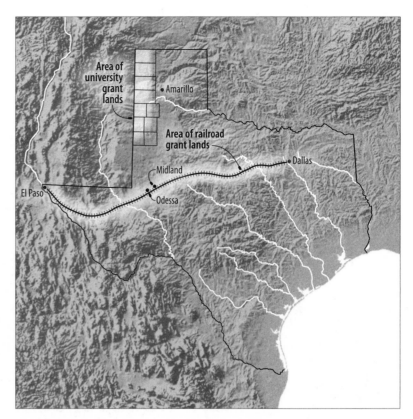

FIG E.03

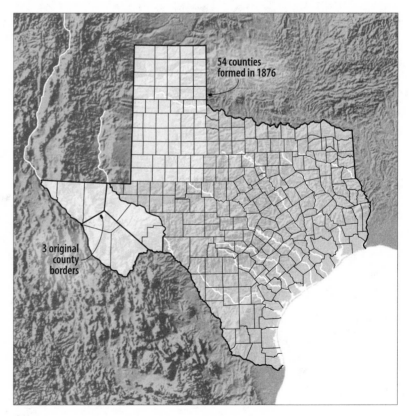

FIG E.04

arily ungovernable in frontier times and to this day lightly inhabited. There, at first, the legislators simply threw up their hands and divided the region into thirds with pie-wedge lines.)[11]

In taking control of its public domain, dividing it into counties, and parceling it out into private ownership, Texas recapitulated the nation-forming acts of the U.S. Congress, but in its own, quite different fashion. In long retrospect, how do the two approaches compare? And what can each tell us about the other?

When comparing a map of Texas counties with a map of American states, the contrast that leaps most vividly to the eye is that whereas states are small in the East and bigger in the West, in Texas (east of the Pecos) counties are roughly the same size everywhere. Texas has a land area of about 262,000 square miles, and if you divide that up into 254 counties, you get an average of just over 1,000 square miles per county—which is pretty much the case, statewide. Part of this uniformity grew out of the same sort of impulse that gave rise to the 6-mile-square New England township, a determination that good governance required a basic political unit of a certain geographic extent. In Texas that unit was the county, and each of the state's counties, no matter how rural, has its grand courthouse, often a bold Victorian pile rising from a grassy town park at the crossing of the county seat's two main streets.

The problem such a uniform county size presents is that in lightly inhabited parts of the state, some counties have less than a thousand souls within their thousand-square-mile borders. Certainly the delegates to the 1876 legislature had a different future in mind when they erected fifty-four counties in the Panhandle. There was undoubtedly an element of logrolling enthusiasm involved (this was the first post-Reconstruction legislative session, when Texans "took back" their government), but there must have been a belief that these new counties would one day be as populous as the counties along the rivers. Texas had by this time been in the business of forming counties for forty years, and the experience up till then had been that new counties soon filled with new families, farms, and towns.

The comparable experience for the U.S. Congress happened earlier, in the first decades of the 1800s, when Ohio, Indiana, and Illinois filled up with new settlement. The nation and its legislators saw a pattern of farms and towns develop that gave every promise of covering those new states with a density of habitation to equal New England. As new states were formed, the pattern of settlement repeated itself—seemed indeed almost self-fulfilling: open up a state to settlement, and in a twinkling, American civilization will spread into it and fill it. By 1839 the journalist John O'Sullivan could write, in his *United States Democratic Review*,

The American people having derived their origin from many other nations, and the Declaration of National Independence being entirely based on the great principle of human equality, these facts demonstrate at once our disconnected position as regards any other nation; that we have, in reality, but little connection with the past history of any of them, and still less with all antiquity, its glories, or its crimes. On the contrary, our national birth was the beginning of a new history, the formation and progress of an untried political system, which separates us from the past and connects us with the future only; and so far as regards the entire development of the natural rights of man, in moral, political, and national life, we may confidently assume that our country is destined to be the great nation of futurity.

Later in the article, O'Sullivan makes a prediction:

The far-reaching, the boundless future will be the era of American greatness. In its magnificent domain of space and time, the nation of many nations is

destined to manifest to mankind the excellence of divine principles; to establish on earth the noblest temple ever dedicated to the worship of the Most High—the Sacred and the True. Its floor shall be a hemisphere—its roof the firmament of the star-studded heavens, and its congregation an Union of many Republics, comprising hundreds of happy millions, calling, owning no man master, but governed by God's natural and moral law of equality, the law of brotherhood—of "peace and goodwill amongst men."[12]

Clearly the manifest destiny of the nation, in 1839, was that it spread to the far Pacific; and as was foretold, that great swath of land does indeed now contain hundreds of millions. But equally clearly, O'Sullivan imagined that population achieved not as it has been, with a crowded periphery and an almost empty center, but rather by a spread of Ohio-like settlement over the whole of the nation's extent.

Anyone with a knowledge of the terrain of the West would instantly recognize the impossibility of such an America, but up until 1839, Americans had shown themselves capable of populating the hills of Pennsylvania, the bayous of Louisiana, and the prairies of Illinois. What was to prevent such a people—with such a destiny—from finding ways to settle the Plains and the deserts and the Rockies?

In 1839 the evidence that would refute this belief was not yet in, and so it seemed reasonable to believe that the best way to govern a new region that would come to have Ohio-like density was with Ohio-sized states—something on the order of 40,000 to 50,000 square miles. As you can see in figure E.05, all the states formed up

FIG E.05

to 1839 fall into this range (Illinois a little bigger because of its northward land grab, Michigan bigger from being given the "uninhabitable" Upper Peninsula). The great exception was Missouri, at almost 70,000 square miles; but remember that Missouri was born as the South's last chance to plant its slave culture north of the Compromise line at latitude 36° 30'.

By the 1840s the idea was beginning to dawn that Ohio-like densities might not be achievable everywhere, and Iowa and Wisconsin were given about 50,000 square miles. By the late 1850s and early 1860s, when Minnesota, Kansas, and Nebraska approached statehood, Congress looked at Minnesota's northern reaches and the Plains in the west of Kansas and Nebraska and, in compensation, made all three states around 80,000 square miles.

(For the rest of the states, we can discount those whose sizes were set not so much by congressional judgment as by external factors. Florida had had its borders since Spanish colonial

times. Oklahoma, we have seen, was an assemblage of remnants. California drew its own border to encompass the Sierra Nevada. Oregon was stretched between the Adams-Onís Treaty line and the Columbia River, leaving Washington the territory north to the 49th parallel. Nevada, rushed prematurely into statehood to provide governance over the silver rush, was born small but then expanded eastward—twice; in the expansion, the new state got frontage on the Colorado River, and [no coincidence] the Mormon empire got diminished.)

With the exception of Colorado (rushed like Nevada into statehood for governance of a mineral boom), all the states of the mountain West, plus the Dakotas, had their borders set, as territories or states, in 1889. By that time it was abundantly clear that these would be states very different from Ohio, and each was given an average of 100,000 square miles. The "exceptions that prove the rule" are Utah, diminished by surrounding states; the Dakotas, which, until the last minute, were to have been admitted as one state; and Idaho and Montana, which would have been more equal in size if their common border had been the Continental Divide.

As it turned out, even this doubling of states' areas—from 50,000 to 100,000 square miles—would be insufficient to give them populations equivalent to some national average. In 2000 the population of the "Lower 48" was about 280 million, in an area of about 3 million square miles. This means that an "average" state would be home to just under 6 million people in borders enclosing 62,500 square miles—a pretty close description of Missouri. Imagine the rural-urban mix of Missouri spread evenly from San Diego to Maine, and you will envision the hundreds of happy millions in O'Sullivan's great nation of futurity.

Clearly the Congress of 1889 saw that the Plains and mountain West would be thinly populated, but it would probably be surprised at the disparity, today, between the least- and most-populous states. The census of 1790 disclosed that the ratio between the least- and most-populous states—Delaware and Pennsylvania or Virginia—was about 1 to 10. Today the ratio between Wyoming (just under 500,000) and California (34 million) is 1 to 68. (Smug easterners should be aware that the ratio between Vermont and California is 1 to 54.)[13]

But what could Congress have done in 1889? If it had made one huge state out of Idaho, Montana, Wyoming, and both Dakotas, even such a super-state would, today, not quite attain that 6-million-person average. What we have here is a continuation of that pattern begun when the monarchs of England made the first colonial grants, when a process of border making conceived at the center of power produced unexpected results when enacted on an unknown periphery. Both U.S. state making and Texas county making were processes invented out of the experience of people settling on lands uniformly fertile and habitable. Whether applied with rigidity, as by the Texas legislature of 1876, or stretched to accommodate new conditions, as by the Congress of 1889, neither process imagined the disparities that would in fact ensue.

These disparities would be merely academic if they didn't have real consequences. Leave aside the political question of the equal representation of states in the U.S. Senate: no matter how formed, each state has by now evolved a distinctive political culture, and it is arguably a good thing to accord weight to each culture. What is, I think, inarguable is that such wide disparities distort the nation's perception of itself. It is ingrained in Americans from an early age that the states are "equal;" and that concept, so firmly held, makes it difficult to imagine states as radically different from each other. We know that some states are densely populous, others less so: no one really imagines the U.S. population spread as evenly as mayonnaise from sea to sea. Yet to imagine a U.S. map the size of a basketball court, as in figure E.06 on the following page, with one person standing in Wyoming and sixty-eight squeezed into California—such an image strains credulity, yet such is the true reality of our population. Our inability to imagine that reality conditions our politics and national life—probably inevitably, but perhaps also fortunately. This is, after all, one nation, and if we truly saw it as it is, it would be awfully easy to ignore that lone cowboy in Wyoming, or that equally solitary figure in Vermont.

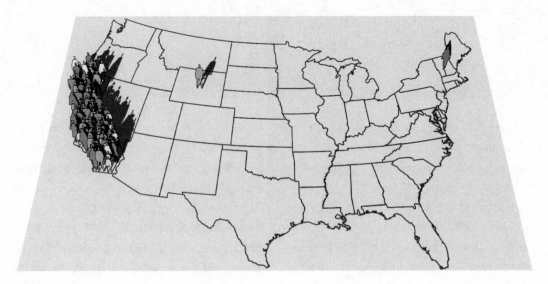

FIG E.06

When it comes to population disparities, Texas—outsize and outlandish in all things—naturally has the topper. In 2000, Harris County, home to Houston, had a population of about 3.4 million; Loving County, at the southeastern corner of New Mexico, had a mere 76 persons.[14] And by the way: lest you still think of Texas as a wide-open, empty place, be aware that with 21 million people on 267,000 square miles, Texas is actually *denser* than "average" Missouri.

TEXAS GRANTS ALL ITS PUBLIC LANDS; THE UNITED STATES RETAINS A LARGE REMNANT

The land area of Texas can also be expressed in acres, about 170 million of them. Of this total, the republic and then the state of Texas gave away over 86 million acres in payment for public improvements. There were several railroad grants like the one you saw earlier, and the fabulous giveaway for the state capitol. But the greatest portion of the land grants—almost 52 million acres—was set aside to benefit education: over 2 million acres for the university system, the rest for elementary and secondary schools. Most of the grade-school land was sold early on and the proceeds put into a permanent fund, but the university held on to its land, much of it in west Texas in the area centered on Midland and Odessa. The grant was vast, but at the turn of the last century it was yielding only about $40,000 a year in rents and grazing fees. Then in May of 1923 two wildcatters drilling test wells on university land brought in the Santa Rita oil well. By law, proceeds from the oil field had to go into a permanent fund, and by 1925 that fund was growing by $2,000 a day.

Texas had been in the business of swapping land for social goods from the founding of the republic until 1900, by which time all of its public domain had been sold into private hands, granted for services, or assigned for education.[15] Did Texas get "value for money" from this policy? Thanks largely to oil revenues, Texas has the best-endowed public university system in the nation, with a fund of several billion dollars. The public school fund, gathering interest since the nineteenth century, stands at about double the university fund. And the investment in roads and railroads (and all those courthouses!) spread development deep into the interior of the state.

But the salient fact about the Texas public domain is that from the first, it was the state's intention to dispose of all its land, and this was accomplished before the turn of the twentieth century. Contrast this to the U.S. Public Domain, where, at the turn of the twenty-first century, millions of acres are still held by the federal government. The stark contrast is explained, of course, by the fact that Congress rejected, early on, the idea of apportioning the Public Domain into huge tracts, and committed itself to sales and grants that could not easily aggregate into large holdings. The checkerboard pattern of grants to the transcontinental railroads is an example of this determination. Even when the last western states were granted four Sections per Township for education, those Sections—numbers 2, 16, 32, and 36—were scattered across the Township.

The determination to "sell small" meant that no one was offered the bargain accepted in the XIT Ranch deal: take on millions of acres of desert waste in order to get the thousands of acres that could be turned to productive use. The result was that as the nineteenth century neared its end, much of the intermountain West—from the Rockies across the basin-and-range region through the Sierra Nevada—was still in federal hands, "on offer" for settlement but finding few takers.

This meant that much of the U.S. Public Domain, unlike that of Texas, was still in public ownership at the time when the idea of *conservation* arose. It was thought in the very first U.S. land ordinances that certain lands might be so intrinsically valuable that no one should be allowed to own them: remember that if a Section were found to contain a salt lick, that Section would be reserved from development, held by the government for the benefit of all. In 1864 Californians decided that their Yosemite Valley was such a unique place that it should be owned in perpetuity by the state so as to be forever available in its natural condition for the enjoyment of the people. In 1872 the federal government decided that the Yellowstone region was equally precious and put it in permanent federal ownership as a national park. By 1900 four more parks had been added, and in 1916 the National Park Service would be created to administer them.[16]

With Yellowstone was broached the idea that the federal government might be the natural steward of lands so valuable to the whole nation that they could not be allowed to pass into private ownership, or indeed ownership by any of the states. The great parks clearly merited such special status, but equally deserving (it was argued) were the forests of the Rocky Mountains. The trees were a resource of obvious value to the whole nation and thus should not be exploited but rather conserved, harvested, and replenished in a manner that could be continued in perpetuity.[17] At the time it was believed that the federal government was the only entity sufficiently insulated from market pressures to accomplish this perpetual stewardship. By the start of Teddy Roosevelt's administration, 50 million acres of woodland had been placed into stewardship as national forests, and TR would add a further 100 million acres, establishing the National Forest Service in 1905.[18]

As the twentieth century wore on, more national parks and forests were set aside, but the Public Domain itself was still posted with a figurative For Sale sign. Any Quarter Section in the Public Domain could be bought (or, after 1862, claimed and homesteaded) unless that land had been specifically designated as "reserved from selection." (You can see in figure E.07 how the demand for land rose and fell in the settlement period.) By the 1930s the number of people applying to own public land had dwindled, and it began to seem plausible that the remaining land might never find takers. So gradually, in a series of executive actions and especially in the Taylor Grazing Act of 1934, the For Sale sign was taken down. None of the Quarter Sections of the Public Domain could be transferred from federal to private hands unless specifically designated as "on offer."[19] You could still buy federal land, but you'd be in the same position as with a private landowner: first you'd have to convince the owner to let the land go, then offer a price that would make selling the land more attractive than holding on to it. The Public Domain that had been "open" since 1784 was now "closed." The closing of the Public Domain was neatly symbolized when in 1936 the General Land Office became the Bureau of Land Management.[20] There would never again be a "land-office business" for federal lands.

The result of these decisions is that several states now find the bulk of their lands locked in federal ownership, largely exempt from taxation and from any development plans the states might formulate. This situation quite naturally arouses resentment in some quarters. Part of the resentment comes from today's free-floating

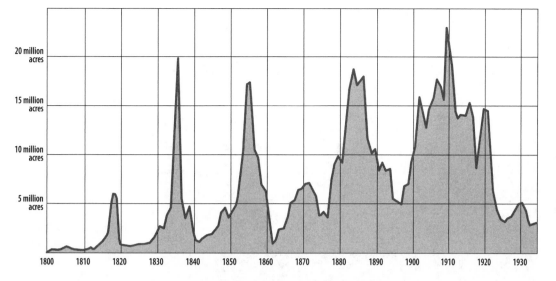

FIG E.07

This chart tracks the number of acres *entered* each year into the records of the General Land Office; the number of acres actually *patented* (transferred from federal to private ownership) is smaller—in some years significantly so. Note that the year in which the most land was entered was 1910; in a few pages, you'll see why.

Drawn from data in Hibbard, *Public Land Policies.*

animus toward all things "federal," which runs the gamut from grumbles on April 15 to fears of black helicopters; but there's also a serious philosophical argument behind the resentment.

As a pointed instance of federal land ownership, take the national forests. Montana, Michigan, and Maine all have within their borders great stretches of woodlands arguably important to the nation, yet the forests of Michigan and Maine are under state jurisdiction, while much of Montana's woodland is in federal hands. Why, apart from historical happenstance, should this be so? If timberlands are a national resource demanding federal stewardship, then the woods of Michigan and Maine—and the pine forests of the South— ought to be under federal control. But if individual states can be good woodland stewards, then Montana and the other western states should gain control of their national forests.

Of course, in debates over national land policy, an argument of "logical consistency" seldom has sufficient force to trump entrenched habits and economic interests. Still, the argument for most civil rights rulings and legislation has been, at base, "logical consistency"; so who can say how the debate over the ownership of the unappropri-

ated Public Domain will eventually shake out?

"Logical inconsistency" is most visible in those maps that show stretches of western states covered in the green tint of national forests, parks, monuments, and wildlife refuges; but federal ownership of land most often "pinches" in the parts of the Public Domain outside those designated reserves. These are the tracts of land that, up to the 1930s, nobody wanted to take on—not individuals or corporations, not the Park Service or the Forest Service or the Bureau of Indian Affairs or the U.S. military. But it's been seventy years since the closing of the Public Domain, and in that time the interstates have been flung across the continent, the suburbs and then the exurbs have spread, new technologies of communication and of energy extraction and use have been developed, and new cultural habits have arisen. In some places in the West, people are imagining ways to make use of the land nobody wanted. It is in these places that the closing of the Public Domain sometimes rankles; but even if the land were on offer, one more consequence of the determination to "sell small" would persist.

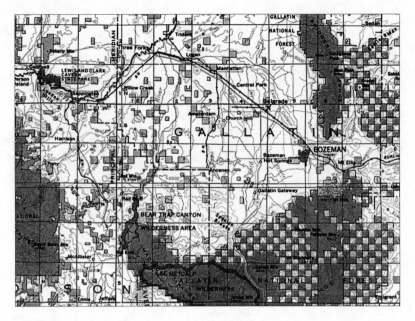

FIG E.08

A detail from a Bureau of Land Management map of Montana. In addition to the "school sections," you can see remnants of alternate-section railroad grants.

Bureau of Land Management, United States Department of the Interior.

You can see the vast forest-and-wilderness holdings of the federal government tinted in green on most any road map, but if you get hold of a really detailed map like the one in figure E.08, you'll often find the land color-coded by ownership. Usually white stands for land in private hands, green is for national forests and reserves, blue is for state-owned lands, and yellow is for "public lands," the unchosen remnant of the Public Domain administered by the Bureau of Land Management. In such a map, what immediately strikes the eye is the scatter-shot nature of landownership.

On Montana maps, for example, you'll repeatedly see Townships in which Sections 16 and 36 are blue—state owned—a legacy of Montana's enabling act of statehood, which granted those Sections to the state for education. A swath of land covered in a checkerboard pattern of white private land and tinted state or federal land is the sure tip-off of an old railroad grant (see figure E.09). And as the eye ranges slowly over the map and encounters a stray Section of yellow BLM land, the mind has to wonder, "Why did nobody ever want that land?" Encountering a blue Half Section of state land raises the question, "Could that be a foreclosure? Some farm family gone bust in the Depression?"

Bear in mind that the map's color coding only hints at the true complexity of landownership, for all that contiguous private land was itself claimed or purchased only by Half Sections. The real legacy of the federal policy of "selling small"—which stands in contrast to the Texas example—is the interwoven distribution of land-ownership. One result of this pattern is a kind of check (not a checkmate) on overlarge plans for the land. By doling out enough money, it's relatively easy for a person or corporation to buy up a large quantity of land, but because of the multiplicity of ownership it's often hard to assemble an unbroken "empire" of acreage. In fact, groups

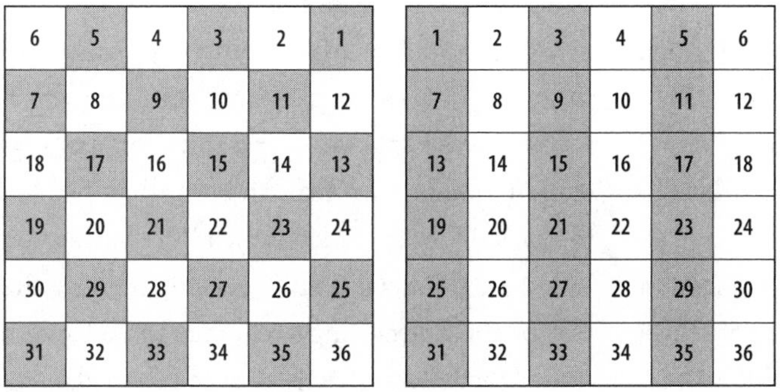

6	5	4	3	2	1
7	8	9	10	11	12
18	17	16	15	14	13
19	20	21	22	23	24
30	29	28	27	26	25
31	32	33	34	35	36

1	2	3	4	5	6
7	8	9	10	11	12
13	14	15	16	17	18
19	20	21	22	23	24
25	26	27	28	29	30
31	32	33	34	35	36

FIG E.09

One advantage of a boustrophedonic numbering system is shown here. If land was to be granted in a checkerboard pattern, the land office merely specified "Even-Numbered (or Odd-Numbered) Sections."

like the Nature Conservancy leverage this interwoven landownership to block big plans. They call it "checkerboarding": buying up just enough small tracts, scattered over a sufficient area, to make a planned development difficult.

So the federal policy of selling and granting only by Half Section has on occasion acted to "put some sand in the gears" if a powerful individual or corporation wanted to impose a single idea of development on a whole region. What that policy could not impede—what it in fact enabled—was the phenomenon of thousands of nonpowerful individuals each arriving independently at the same idea of how to develop a region. Gold and silver rushes are the best known examples of this: valley after valley east of Sacramento stripped bare by thousands of gold prospectors following a single dream. But the most poignant example, and one abetted directly by the Rectangular Survey, occurred in eastern Montana in the years just before and during World War I.

At the turn of the twentieth century, eastern Montana was still pretty much open country. Families had spread westward from Minnesota, but halfway across the Dakotas, farmers began to find insufficient rain to grow their accustomed crops of wheat, oats, and alfalfa. A chance collision of events would change that.

The collision began with the 1902 publication of *Campbell's Soil Culture Manual* (price $2.50).[21] Hardy W. Campbell was a self-taught agricultural scientist who took the experiences of farmers in regions like Utah and the Dakotas, added some insights and techniques of his own, and produced the "Campbell System" of dry farming. His system relied on leveraging three facts about crops, soil, and water:

- *Certain crops like wheat and especially alfalfa have deep roots that can tap water far below the surface: they can grow and even thrive in areas of little rain as long as the subsoil is damp.*
- *In the northern Plains, the scant rain falls mostly during the growing season, but the air is so dry that rain can evaporate into the air rather than percolate down through the soil. Ordinary cultivation breaks the top of the soil into clods, which increases surface area and hastens evaporation; but if the surface of the soil is smoothed, much more water will percolate downward.*
- *Rather than being broadcast over the land, seeds can be drilled into the soil, down past the dry surface layers and into moisture where they can germinate and grow.*[22]

The ideal dry-farm field had a smooth top layer of soil, to lessen evaporation; below that was a layer of broken-up soil, to impede water rising to the surface through capillary action. Below that was the moisture-holding layer into which the seeds would be drilled—but only on alternating years. To soak up and store sufficient water, the field was kept bare of moisture-robbing vegetation for an entire year, and only then planted to crops. During the fallow year, the farmer followed a complicated procedure of opening the soil to rains and closing it to evaporation. He continued that process after the seeds were planted, stopping only when the new crop broke the surface. After harvest, the stubble was plowed under immediately, to put any water in the plant stalks back in the ground, and the fallowing cycle begun again.[23]

Beginning in 1906 there would be an annual Dry-Farming Congress in some western city.[24] These expositions—and Campbell's manual—began to raise the possibility that even dry eastern Montana could be made to bloom. In 1909 the exposition came to Billings, at least in part to mark the coming of a new railroad.[25]

The Chicago, Milwaukee & Puget Sound Railway was more familiarly known as the Milwaukee Road, and by 1909 it stretched, as its full name promised, all the way to Seattle (its route is shown in figure E.10). To get business, the line's owners needed farmers along the route who would ship out their crops, and town merchants who would ship in goods to sell. So the Milwaukee Road established towns—one about every 10 miles on its route through eastern Montana—and placed advertisements in newspapers in eastern cities and even in Europe, trumpeting the land and the successes that dry farming had achieved on it.[26] In the ads the potential farmer would learn that for a mere $22.50 the Mil-

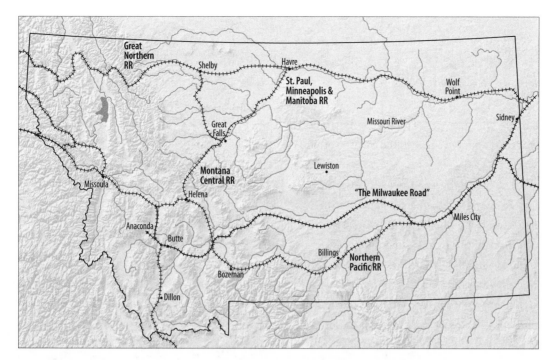

waukee Road would ship his family and all his belongings—even livestock—all the way from the East Coast to Montana. The other already established railroads followed suit, and soon the final western land boom was on.[27]

The decade that bridged the turn of the twentieth century had been dry in the northern Plains, but starting about 1906—that year of the first Dry-Farming Congress—the drought broke. Farmers who tried the Campbell method reaped abundant harvests.[28] There had been a recurrent phrase on the Plains, spoken hopefully among farmers and boastfully among land promoters:

Rain Follows the Plow.[29]

The very act of turning under the grasslands and releasing the subsurface moisture would alter the weather of the Plains. By about 1910, the truth of the phrase seemed confirmed, and tens of thousands came west on the railroads and in the new automobiles. To their delight they found that not only could they grow wheat and oats and

FIG E.11

In Montana today, the dry-farming system has been refined: the fallow and planted fields are in long alternating strips, oriented usually north-south. The planted strips partially shield the fallow ones from the prevailing western winds.

Airphoto—Jim Wark.

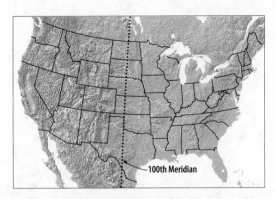

barley, but there was a lucrative international market for their crops with the coming of World War I.[30]

Then, just as the war drew to a close, the rains failed—first along the Canadian border, then, by 1919, across the whole state. The drought continued through the 1920s and on into the Depression. Eventually sixty thousand people would abandon Montana. The legacy of the last great land boom was thousands of hopes dashed, and millions of acres of grass plowed under, never to return.[31]

It was the Montana land bust that finally convinced the nation that the remaining Public Domain contained no undiscovered farmland, only rangeland and forests and deserts. The land bust also forced people to admit that rain did not follow the plow (see figure E.11), that the West was irrevocably and eternally an arid place.

Aridity is, of course, a relative term: there is eastern Montana and then there's Death Valley. But with the exception of the mountains and the Pacific coast, once you get past the Mississippi, the land gets drier the farther west you go. For a farmer, though, one question distinguishes true aridity from mere dryness: "Can I grow a crop, year after year, with only the rain that falls from the sky?"

Crops that can get by with scant water, like wheat, still need about 20 inches of rainfall to thrive; and by a cartographic coincidence that no one could have foreseen, a line of longitude 100° west of Greenwich marks where rainfall drops below that magic number (see figures E.12–14). Not in every year, of course: with dry and wet years the 20-inch line swings east and west of the meridian. And south of the Texas Panhandle,

FIGS E.12–14

The railroad town of Cozad, Nebraska, was placed specifically to sit astride the 100th meridian. (Don't scoff: grander cities have been founded on flimsier pretexts.) Cozad's main avenue, Meridian Street, is set right on the longitude line. A seam in the paving runs down the center of the street: walk north on that seam (there's not much danger from traffic, on even the busiest days), and you'll have your right foot in the East and your left in the West.

weather from the Gulf of Mexico pushes suffi-cient rainfall farther inland. But in the Plains, the earliest explorers and later the first emigrants to Oregon noticed a change at about the 100th meridian. Before they crossed that line, they were moving through grass that could grow higher than their eyes—for many, a frightening experi-ence. Beyond the line the grass thinned and grew shorter: they could see to the far horizon (which held its own fears). They had emerged, literally, into a new country.

In ancient times, the line between tall-grass prairie and short-grass prairie swung eastward and westward with times of drought or rain. The buffalo could shift their grazing patterns to fol-low the swing, and the Plains Indians, being nomadic, could sustain themselves by shifting with them. Nomadism came to an end when farms were carved from the grasslands. Now a Plains dweller was rooted to one place, and his entire livelihood, year to year, depended on whether that line of sufficient rain would swing east or west of his farmstead.

That immovable farmstead was inevitably a rectangular parcel of land, surrounded by similar farms, all of them also rectangular—which raises the question on which I'll end this book:

After about 1796, when the Rectangular Survey was resumed, Congress would never again tam-per with the system of 36 Sections to a 6-mile-square Township. Every time complaints about the system came from the hinterland, Con-gress responded not by changing the tracts the Survey marked on the land but by modifying how settlers could choose from among those marked tracts.

Early on, it was apparent that no family could successfully manage a full, mile-square Section as a farmstead, so in 1800 (as shown in figure E.15 on the following page) Congress allowed set-tlers to buy a Half Section—but only one made by a north-south line through the center of a full Section. Even that was too big for farm families, so in 1804 Congress allowed land to be sold by the Quarter Section—but only if achieved by drawing an east-west line across a north-south Half Section.

Under pressure from the frontier for more flexibility in laying out farmsteads, Congress in 1820 allowed farmers to assemble the man-dated Quarter Sections by grouping together two adjacent Half-Quarter Sections. Farmers asked for still more flexibility, and in 1832 Congress made its final concession: a 160-acre, Quarter-Section farmstead could be assembled by group-ing together four Quarter-Quarter Sections of 40 acres each—as long as the whole farmstead was contiguous.[32] This allowed for a multiplicity of possible farmstead shapes, and gave rise to that great Americanism "the back forty," the 40-acre Quarter-Quarter Section farthest from the farm-er's house.[33]

Congress's reluctance about these conces-sions is easily explained. One of the intentions

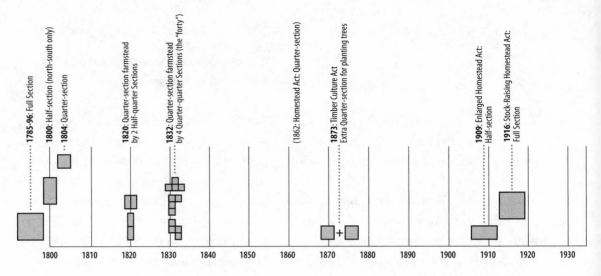

1785-96: Full Section
1800: Half-section (north-south only)
1804: Quarter-section
1820: Quarter-section farmstead by 2 Half-quarter Sections
1832: Quarter-section farmstead by 4 Quarter-quarter Sections (the "forty")
(1862: Homestead Act: Quarter-section)
1873: Timber Culture Act Extra Quarter-section for planting trees
1909: Enlarged Homestead Act: Half-section
1916: Stock-Raising Homestead Act: Full Section

1800 1810 1820 1830 1840 1850 1860 1870 1880 1890 1900 1910 1920 1930

FIG E.15

behind the Rectangular Survey was that the Public Domain would be taken up without gaps. Congress didn't want to be stuck with tiny tracts of federal land completely islanded in a sea of privately owned farms. The bargain implicit in the rectangles of federal land was that the boundary lines would enclose good land along with bad, and in order to get the good land, the buyer would have to take the bad land off the hands of the government. Congress feared—with some justification—that farmers would use the smaller tracts to wiggle their farmsteads around undesirable land, leaving the government stuck with those isolated parcels.[34]

Allowing the "forty" was as far as Congress was willing to go, and the inventory of parcels the government had on offer remained unchanged until the 1870s. What Congress addressed during this period was the method of purchasing land—again in response to pressure from the West. In 1841, after resisting for decades, it allowed *preemption*, which meant that if a settler established a farmstead in advance of the Rectangular Survey, once the Section lines were drawn around his fields, he would have a "right of first refusal" to buy that Quarter Section at the government's minimum price per acre.[35]

Farmers agitated for more, specifically the right to buy a Quarter Section not with money but with the "sweat equity" of building a home and planting fields. Southern Congressmen resisted the idea: a region covered in Quarter-Section farms would be a territory inimical to slavery. But in 1862 Southerners were absent from Washington, and the Homestead Act was finally passed. A settler could lay claim to any Quarter Section in the Public Domain, pay a small filing fee, and if after five years he (or she!) had met the requirements for making improvements, he would have "proved up" his claim and the land would be his.[36] (See figure E.16 for a satirical take on homesteading.)

After the Civil War, and especially with the coming of the transcontinental railroads, settlement was pushing right up against the 100th meridian, and farmers on the frontier were facing conditions not seen before. Farmers in Indiana and Illinois had found ways to make the Quarter-Section farmstead work, and those methods continued to yield good results as farming pushed into Iowa and even eastern Nebraska and Kansas. Farther to the west the rain diminished, and so did the crop yield per acre. Farmers asked to be able to homestead bigger farms, and Congress responded, somewhat fitfully. Its first response came under the guise of promoting the planting of trees on the Plains, in the Timber Culture Act of 1873. If a farmer homesteading a Quarter Section were to plant a sufficient number of trees on an adjacent Quarter Section, he could "prove up" both tracts.[37] Congress soon found that Plains "farmers" were planting trees for the real purpose of getting grazing land for their cattle. In Washington there was still a strongly held ideal that homesteading was for the benefit of Jeffersonian yeoman farmers and not cattle barons, and so the act was repealed in 1891.[38]

The pressure for larger farms continued, and the Half-Section homestead was reinstituted in the Enlarged Homestead Act of 1909—just in time to allow Montana dry farmers to leave half their land in fallow. By 1916 Congress had finally accepted the idea that some parts of the Public Domain were suited only for grazing, and it adopted the Stock-Raising Homestead Act, allowing a full 640-acre Section to be claimed and proved up.[39] In the thirty years centered on 1900, there were other land acts, some of them in force for only a short time, some valid in only a particular region; but by the time of World War I —and especially after the Montana land bust— Congress began to see the futility of passing legislation that would actively encourage further development on the Plains. The experience of the Dust Bowl in the 1930s convinced everyone that it was time to close the Public Domain.

A HOUSE "TWELVE BY FOURTEEN"

FIG E.16

There is no doubt that some unscrupulous types skirted the Homestead Act's requirements for "proving up" a claim. Bear in mind, though, that just as there is *tax avoidance* (legal) and *tax evasion* (illegal), so there is land-law *manipulation* and land-law *fraud*. And just as has been the case with taxes, Congress abetted the manipulation by layering land law upon land law, all during the settlement period.

U.S. Government Printing Office.

John Wesley Powell's Alternative to the Rectangular Survey

While it's true that Congress never did tamper with the Sections and Townships of the Rectangular Survey, there was one poignant, might-have-been moment when an alternative managed to gain serious consideration. In 1869 John Wesley Powell and his crew became the first people ever to boat down the length of the Colorado River. When they emerged from the canyons on August 30, Powell became a national hero, the Lindbergh of his day. Powell's trip, though, was not a stunt but a serious scientific expedition, and he leveraged his fame to get himself appointed head of several other missions of exploration, the most important of which was the grandly named United States Geographical and Geological Survey of the Rocky Mountain Region, J. W. Powell in Charge.

That survey took Powell back to the area around the Colorado River, and especially to southwestern Utah, which the Colorado and Green rivers traverse. While there, Powell observed not only the topography and geology of the land, but the habits of both the region's native peoples and the Mormon farmers working irrigated farm plots on a cooperative basis. Powell poured all of his experiences in the West into his great work, *A Report on the Lands of the Arid Region of the United States, With a More Detailed Account of the Lands of Utah*, and presented it in 1877.[40]

In the report Powell argued that in large parts of the West, the rectangular Quarter-Section farmstead could never be made to work, and he proposed a wholly new agricultural model. The West would be surveyed not into rectangles but into watersheds, and the farmers in each watershed would govern the agriculture there cooperatively. Their first task would be to build (and then maintain) irrigation channels running parallel to the watershed's major streams. Each farmer would then get a two-part farmstead. The first part would be a 20-acre plot along a stream, which the farmer could irrigate from the shared water channel. Here the farm family would grow vegetables and fruits for their own use. The farm's second part would extend back from the streamside plot—a parcel equivalent to four whole Sections, for grazing cattle to be sold on the national market. The configuration of the streams would determine how many such dual farmsteads could be laid out, but when their boundaries had all been surveyed, the remainder of the watershed would be held by the cooperative as a common grazing ground.[41]

Figures E.17–19 show Powell's system hypothetically applied to the drainage basin of the Beaverhead River, which the Initial Point survey team traversed in 1867.

Despite Powell's lobbying, no action was taken on his report for years. Then in October of 1888, Congress passed a bill with a deeply buried provision directing the Interior Department to investigate which lands in the West were capable of being irrigated, and authorizing the "segregation from sale" of all such irrigable lands. Without truly meaning to, Congress had (it could be argued) asked the department to survey the West into watershed basins, and until that had been done, to "reserve from selection" all lands that might be irrigable.[42]

Congress soon realized what it had done. When Powell's 1877 report was made more generally known, some members thought that four Sections was too much for one farm family to own, while to others the idea of a cooperatively run irrigation system smacked of socialism. (Then as now, any bill could be killed if a legislator could hurl that epithet and make it stick.) On August 30, 1890, the provision for "segregation from sale" was repealed.[43] Until the closing of the Public Domain, farmers would remain free to try their luck at making an Iowa farm work, anywhere in the West.

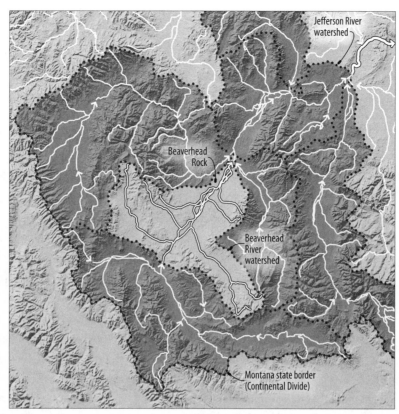

You'll remember Beaverhead Rock from the story of the Initial Point of the Montana Survey; figure E.17 is a view of the rock from the south (Airphoto—Jim Wark). What if Powell's land system had been applied in the watershed of the Beaverhead River? Figure E.18 shows the complexity of the watersheds that eventually flow into the Jefferson River. Figure E.19 uses the Sections of the Survey to give a hint of how land might have been distributed under Powell's system on one part of this watershed. (The 20-acre irrigation farms, not shown, would be strips of land directly alongside the streams.)

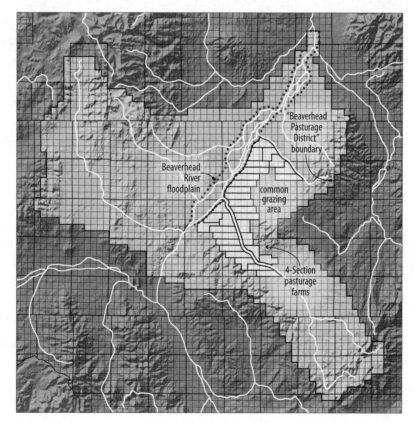

FIG E.19

So has the great Rectangular Survey served the nation well? It's certainly time for a look back. When the Public Domain "closed" seventy years ago, there ceased to be a need for new surveys—just some touching up here and there, some filling in of gaps, some resurveying to find lost corners. In the contiguous states, the Public Land Survey System is now an artifact. What was intended as a way of doing things has now become a *done thing.*

For scholars, mythologizing the survey is a constant temptation. They look at this great American artifact and, out of the habits of their calling, insist that it must mean something, must say something profound about the national character. Me, I'm suspicious when people pull meaning out of something made, when the makers of that thing didn't consciously pour meaning in. I'm more content just to say that the only "meaning" that can be read out of the Rectangular Survey is what it embodied to its originators: fairness.

It was a system drawn up to no one's obvious advantage, and its operations required no one's judgment. It was

a machine that would go of itself.

That last is a phrase spoken about the U.S. Constitution by James Russell Lowell in 1888, and it describes a peculiarly American concept of fairness. A group of free individuals negotiates among themselves, and they agree to adopt a procedure because it appears to favor none of them in particular. They then apply the procedure to a condition not foreseen in the original negotiations. Whatever results that procedure might then produce—*that* result, all will agree, constitutes fairness to all parties. Deciding farm boundaries by drawing straight lines across the landscape might confer benefit on some farmers and hardship on others, but those outcomes will not have been caused by a person or power, but rather by the impartial operations of a system blind to everyone's interests.

Even if the results produced by the Rectangular Survey are fair in this procedural sense, it can't be said that those results were always wise. The Montana land bust is only one example of an unwise use of the land facilitated by the "fair" application of the survey system. Nonetheless, the mistakes made under the Rectangular Survey have a saving quality that its originators could not have foreseen but which has become apparent in the two centuries since their debates in Congress. When mistakes are made *rectangularly,* they are more likely to be revocable. Let me explain (and end) by drawing a contrast.

As I write this, I am about to enter my fifty-ninth year. If I had been born in 1840, and it was now 1899, I would have come to consciousness in time to witness virtually the entire settlement of the West. As I started school, I would have heard about wagons heading overland to Oregon, then a little later, about the gold rush to California. If I survived the Civil War, I would have spent my late twenties hearing news of the transcontinental railroad. In my thirties I would have read about the Indian wars on the Great Plains, and in my forties I would have seen those same lands grazed by cattle and planted to wheat.

Three months after my fifty-third birthday, if I had been in Chicago for the great Columbian Exposition, I might have heard Frederick Jackson

Turner deliver that famous lecture, "The Significance of the Frontier in American History," in which he reflected on the finding in the 1890 census that settlement had so overspread the continent that "there can hardly be said to be a frontier line."

An entire continent settled in the space of a single lifetime, and all of it—*all of it*—done according to the lines laid down by the Rectangular Survey. Not all of the country would have been "gridded up" by my fifty-third birthday in 1893, but much of it would have been, and the survey system itself would have predicted the shape of the rest. Americans had conceived an efficient system for apportioning the land and then had set that system in motion. Once set in motion, manic American energy applied the system pervasively, a can-do American spirit pushed past obstacles, and American know-how invented ever-new ways to get the job done faster. By the time of Turner's lecture, I would have been living in a new, Rectangular America, and I would have seen it happen in the span of my lifetime.

The real me also lives in a new America. This new country also came into being with manic energy, and it too has happened in the span of a lifetime. I wasn't born in 1840, but in 1947. As the actual me started school, the pioneers were not heading for Oregon; the pioneering I experienced was my family's five-mile move to a brand-new house in the suburbs ($13,250 for 1,200 square feet, and we could imagine nothing grander). Our house faced onto what we called a court, but which would later gain the classier designation of *cul-de-sac*. Our cul-de-sac and several others like it fed into a collector street which looped around our subdivision and then con-

nected to a two-lane highway. That highway was how we got to a bigger, four-lane highway, which was our route to the new interstate.

If you are an American of the twenty-first century, you know this system intimately. It's been compared to a tree, with houses as leaves, cul-de-sacs as twigs, the highways as branches and limbs, and the interstate as the trunk. What continually amazes me is that when I was a child, this trunk-to-twig system of apportioning a landscape existed almost nowhere, yet now, on the brink of my fifty-ninth year, it is pervasive. It is the America we now live in. May I quote myself?

Americans had conceived an efficient system for apportioning the land and then had set that system in motion. Once set in motion, manic American energy applied the system pervasively, a can-do American spirit pushed past obstacles, and American know-how invented ever-new ways to get the job done faster.

Like the Rectangular Survey, the trunk-to-twig system is supremely rational, especially in its handling of automobile traffic: in any given region there'll be thousands of cars on the interstate, hundreds on the highways, dozens on the collector streets, and only the few cars of the actual residents in each cul-de-sac—perfectly safe for children to play in the street. Like the Rectangular Survey, the trunk-to-twig system was conceived by the finest minds of the time, and it seemed to favor no particular group or interest (black people were initially kept out of the suburbs by means other than the road system). But the most salient similarity is that both systems were enacted with *speed*—such speed that vast areas were apportioned before any consequences could be discerned; and even when negative reports began to trickle in, the system continued, so great was its momentum.

The difference between the two fifty-year land booms is, of course, that the first was directed at turning wilderness into farmland, while the second turned farmland into suburbs. But it's that distinction that makes the second apportionment so irrevocable.

Under any system of land apportionment, the boundaries of farms can be changed without too much difficulty; the Rectangular Survey's aliquot principle—and more especially the aliquot habit that grew out of it—simply makes those changes easier and less contentious. The trunk-to-twig apportionment, by contrast, is based on *roads*, and that makes all the difference, because roads are almost impossible to change. Streets themselves may be just a few inches of asphalt, but in their rights-of-way are buried pipes for water and sewage, along with buried or suspended cables for electricity and for so much else these days. Major highways might be raised on viaducts or depressed in trenches, and they always require overpasses and interchanges. You know how rarely major highways are ever removed, but even residential streets are rarely changed.

There is almost no decision taken by a citizenry that is more permanent than the laying down of streets. (The streets at the center of my city of Boston do not—contrary to legend—follow cow paths, but the alignments of those streets haven't changed in almost four hundred years.) Of course, what faces onto those streets will change, sometimes drastically, with the passage of time: the Empire State Building replaced a hotel, which itself replaced some grand houses, all within the same streets. Once a people lays down a pattern of streets, however, that is the pattern within which the city will have to grow.

We know how a city evolves when its streets are laid out in a grid, as in so many American cities, and when they are a net, as is common in Europe. We know from the history of such cities that when they spread far enough, they develop multiple centers, not just one "downtown" (think of New York or Los Angeles—or London); but we also know that in vibrant cities, those centers migrate (think how the prime shopping districts of those three cities have moved in the last century—and in recent decades, fragmented). That movement is relatively easy in a grid or a net: the blocks are pretty much the same everywhere, and so centers can die off over here and spring up over there.

Now, though, even middling American cities are spread out, and so they are naturally developing multiple centers. What we don't know from history is how those centers can evolve when a city is apportioned in a trunk-to-twig pattern. Unlike a net or a grid, a trunk-to-twig pattern irrevocably marks certain places as the most important—the interchanges of the great, funneling highways; and marks other places as not important—all the cul-de-sacs at the "twig" ends of the system. In these new spread-out cities, the first centers naturally have grown up at the interchanges, but history suggests that the first centers of a great city will not be—probably *should* not be—the centers it needs later in its life. In a trunk-to-twig system, how can new centers arise anywhere other than where the system mandates them?

We do know, from history, that new centers can indeed arise in spread-out metropolitan areas. Melrose Avenue in Los Angeles was for sixty years a nondescript street in the endless LA grid, lined with tiny houses and little shops and corner gas stations. It dozed in obscurity until suddenly, in the 1980s, it became *the* chic shopping district for the region. The little shops were renovated and enlarged, gas stations became cafes with outdoor dining, tiny houses were tricked out as fantasy worlds or replaced entirely. It's impossible to imagine Melrose Avenue happening on a cul-de-sac.

But who can say? Americans found ways to live and prosper in the first apportionment system invented out of whole cloth and pushed blindly across the continent. And in places like Chicago and Los Angeles, they even found ways to redivide the rectangles of the survey into blocks and streets, and make real cities. Who can foresee what Americans might make out of this second system, as much an invention as the first, and pursued with equal blindness?

When faced with the question, "How will we apportion the land?" the best that any people can do is to consult their experience, critique it in the light of current knowledge, and take their best shot. That's what the Continental Congress did in 1785. We might wish that those founding designers of the Rectangular Survey had mandated a midcourse correction; but at the moment when that correction might have been made— around 1880—America was as incapable of consensus as it is today. Then as now, the nation had too much invested—economically, ideologically, and even culturally—to turn from the ongoing system and find another.

Perhaps the truest verdict on the Rectangular Survey as a method of apportioning land would be, "Well, you could have done worse." As a sys-

tem for appropriating land and putting it into the hands of a populace, we have only to look to the sad experience of Latin America or southern Africa for a counterexample of what might have been done instead. As a system for ensuring the sustainable cultivation of the appropriated land, probably no system could have impeded the growth of monoculture crops on an ever-larger scale. As a template for turning farmsteads into cityscapes—well, the designers of the Rectangular Survey never imagined that their system would be put to such uses, and yet Americans took that system into their hands and produced cities and towns of which it might truly be said, "You could have done worse."

Will Americans, a lifetime from now, be able to say the same about the trunk-to-twig method of apportioning land? I'm hopeful, but I have doubts. I'm not so certain that the spirit that gave rise to the rectangular idea is the direct equivalent of the spirit that gave us trunks and twigs.

My parents were persuaded to buy that house on the cul-de-sac in part by a whole world of advertisements that emphasized *safety for children*: no cars will speed past your front yard and knock your kids off their bikes! And indeed, my brother and I were able to ride our bicycles all over Azalea Acres with absolute abandon, never once, in our entire childhood, encountering a runaway automobile.

And yet, I cannot help but feel that many of my parents' generation read the selling-point of "safety from strange autos" as really meaning "safety from strange people." And some part of me suspects that this unspoken (indeed, in today's climate, unspeakable) idea still impels people toward houses on cul-de-sacs. The rectangular idea—as a way to draw borders and as

FIG E.20   Airphoto—Jim Wark.

FIG E.21   Airphoto—Jim Wark.

a way to apportion the land—springs, I'm compelled to suspect, from a different impulse.

I'm far from certain of this, and as a child of the 'burbs, I'm probably incapable of objectivity. So let me end these reflections not with my own words but with those of another—and with the images in figures E.20 and E.21, two suburbs from opposite ends of the twentieth century.

There is a piece of American doggerel that gives voice to the particular ethos of living on one of the arrow-straight roads wedged into the Rectangular Survey—and by extension, of living on any street that extends far from your house. These verses by Sam Walter Foss (1858–1911) are more than a little wince-making today: nobody speaks this way anymore. But if you can get past the language, you might wince again to realize how few Americans *feel* this way anymore.

The House by the Side of the Road

There are hermit souls that live withdrawn
In the peace of their self-content;
There are souls, like stars, that dwell apart,
In a fellowless firmament;
There are pioneer souls that blaze their paths
Where highways never ran;—
But let me live by the side of the road
And be a friend to man.

Let me live in a house by the side of the road,
Where the race of men go by—
The men who are good and the men who are bad,
As good and as bad as I.
I would not sit in the scorner's seat,
Or hurl the cynic's ban;—
Let me live in a house by the side of the road
And be a friend to man.

I see from my house by the side of the road,
By the side of the highway of life,
The men who press with the ardor of hope,
The men who are faint with the strife.
But I turn not away from their smiles nor their tears—
Both parts of an infinite plan;—
Let me live in my house by the side of the road
And be a friend to man.

I know there are brook-gladdened meadows ahead
And mountains of wearisome height;
That the road passes on through the long afternoon
And stretches away to the night.
But still I rejoice when the travelers rejoice,
And weep with the strangers that moan,
Nor live in my house by the side of the road
Like a man who dwells alone.

Let me live in my house by the side of the road
Where the race of men go by—
They are good, they are bad, they are weak, they are strong,
Wise, foolish—so am I.
Then why should I sit in the scorner's seat
Or hurl the cynic's ban?—
Let me live in my house by the side of the road
And be a friend to man.

Notes

PART I

1 McIntyre, *Land Survey Systems*, 3.
2 Every British student knows this chronology of British monarchs; an American like myself has to rely instead on *Encyclopaedia Britannica*, 2007 deluxe edition.

CHAPTER ONE

1 Blum, *National Experience*, 20.
2 Paullin, *Atlas of Geography*, 25.
3 Ibid., 26.
4 Ibid.
5 Ibid., 26–28, 72–74.
6 Ibid., 29.
7 Pattison, *Beginnings of Survey*, 6.
8 Paullin, *Atlas of Geography*, 73.
9 Ibid., 29, 73.
10 Ibid., 29.
11 Douglas, *Boundaries of States*, 118.
12 Paullin, *Atlas of Geography*, 28.
13 Ibid., 29.
14 Ibid., 84–85.
15 Ibid.
16 Douglas, *Boundaries of States*, 121.
17 Descriptions of surveying techniques are drawn from Elgin, *2000 Celestial Ephemeris*; Pence, *Surveying Manual*; and Stewart, *Public Land Surveys*.
18 Paullin, *Atlas of Geography*, 77–78.
19 Abernethy, *Western Lands*, 91.
20 Paullin, *Atlas of Geography*, 77–78.
21 White, *Initial Points*, 2–3.
22 Ibid., 3.
23 Paullin, *Atlas of Geography*, 26–27.
24 Ibid., 82–83.
25 Ibid., 83–84.

CHAPTER TWO

1 Paullin, *Atlas of Geography*, 22–23.
2 Ibid.
3 Ibid., 33.
4 Blum, *National Experience*, 129.
5 Rakove, *National Politics*, 155–56.
6 Paullin, *Atlas of Geography*, 73.
7 Pattison, *Beginnings of Survey*, 6.
8 Rakove, *National Politics*, 158.
9 Ibid., 187.
10 Ibid., 187–89.
11 Ibid., 212.
12 Ibid., 285–86.
13 Douglas, *Boundaries of States*, 113.
14 Pattison, *Beginnings of Survey*, 121.

15 Treat, *National Land System*, 164.
16 Pattison, *Beginnings of Survey*, 6, 10.
17 Abernethy, *Western Lands*, 244.
18 Ibid.
19 Pattison, *Beginnings of Survey*, 19.
20 Abernethy, *Western Lands*, 244.
21 Ibid., 246.
22 Rakove, *National Politics*, 287.
23 Ibid., 334.
24 Abernethy, *Western Lands*, 272.
25 Ibid., 272.
26 Pattison, *Beginnings of Survey*, 3.
27 Blum, *National Experience*, 128.
28 Paullin, *Atlas of Geography*, 54.
29 Blum, *National Experience*, 138.
30 Ibid., 180.
31 Pattison, *Beginnings of Survey*, 8.
32 Douglas, *Boundaries of States*, 66.
33 Pattison, *Beginnings of Survey*, 8.
34 Douglas, *Boundaries of States*, 66.
35 Ibid., 186.
36 Ibid., 182.
37 Ibid.
38 Paullin, *Atlas of Geography*, 83.
39 Douglas, *Boundaries of States*, 164.
40 Ibid., 154.
41 Ibid.

CHAPTER THREE

1 Douglas, *Boundaries of States*, 21.
2 Paullin, *Atlas of Geography*, 58.
3 Douglas, *Boundaries of States*, 9.
4 Paullin, *Atlas of Geography*, 60.
5 Douglas, *Boundaries of States*, 29.
6 Adams, *First Jefferson Administration*, 12.
7 Paullin, *Atlas of Geography*, 63.
8 Douglas, *Boundaries of States*, 29.
9 Ford, *Jefferson Writings*, 262.
10 Ibid., 243.
11 Adams, *First Jefferson Administration*, 72.
12 Ambrose, *Undaunted Courage*, 222.
13 Ibid., 386.
14 Paullin, *Atlas of Geography*, 22.
15 Douglas, *Boundaries of States*, 33–34.
16 Ibid., 60.
17 Paullin, *Atlas of Geography*, 15.
18 Utley, *Wild and Perilous*, 22.
19 Douglas, *Boundaries of States*, 35.
20 Ibid., 166–67.
21 Paullin, *Atlas of Geography*, 60.

22 Douglas, *Boundaries of States*, 13.
23 Blum, *National Experience*, 287.
24 Douglas, *Boundaries of States*, 35.
25 Blum, *National Experience*, 197.
26 Ibid.
27 Douglas, *Boundaries of States*, 36.
28 Utley, *Wild and Perilous*, 86.
29 Ibid., 200.
30 Paullin, *Atlas of Geography*, 58.
31 Douglas, *Boundaries of States*, 17.
32 Ibid., 19–20.
33 Blum, *National Experience*, 291–92.
34 Paullin, *Atlas of Geography*, 37.
35 Blum, *National Experience*, 292.
36 Douglas, *Boundaries of States*, 21.
37 Donaldson, *Public Domain*, 7.
38 Blum, *National Experience*, 286.
39 Douglas, *Boundaries of States*, 37.
40 Blum, *National Experience*, 286.
41 Ibid., 290.
42 Polk, *Polk's Folly*, 224.
43 Paullin, *Atlas of Geography*, 64–65.
44 Blum, *National Experience*, 295.
45 Paullin, *Atlas of Geography*, 64–65.
46 Ibid., 65–66.
47 Blum, *National Experience*, 254.
48 Paullin, *Atlas of Geography*, 66.
49 Ibid.
50 Blum, *National Experience*, 313.
51 Paullin, *Atlas of Geography*, 66.
52 Blum, *National Experience*, 411.
53 Douglas, *Boundaries of States*, 42.
54 Ibid.
55 Paullin, *Atlas of Geography*, 69–70.

CHAPTER FOUR

1 Johnson, *American People*, 103.
2 The Vandalia story is drawn from a chapter in Abernethy, *Western Lands*, 40–57.
3 Boyd, *Jefferson Papers*, 6:582–83.
4 Ibid., 583.
5 Ibid., 584.
6 Ibid.
7 Ibid., 585.
8 Ibid., 607–8.
9 Ibid., 609.
10 Abernethy, *Western Lands*, 356.
11 Boyd, *Jefferson Papers*, 6:174.
12 Peterson, *Jefferson Writings*, 394–96.
13 Boyd, *Jefferson Papers*, 6:174.
14 Ibid.
15 *Continental Congress Journals*, 26:247.
16 Boyd, *Jefferson Papers*, 6:614.
17 Ibid.
18 Ibid., 611.

19 *Continental Congress Journals*, 26:279.
20 Pattison, *Beginnings of Survey*, 31.
21 Ibid.
22 Ibid., 32–33.
23 Abernethy, *Western Lands*, 291–94; Paullin, *Atlas of Geography*, 24–25.
24 Treat, *National Land System*, 14.
25 Paullin, *Atlas of Geography*, 24–25.
26 Ibid., 33–34.
27 *Continental Congress Journals*, 32:274–85, 312–21, 332–43.

CHAPTER FIVE

1 Douglas, *Boundaries of States*, 258.
2 Ibid., 164.
3 Ibid.
4 Ibid., 69.
5 Ibid., 186–87.
6 Ibid., 187.
7 Ibid., 187–88.
8 Ibid., 188–89.
9 Ibid., 166.
10 Ibid., 195.
11 Ibid., 193.
12 Ibid., 166–68.
13 Ibid., 200.
14 Ibid., 191.
15 Ibid., 164.
16 Ibid., 193.
17 Ibid., 162.
18 Ibid., 178.
19 Ibid., 183–86.
20 Hibbard, *Public Land Policies*, 145.
21 Blum, *National Experience*, 219.
22 Douglas, *Boundaries of States*, 178.
23 Ibid.
24 Ibid.
25 Ibid., 195.
26 Ibid., 202.
27 Ibid., 189, 196.
28 Ibid., 196–98.
29 Ibid., 203–6.
30 Ibid., 161.
31 Ibid., 204–5.
32 Ibid., 199–201.
33 Ibid., 200.
34 Ibid., 206.
35 Ibid., 241.
36 Ibid., 171–72.
37 Ibid., 242–43.
38 Ibid., 226–27.
39 Ibid., 231.
40 Ibid., 239–40.
41 Ibid., 211.
42 *Historical Atlas of U.S.*, 104–5.
43 Douglas, *Boundaries of States*, 207.

44 Ibid., 241–42.

45 Blum, *National Experience*, 334.

46 Douglas, *Boundaries of States*, 215.

47 Ibid., 224.

48 Ibid., 208.

49 Ibid., 234.

50 Uzes, *Chaining the Land*, 73.

51 Douglas, *Boundaries of States*, 234.

52 Ibid., 233–34.

53 Ibid., 237.

54 Ibid., 144–45.

55 Ibid., 234.

56 Malone, *Montana: A History*, 92–96.

57 Douglas, *Boundaries of States*, 219–20.

58 Ibid., 145.

59 Ibid., 234.

60 Ibid., 211.

61 Ibid., 222.

62 Ibid., 220.

63 Ibid., 224.

64 Ibid., 212.

65 Ibid., 240, 220, 210.

66 Iseminger, *Quartzite Border*, 7–13.

67 Douglas, *Boundaries of States*, 238, 222.

68 Ibid., 217.

69 Ibid.

70 Ibid., 231–32.

71 Ibid., 217.

72 Ibid., 219, 229, 233.

73 Ibid., 219.

74 Douglas, *Boundaries of States*, 229, 233.

PART THREE

1 Boyd, *Johnson Papers*, 6:585.

2 Pattison, *Beginnings of Survey*, 3.

3 Boyd, *Johnson Papers*, 7:2.

4 *Continental Congress Journals*, 28:114, 28, 165, 381.

CHAPTER SIX

1 Ford, *Jefferson Writings*, 174.

2 *Continental Congress Journals*, 26:324–25.

3 Hibbard, *Public Land Policies*, 118.

4 *Continental Congress Journals*, 26:325–30.

5 Ibid., 26:326.

6 Pattison, *Beginnings of Survey*, 13.

7 Ibid., 14.

8 *Continental Congress Journals*, 28:251.

9 Ibid., 28:252.

10 Ibid., 28:253.

11 Ibid., 28:254–55.

12 Ibid., 28:256.

13 Ibid., 28:299, 377.

14 Ibid., 28:293–95.

15 Ibid., 28:326.

16 Ibid., 28:328–29.

17 Ibid., 28:335–36.

18 Ibid., 28:337–39.

19 Ibid., 28:342.

20 Ibid., 28:381.

21 Ibid., 28:378.

22 Ibid., 28:376.

23 Peterson, *Jefferson Writings* (in *Notes on the State of Virginia*), 296.

24 Richeson, *English Land Measuring*, 37.

25 Orwin, *Open Fields*, 32–36, 55–58.

26 Richeson, *English Land Measuring*, 84.

27 Mathews, *Expansion of New England*, 84.

28 Ibid., 17.

29 Ibid., 111.

CHAPTER SEVEN

1 Pattison, *Beginnings of Survey*, 123n3.

2 Ibid., 119.

3 Descriptions of surveying techniques are drawn from Elgin, *2000 Celestial Ephemeris*; Pence, *Surveying Manual*; and Stewart, *Public Land Surveys*.

4 Pattison, *Beginnings of Survey*, 122.

5 Ibid., 117, 123.

6 Ibid., 129–30.

7 White, *Rectangular Survey*, 9–12.

8 White, *Initial Points*, 10–11.

9 Pattison, *Beginnings of Survey*, 130–32.

10 Stewart, *Public Land Surveys*, 18.

11 Pattison, *Beginnings of Survey*, 133.

12 Ibid., 138.

13 White, *Initial Points*, 4.

14 Pattison, *Beginnings of Survey*, 138.

15 Treat, *National Land System*, 44–45.

16 Pattison, *Beginnings of Survey*, 155–56.

17 Ibid., 140.

18 Treat, *National Land System*, 44–45.

19 Pattison, *Beginnings of Survey*, 140.

20 Ibid., 140–42.

21 Ibid., 185.

22 Ibid., 170.

23 Ibid.

24 Treat, *National Land System*, 48.

25 Hurt, *Ohio Frontier*, 157.

26 Treat, *National Land System*, 48.

27 Ibid., 50–52.

28 Ibid., 53–54.

29 Pattison, *Beginnings of Survey*, 174.

30 Ibid., 176.

31 Ibid., 177.

32 *Historical Atlas of U.S.*, 99.

33 Pattison, *Beginnings of Survey*, 173, 178.

34 Ibid., 181–82.

35 Ibid., 185.

36 Ibid., 181.

37 Ibid., 179–80.

38 Ibid., 181–82.
39 Treat, *National Land System*, 70–72.
40 Pattison, *Beginnings of Survey*, 188.
41 White, *Initial Points*, 16.
42 Treat, *National Land System*, 357–58.
43 Pattison, *Beginnings of Survey*, 192.
44 Ibid., 193.
45 Ibid., 194–98.
46 Stewart, *Public Land Surveys*, 43.
47 Pattison, *Beginnings of Survey*, 194–98.
48 Stewart, *Public Land Surveys*, 23.
49 White, *Initial Points*, 16.
50 Pattison, *Beginnings of Survey*, 202.
51 White, *Initial Points*, 16.
52 Pattison, *Beginnings of Survey*, 202.
53 Treat, *National Land System*, 334–35.
54 Stewart, *Public Land Surveys*, 119.
55 Pattison, *Beginnings of Survey*, 159.
56 Ibid., 148.
57 Ibid., 146.

CHAPTER EIGHT

1 Treat, *National Land System*, 94–95.
2 Stewart, *Public Land Surveys*, 27–28.
3 Ibid., 26.
4 Treat, *National Land System*, 167–68.
5 White, *Initial Points*, 28–30.
6 Pattison, *Beginnings of Survey*, 182.
7 White, *Initial Points*, 30–31.
8 Ibid., 30–32.
9 Ibid., 40–43.
10 Stewart, *Public Land Surveys*, 27.
11 Ibid., 33–34.
12 McEntyre, *Land Survey Systems*, 50.
13 Stewart, *Public Land Surveys*, 94.
14 Ibid.
15 McEntyre, *Land Survey Systems*, 129.

CHAPTER NINE

1 Stewart, *Public Land Surveys*, 95.
2 Ibid., 87.
3 Interview with Norman Caldwell, August 13, 1999, Lansing, Michigan.
4 Burt, *They Left Their Mark*, 2–4, 13.
5 Stewart, *Public Land Surveys*, 88.
6 Burt, *They Left Their Mark*, 23.
7 Stewart, *Public Land Surveys*, 88.
8 Burt, *They Left Their Mark*, xv.
9 My description of how to set up the solar compass is a shortened version of that recommended in Burt, *Key to Solar Compass*.
10 Burt, *They Left Their Mark*, 32–34.
11 Ibid., 35, 50.
12 Ibid., 98.
13 Ibid., 100, 137.

14 Stewart, *Public Land Surveys*, 120–21.
15 Rolvaag, *Giants in the Earth*, passim.
16 White, *Initial Points*, 17.
17 Ibid., 19.
18 Ibid., 22–23.
19 Ibid., 23.
20 Ibid., 32.
21 Ibid., 41.
22 Ibid., 50–52.
23 Ibid., 60–61.
24 Ibid., 74.
25 Ibid., 76.
26 Ibid.
27 Ibid., 84–85.
28 Ibid., 91–92.
29 Ibid., 99, 107.
30 Ibid., 92.
31 Ibid., 113
32 Ibid., 118, 120–23.
33 Ibid., 126–27.
34 Ibid., 128–31.
35 Ibid., 156–57.
36 Ibid., 136–37.
37 Ibid., 142–44.
38 Ibid., 148–49.
39 Ibid., 170, 173.
40 Ibid., 179–80, 182.
41 Ibid., 228.
42 Ibid., 234–35.
43 Ibid., 246, 250–51.
44 Ibid., 262–65.
45 Ibid., 273.
46 Ibid., 276–78.
47 Ibid., 295.
48 Ibid., 296.
49 Ibid., 296–97.
50 Ibid., 309–12.
51 Ibid., 338–39.
52 Ibid., 351–52.
53 Ibid., 377.
54 Ibid., 391–95, 407.
55 Ibid., 413–15, 423.
56 Ibid., 430.
57 Ibid., 433, 439.
58 Ibid., 445–46.
59 Ibid., 450.
60 Ibid., 450–53.
61 Ibid., 460–68.
62 Ibid., 476–79, 490–95.
63 Ibid., 500–503.
64 Ibid., 519.
65 Ibid., 523–24.
66 Ibid., 534–38.
67 Ibid., 544–45.
68 *Dominion Survey*, 44.

CHAPTER TEN

1   McRae, *Montana Handbook*, 14, 16.
2   White, *Initial Points*, 362.
3   Ibid., 363.
4   Ambrose, *Undaunted Courage*, 258–60.
5   White, *Initial Points*, 365.
6   Ibid.
7   Ibid.
8   Burt, *Key to Solar Compass*, 79.
9   Ibid., 79–80.
10 Ibid., 81.
11 Ibid., 83.
12 Ibid.
13 White, *Initial Points*, 367–68.
14 Ibid., 368.
15 Ibid., 366.
16 Ibid., 368.
17 Ibid., 444–46.
18 Malone, *Montana: A History*, 175.
19 Ibid., 177.
20 Ibid.
21 Ibid.
22 Ibid., 180–81.

EPILOGUE

1   Arneson, "Measurement in Texas," in *One League to Each Wind*, 8.
2   Ibid., 13.
3   Ibid., 47.
4   Powell, "Texas and Pacific Surveys," in *One League to Each Wind*, 171–73.
5   Mabry, "Survey in Panhandle," in *One League to Each Wind*, 116.
6   Ibid.
7   Handbook of Texas Online, s.v "Capitol."
8   Handbook of Texas Online, s.v "XIT Ranch."
9   Handbook of Texas Online, s.v "Capitol."
10 Gournay, *Texas Boundaries*, 95–96.
11 Ibid., 88–91.
12 Cornell University Library, Making of America, "Great Nation of Futurity."
13 State areas and populations are from *Roadmaster Road Atlas*, passim.
14 Ibid.
15 Handbook of Texas Online, s.v "Permanent University Fund."
16 *Historical Atlas of U.S.*, 109.
17 Peffer, *Closing of Public Domain*, 17.
18 Ibid., 57–66.
19 Ibid., 220, 257.
20 Ibid., 306.
21 Raban, *Bad Land*, 30.
22 MacDonald, *Dry-Farming*, 43, 72, 135.
23 Ibid., 152–60.
24 Ibid., 37.
25 Malone, *Montana: A History*, 241.

26 Raban, *Bad Land*, 19.
27 Malone, *Montana: A History*, 240.
28 Ibid., 242.
29 Stegner, *Beyond the Hundredth Meridian*, 3.
30 Malone, *Montana: A History*, 242.
31 Ibid., 283.
32 Hibbard, *Public Land Policies*, 67–75.
33 Johnson, *Order upon the Land*, 66–72.
34 Hibbard, *Public Land Policies*, 331.
35 Ibid., 228.
36 Ibid., 331.
37 Ibid., 414.
38 Ibid., 418–20.
39 Ibid., 393, 398.
40 Stegner, *Beyond the Hundredth Meridian*, 210.
41 Ibid., 225–27.
42 Ibid., 303.
43 Peffer, *Closing of Public Domain*, 19.

Bibliography

The first serious notes for this book date from 1997. They were taken in a spiral notebook in a carrel borrowed from some absent graduate student in the (then not air-conditioned) Widener Library at Harvard University. The Widener is truly one of the great repositories of humankind: in a set-aside alcove of the stacks, I passed an afternoon perusing Teddy Roosevelt's personal book collection! In the short decade since then, many of the materials I spent hours hand-copying (or photocopying, page by page) have become available online. A boon to all—but I wonder what will be lost when future scholars no longer know the pleasure of sitting cross-legged on the floor between bookshelves with some century-old volume spread across their knees.

Nonetheless, I feel compelled to point out, to those future scholars, one online resource that I found absolutely invaluable in my research. It's the Avalon Project at Yale Law School; and if you want the exact wording of the founding charters of the American colonies—or the text of the United Nations resolutions on Israel and Palestine—it's there. Many thanks, from a scholar fortunate to live into these miraculous times, to www.yale.edu/lawweb/avalon.

Abernethy, Thomas Perkins. *Western Lands and the American Revolution*. New York: Appleton-Century Co., 1937.

Adams, Henry. *History of the United States of America during the First Administration of Thomas Jefferson*. New York: Charles Scribner's Sons, 1909.

Ambrose, Stephen E. *Undaunted Courage: Meriwether Lewis, Thomas Jefferson, and the Opening of the American West*. New York: Simon & Schuster, 1996.

Arneson, Edwin P. "The Early Art of Terrestrial Measurement and Its Practice in Texas." In *One League to Each Wind*, edited by Sue Watkins. Austin: Texas Surveyors Association, 1964.

Blum, John M., et al. *The National Experience, Part One: A History of the United States to 1877*. Fort Worth, TX: Harcourt Brace College Publishers, 1993.

Boyd, Julian P. *The Papers of Thomas Jefferson*. Princeton, NJ: Princeton University Press, 1952–54.

Burt, John S. *They Left Their Mark: William Austin Burt and His Sons, Surveyors of the Public Domain*. Rancho Cordova, CA: Landmark Enterprises, 1985.

Burt, William A. *A Key to the Solar Compass and Surveyor's Companion*. New York: D. Van Nostrand, 1873.

Continental Congress. *Journals of the Continental Congress, 1774–1789*. Edited by Worthington C. Ford et al. 34 vols. Washington, DC, 1904–37.

Cornell University Library. Making of America, "Great Nation of Futurity." http://cdl.library.cornell.edu.

Donaldson, Thomas. *The Public Domain: Its History, with Statistics*. Washington, DC: Government Printing Office, 1884.

Douglas, Edward M. *Boundaries, Areas, Geographic Centers and Altitudes of the Several States and the United States*. Washington, DC: Government Printing Office, 1932.

Ebeling, Walter. *The Fruited Plain: The Story of American Agriculture*. Berkeley and Los Angeles: University of California Press, 1979.

Elgin, Richard L., et. al. *2000 Celestial Observation Handbook and Ephemeris*. Overland Park, KS: Sokkia Corporation, 1999.

Ford, Amelia Clewley. *Colonial Precedents of Our National Land System as It Existed in 1800*. Madison: University of Wisconsin Press, 1910.

Ford, Paul Leicester. *The Writings of Thomas Jefferson*. New York: G. P. Putnam's Sons, 1897.

Goode, Richard E. *Survey of the Boundary Line between Idaho and Montana from the International Boundary to the Crest of the Bitterroot Mountains*. Washington, DC: Government Printing Office, 1900.

Gournay, Luke. *Texas Boundaries: Evolution of the State's Counties*. College Station: Texas A&M University Press, 1995.

Handbook of Texas Online. http://www.tsha.utexas/handbook.

Hibbard, Benjamin Howard. *A History of the Public Land Policies*. 1924. Reprint, Madison: University of Wisconsin Press, 1965.

Historical Atlas of the United States. Washington, DC: National Geographic Society, 1993.

Hodgman, F. A Manual of Land Surveying. 1913. Reprinted by Michigan Society of Professional Surveyors, 1994.

Hurt, R. Douglas. The Ohio Frontier: Crucible of the Old Northwest, 1720–1830. Bloomington: Indiana University Press, 1996.

Iseminger, Gordon L. The Quartzite Border: Surveying and Marking the North Dakota-South Dakota Boundary, 1891–1892. Sioux Falls, SD: The Center for Western Studies, 1988.

Jensen, Merrill. "The Creation of the National Domain, 1781–1784." Mississippi Valley Historical Review 27, no. 3 (December 1939).

Johnson, Hildegard Binder. Order upon the Land: The U.S. Rectangular Land Survey and the Upper Mississippi Country. New York: Oxford University Press, 1976.

Johnson, Paul. A History of the American People. New York: HarperCollins Publishers, 1997.

Mabry, W. S. "Early Survey in the Texas Panhandle." In One League to Each Wind, edited by Sue Watkins. Austin: Texas Surveyors Association, 1964.

MacDonald, William. Dry-Farming: Its Principles and Practice. New York: Century Company, 1909.

Malone, Michael P., Richard B. Roeder, and William L. Lang. Montana: A History of Two Centuries. Seattle: University of Washington Press, 1976.

Manual of Instructions for the Survey of Dominion Lands. Ottawa, ON: Department of the Interior, 1918.

Mathews, Lois Kimball. The Expansion of New England: The Spread of New England Settlement and Institutions to the Mississippi River, 1620–1865. Boston: Houghton Mifflin Company, 1909.

McEntyre, John G. Land Survey Systems. New York: John Wiley & Sons, 1978.

McRae, W. C., and Judy Jewell. Montana Handbook. Chico, CA: Moon Publications, 1999.

Orwin, Charles Stewart, and C. S. Orwin. The Open Fields. Oxford: Clarendon Press, 1954.

Pattison, William D. Beginnings of the Rectangular Survey System, 1784–1800. Chicago: University of Chicago Press, 1957.

Paullin, Charles O. Atlas of Historical Geography of the United States. Washington, DC: Carnegie Institution and American Geographical Society, 1932.

Peffer, E. Louise. The Closing of the Public Domain: Disposal and Preservation Policies, 1900–1950. Palo Alto, CA: Stanford University Press, 1951.

Pence, William D., and Milo S. Ketchum, Surveying Manual: A Manual of Field and Office Methods for the Use of Students in Surveying. New York: McGraw Hill, 1932.

Peterson, Merrill D. Thomas Jefferson: Writings. New York: Library of America, 1994.

Polk, William R. Polk's Folly: An American Family History. New York: Doubleday, 2000.

Powell, W. J. "Re-Surveying of the Texas and Pacific Surveys West of the Pecos River." In One League to Each Wind, edited by Sue Watkins. Austin: Texas Surveyors Association, 1964.

Raban, Jonathan. Bad Land: An American Romance. New York: Pantheon Books, 1996.

Rakove, Jack N. The Beginnings of National Politics: An Interpretive History of the Continental Congress. New York: Alfred A. Knopf, 1979.

Richeson, A. W. English Land Measuring to 1800: Instruments and Practices. Cambridge, MA: MIT Press, 1966.

Roadmaster 2005 Standard Road Atlas. Mountville, PA: MapQuest. com, 2005.

Rolvaag, O. E. Giants in the Earth. 1925. Reprint, New York: HarperCollins Publishers, 1991.

Stegner, Wallace. Beyond the Hundredth Meridian: John Wesley Powell and the Second Opening of the West. 1954. Reprint, New York: Penguin Books, 1992.

Stewart, Lowell O. Public Land Surveys: History Instructions Methods. Ames, IA: Collegiate Press, 1935.

Treat, Payson Jackson. The National Land System, 1785–1820. New York: E. B. Treat & Company, 1910.

Utley, Robert M. A Life Wild and Perilous: Mountain Men and the Paths to the Pacific. New York: Henry Holt and Company, 1997.

Uzes, Francois D. Chaining the Land: A History of Surveying in California. Sacramento, CA: Landmark Enterprises, 1985.

White, C. Albert. A History of the Rectangular Survey System. Washington, DC: Government Printing Office, 1978.

White, C. Albert. Initial Points of the Rectangular Survey System. Westminster, CO: Professional Land Surveyors of Colorado, 1996.

Illustration Credits

The charts and graphic depictions of surveying techniques were prepared in Adobe Illustrator. The maps were prepared in Adobe Illustrator also, with the terrain imagery processed in Adobe Photoshop. The terrain imagery, the national hydrography, and the national borders came from MapArt US Terrain by Cartesia Software. All Public Land Survey System boundaries and local hydrography and railroads came from 1:2,000,000-Scale Digital Line Graph Data from the U.S. Geological Survey of the U.S. Department of the Interior. The images of the "horizon plane" earth and the abstracted solar compass were prepared in Nemetschek VectorWorks, rendered in Nemetschek RenderWorks, and processed in Adobe Photoshop. The night sky image in figure 1.29 was prepared in Starry Night Backyard by Sienna Software. The standing figures in figure E.06 were prepared in SketchUp, now known as Google SketchUp.

All of the illustrations in this book are by the author, with the following exceptions:

Dedication Figure: Courtesy of C. Albert White

Fig. 1: Courtesy of Joseph Nonneman

Fig. 2: Courtesy of Bank of America

Fig. 3: U.S. Government Printing Office

Fig. 1.31: Courtesy of Physics Department, University of California, Berkeley

Fig. 3.21: Library of Congress, Geography and Map Division

Fig. 4.03: Library of Congress, Geography and Map Division

Fig. 6.10: Redrawn from Orwin, The Open Fields

Figs. 6.11, 6.12: Courtesy of the Museum of English Rural Life

Fig. 6.13: Courtesy of Helen French

Fig. 7.01: National Museum of American History, ©2006 Smithsonian Institution

Fig. 7.17: Courtesy Nebraska Historical Society

Fig. 7.22: Library of Congress, Geography and Map Division

Fig. 7.32: Bureau of Land Management, United States Department of the Interior

Fig. 8.03: Bureau of Land Management, United States Department of the Interior

Fig. 9.16: U.S. Government Printing Office

Fig. 9.17: Gurley Precision Instruments

Fig. 9.18: U.S. Government Printing Office

Fig. 9.46: Courtesy of C. Albert White

Fig. 9.47: Canada Department of the Interior

Fig. 10.01: Sam Abell/National Geographic Image Collection

Figs. 10.04, 10.05: Bureau of Land Management, United States Department of the Interior

Fig. E.02: Redrawn from Gournay, Texas Boundaries

Fig. E.07: Drawn from data in Hibbard, Public Land Policies

Fig. E.08: Bureau of Land Management, United States Department of the Interior

Fig. E.11: Airphoto—Jim Wark

Fig. E.16: U.S. Government Printing Office

Fig. E.17: Airphoto—Jim Wark

Figs. E.20, E.21: Airphoto—Jim Wark

Index

INDEX OF TOPICS